IET ENERGY ENGINEERING 106

Surface Passivation of Industrial Crystalline Silicon Solar Cells

Other volumes in this series:

Volume 1	**Power Circuit Breaker Theory and Design** C.H. Flurscheim (Editor)
Volume 4	**Industrial Microwave Heating** A.C. Metaxas and R.J. Meredith
Volume 7	**Insulators for High Voltages** J.S.T. Looms
Volume 8	**Variable Frequency AC Motor Drive Systems** D. Finney
Volume 10	**SF_6 Switchgear H.M.** Ryan and G.R. Jones
Volume 11	**Conduction and Induction Heating** E.J. Davies
Volume 13	**Statistical Techniques for High Voltage Engineering** W. Hauschild and W. Mosch
Volume 14	**Uninterruptible Power Supplies** J. Platts and J.D. St Aubyn (Editors)
Volume 15	**Digital Protection for Power Systems** A.T. Johns and S.K. Salman
Volume 16	**Electricity Economics and Planning** T.W. Berrie
Volume 18	**Vacuum Switchgear** A. Greenwood
Volume 19	**Electrical Safety: A guide to causes and prevention of hazards** J. Maxwell Adams
Volume 21	**Electricity Distribution Network Design, 2nd Edition** E. Lakervi and E.J. Holmes
Volume 22	**Artificial Intelligence Techniques in Power Systems** K. Warwick, A.O. Ekwue and R. Aggarwal (Editors)
Volume 24	**Power System Commissioning and Maintenance Practice** K. Harker
Volume 25	**Engineers' Handbook of Industrial Microwave Heating** R.J. Meredith
Volume 26	**Small Electric Motors** H. Moczala *et al.*
Volume 27	**AC-DC Power System Analysis** J. Arrillaga and B.C. Smith
Volume 29	**High Voltage Direct Current Transmission, 2nd Edition** J. Arrillaga
Volume 30	**Flexible AC Transmission Systems (FACTS)** Y.-H. Song (Editor)
Volume 31	**Embedded Generation** N. Jenkins *et al.*
Volume 32	**High Voltage Engineering and Testing, 2nd Edition** H.M. Ryan (Editor)
Volume 33	**Overvoltage Protection of Low-Voltage Systems, Revised Edition** P. Hasse
Volume 36	**Voltage Quality in Electrical Power Systems** J. Schlabbach *et al.*
Volume 37	**Electrical Steels for Rotating Machines** P. Beckley
Volume 38	**The Electric Car: Development and future of battery, hybrid and fuel-cell cars** M. Westbrook
Volume 39	**Power Systems Electromagnetic Transients Simulation** J. Arrillaga and N. Watson
Volume 40	**Advances in High Voltage Engineering** M. Haddad and D. Warne
Volume 41	**Electrical Operation of Electrostatic Precipitators** K. Parker
Volume 43	**Thermal Power Plant Simulation and Control** D. Flynn
Volume 44	**Economic Evaluation of Projects in the Electricity Supply Industry** H. Khatib
Volume 45	**Propulsion Systems for Hybrid Vehicles** J. Miller
Volume 46	**Distribution Switchgear** S. Stewart
Volume 47	**Protection of Electricity Distribution Networks, 2nd Edition** J. Gers and E. Holmes
Volume 48	**Wood Pole Overhead Lines** B. Wareing
Volume 49	**Electric Fuses, 3rd Edition** A. Wright and G. Newbery
Volume 50	**Wind Power Integration: Connection and system operational aspects** B. Fox *et al.*
Volume 51	**Short Circuit Currents** J. Schlabbach
Volume 52	**Nuclear Power** J. Wood
Volume 53	**Condition Assessment of High Voltage Insulation in Power System Equipment** R.E. James and Q. Su
Volume 55	**Local Energy: Distributed generation of heat and power** J. Wood
Volume 56	**Condition Monitoring of Rotating Electrical Machines** P. Tavner, L. Ran, J. Penman and H. Sedding
Volume 57	**The Control Techniques Drives and Controls Handbook, 2nd Edition** B. Drury
Volume 58	**Lightning Protection** V. Cooray (Editor)
Volume 59	**Ultracapacitor Applications** J.M. Miller
Volume 62	**Lightning Electromagnetics** V. Cooray
Volume 63	**Energy Storage for Power Systems, 2nd Edition** A. Ter-Gazarian
Volume 65	**Protection of Electricity Distribution Networks, 3rd Edition** J. Gers
Volume 66	**High Voltage Engineering Testing, 3rd Edition** H. Ryan (Editor)
Volume 67	**Multicore Simulation of Power System Transients** F.M. Uriate
Volume 68	**Distribution System Analysis and Automation** J. Gers
Volume 69	**The Lightening Flash, 2nd Edition** V. Cooray (Editor)
Volume 70	**Economic Evaluation of Projects in the Electricity Supply Industry, 3rd Edition** H. Khatib
Volume 72	**Control Circuits in Power Electronics: Practical issues in design and implementation** M. Castilla (Editor)
Volume 73	**Wide Area Monitoring, Protection and Control Systems: The enabler for smarter grids** A. Vaccaro and A. Zobaa (Editors)
Volume 74	**Power Electronic Converters and Systems: Frontiers and applications** A.M. Trzynadlowski (Editor)
Volume 75	**Power Distribution Automation** B. Das (Editor)
Volume 76	**Power System Stability: Modelling, analysis and control** B. Om P. Malik
Volume 78	**Numerical Analysis of Power System Transients and Dynamics** A. Ametani (Editor)
Volume 79	**Vehicle-to-Grid: Linking electric vehicles to the smart grid** J. Lu and J. Hossain (Editors)
Volume 81	**Cyber-Physical-Social Systems and Constructs in Electric Power Engineering** S. Suryanarayanan, R. Roche and T.M. Hansen (Editors)
Volume 82	**Periodic Control of Power Electronic Converters** F. Blaabjerg, K.Zhou, D. Wang and Y. Yang
Volume 86	**Advances in Power System Modelling, Control and Stability Analysis** F. Milano (Editor)
Volume 87	**Cogeneration: Technologies, Optimisation and Implementation** C.A. Frangopoulos (Editor)
Volume 88	**Smarter Energy: From smart metering to the smart grid** H. Sun, N. Hatziargyriou, H.V. Poor, L. Carpanini and M.A. Sánchez Fornié (Editors)
Volume 89	**Hydrogen Production, Separation and Purification for Energy** A. Basile, F. Dalena, J. Tong and T.N.Veziroğlu (Editors)
Volume 90	**Clean Energy Microgrids** S. Obara and J. Morel (Editors)
Volume 91	**Fuzzy Logic Control in Energy Systems with Design Applications in MATLAB®/Simulink®** İ.H. Altaş
Volume 92	**Power Quality in Future Electrical Power Systems** A.F. Zobaa and S.H.E.A. Aleem (Editors)
Volume 93	**Cogeneration and District Energy Systems: Modelling, analysis and optimization** M.A. Rosen and S. Koohi-Fayegh
Volume 94	**Introduction to the Smart Grid: Concepts, technologies and evolution** S.K. Salman
Volume 95	**Communication, Control and Security Challenges for the Smart Grid** S.M. Muyeen and S. Rahman (Editors)
Volume 97	**Synchronized Phasor Measurements for Smart Grids** M.J.B. Reddy and D.K. Mohanta (Editors)
Volume 98	**Large Scale Grid Integration of Renewable Energy Sources** A. Moreno-Munoz (Editor)
Volume 100	**Modeling and Dynamic Behaviour of Hydropower Plants** N. Kishor and J. Fraile-Ardanuy (Editors)
Volume 101	**Methane and Hydrogen for Energy Storage** R. Carriveau and D.S.-K. Ting
Volume 104	**Power Transformer Condition Monitoring and Diagnosis** A. Abu-Siada (Editor)
Volume 107	**Bifacial Photovoltaics: Technology, applications and economics** J. Libal and R. Kopecek (Editors)
Volume 108	**Fault Diagnosis of Induction Motors** J. Faiz, V. Ghorbanian and G. Joksimović
Volume 110	**High Voltage Power Network Construction** K. Harker
Volume 111	**Energy Storage at Different Voltage Levels: Technology, integration, and market aspects** A.F. Zobaa, P.F. Ribeiro, S.H.A. Aleem and S.N. Afifi (Editors)
Volume 112	**Wireless Power Transfer: Theory, technology and application** N.Shinohara
Volume 115	**DC Distribution Systems and Microgrids** T. Dragičević, F.Blaabjerg and P. Wheeler
Volume 117	**Structural Control and Fault Detection of Wind Turbine Systems** H.R. Karimi
Volume 119	**Thermal Power Plant Control and Instrumentation: The control of boilers and HRSGs, 2nd Edition** D. Lindsley, J. Grist and D. Parker
Volume 123	**Power Systems Electromagnetic Transients Simulation, 2nd Edition** N. Watson and J. Arrillaga
Volume 124	**Power Market Transformation** B. Murray
Volume 126	**Diagnosis and Fault Tolerance of Electrical Machines, Power Electronics and Drives** A.J.M. Cardoso
Volume 128	**Characterization of Wide Bandgap Power Semiconductor Devices** F. Wang, Z. Zhang and E.A. Jones
Volume 130	**Wind and Solar Based Energy Systems for Communities** R. Carriveau and D.S.-K. Ting (Editors)
Volume 131	**Metaheuristic Optimization in Power Engineering** J. Radosavljević
Volume 132	**Power Line Communication Systems for Smart Grids** I.R.S. Casella and A. Anpalagan
Volume 905	**Power System Protection, 4 volumes**

Surface Passivation of Industrial Crystalline Silicon Solar Cells

Edited by
Joachim John

The Institution of Engineering and Technology

Published by The Institution of Engineering and Technology, London, United Kingdom

The Institution of Engineering and Technology is registered as a Charity in England & Wales (no. 211014) and Scotland (no. SC038698).

First published 2018

The Institution of Engineering and Technology
Michael Faraday House
Six Hills Way, Stevenage
Herts, SG1 2AY, United Kingdom

www.theiet.org

British Library Cataloguing in Publication Data
A catalogue record for this product is available from the British Library

ISBN 978-1-78561-246-6 (hardback)
ISBN 978-1-78561-247-3 (PDF)

Typeset in India by MPS Limited

Contents

About the editor

Joachim John was born in Freiburg, Germany, on November 8, 1966.

He received his diploma degree in physics at the Albert-Ludwigs-University in Freiburg in 1993.

He received his Ph.D. in physics from the Federal Institute of Technology (ETH) in Zürich, Switzerland, in 1997.

From 1990 to 1993, he joined the Fraunhofer Institute for Physical Measurement Techniques in Freiburg, Germany, doing research and development on semiconductor mid-infrared (IR) laser.

From 1993 to 1998, he worked first as a junior scientist and later as a senior scientist on photovoltaic mid-IR sensor arrays at the Federal Institute of Technology (ETH) in Zürich, Switzerland.

In 1998, he joined the Interuniversity Micro Electronic Centre (IMEC) in Leuven, Belgium, where he is presently a Management Team member and Program Manager in the PV department.

Joachim John is the session chairperson, scientific committee member, and organizing committee member of the EU PVSEC, IEEE PVSC, Si-PV/n-PV, and the SNEC.

He is a paper reviewer for *Progress in Photovoltaics*, *SOLMAT*, *IEEE*, *Renewable Energy* and *Electrochemical Society*.

Since 2014, he is an expert team member of the PV Technology Platform of the European Commission. He has published more than 200 papers, gave more than 40 invited presentations, and holds five patents.

In 2014, he was awarded with the Otto-von-Guericke medal of the University of Magdeburg by the minister-president of Saxony-Anhalt Dr. Reiner Haseloff.

Dr. John is currently working on the industrial crystalline silicon solar cell process improvement and device physics pushing toward higher conversion efficiencies and lower cost.

Authors' biographies

Prof. Armin Aberle serves as the CEO of the Solar Energy Research Institute of Singapore (SERIS) at the National University of Singapore (NUS). In parallel, he is a tenured full professor in the university's Department of Electrical and Computer Engineering. He holds B.Sc./M.Sc., Ph.D., and Dr. habil degrees in physics from German universities. His research focus is on photovoltaic materials, devices, and modules. His R&D work covers the full spectrum from fundamental materials research to the industrial evaluation of advanced or novel PV technologies at the pilot line level and their transfer to industry. In his 31-year career in the PV sector, he has (co)authored more than 450 papers (which have received more than 10,000 citations, h-index 45) and has been supervising more than 70 Ph.D. students. He is an editor of *IEEE Transactions on Electron Devices* and an Editorial Board member of *Progress in Photovoltaics* and other scientific journals. In the 1990s, he established and led the Silicon Photovoltaics Department at the Institute for Solar Energy Research (ISFH) in Hamelin, Germany. He then worked for 10 years in Sydney, Australia, as a professor for photovoltaics at the University of New South Wales (UNSW). In 2008, he joined NUS to establish SERIS, as the Deputy CEO and the Director of the Silicon PV Department. Since 2012, he serves as the SERIS CEO. He is presently the director of the institute's Silicon Materials & Cells Cluster and the Novel PV Concepts Cluster.

Lachlan E. Black was born in Canberra, Australia, in 1985. He received his B.Eng. (with first class honors) and B.A. degrees in 2009, and the Ph.D. degree in engineering in 2015, from The Australian National University (ANU). From 2009 to 2010, he worked as an intern at the Energy Research Centre of the Netherlands (ECN), and for part of 2014, he was a visiting researcher at the Interuniversity Microelectronics Centre (IMEC) in Belgium. His Ph.D. research focused on the surface passivation of crystalline Si by Al_2O_3 prepared by atmospheric pressure chemical vapor deposition and the electrical characterization of the Si–Al_2O_3 interface. Since 2015, he has been a postdoctoral researcher with the Department of Applied Physics at the Eindhoven University of Technology in the Netherlands, where he has worked on the application of novel thin film materials prepared by atomic layer deposition for surface passivation of III/V semiconductor nanowires and crystalline silicon solar cells.

Dr. Black has authored or co-authored more than 30 peer-reviewed articles, conference proceedings, and one patent. His current research interests focus on the investigation of novel thin film materials for surface passivation and contacting of silicon solar cells and III/V semiconductor devices, as well as the physics, characterization, and modeling of recombination processes in semiconductors more broadly.

Emanuele Cornagliotti was born in Torino, Italy, in 1981. He received his M.Sc. in Electronic Engineering in 2006 from the Politecnico of Torino and his Ph.D. in Electrical and Electronic Engineering in 2011 from KU Leuven, Belgium. Since 2007 he has been a researcher at IMEC Belgium, working on the development of industrial crystalline silicon solar cells (p-PERC and n-PERT), with a focus on process integration and surface passivation. Since 2017, he is a researcher at the Kaneka European Photovoltaics Laboratory, in Leuven, Belgium. He is the author or co-author of more than 80 papers in the field of crystalline silicon solar cells and semiconductor process technology.

Dr. Kristopher O. Davis is an assistant professor in Materials Science and Engineering at the University of Central Florida (UCF) and a member of the Resilient, Intelligent and Sustainable Energy Systems (RISES) faculty cluster at UCF. His research is focused on the development of new materials, manufacturing processes, and characterization techniques to improve photovoltaic (PV) technologies and speed up the adoption of PV as an energy source. This work is currently supported through multiple awards from the U.S. Department of Energy and from industry.

Previously, Dr. Davis was a research engineer at the Florida Solar Energy Center (FSEC) and the c-Si Metrology Program Manager for the U.S. Photovoltaic Manufacturing Consortium (PVMC), a DOE-funded initiative in collaboration with SEMATECH. As a program manager in the PVMC, Dr. Davis worked with a number of companies across the US PV supply chain to better understand the challenges facing the manufacturing sector and develop metrology solutions to these problems through collaborative R&D projects.

Dr. Davis received his B.S. in Electrical Engineering, M.S. in Optics, and Ph.D. in Optics, all from the UCF. He is a peer reviewer for notable journals (e.g., *Scientific Reports*, *Journal of Materials Research*, *Solar Energy Materials & Solar Cells*) and for national laboratories. Since 2012, he has co-authored over 20 peer-reviewed journal publications related to PV.

Dr. Thorsten Dullweber studied physics at the University of Hanover, Germany, and received his physics diploma degree in 1996 with a thesis on ordering phenomena of atomic mono layers on single crystal surfaces. Afterwards, Thorsten joined the Institute of Physical Electronics (IPE) at the University of Stuttgart where he experimentally investigated band gap grading in $Cu(In,Ga)Se_2$ solar cells in the research group of Hans-Werner Schock. Thorsten received his Ph.D. degree in electrical engineering in 2002 from the University of Stuttgart. From 2001 to 2009, he worked in the microelectronics industry, first as an R&D engineer at Siemens AG and later as a technology project leader at Infineon Technologies AG developing DRAM memory chips with CMOS transistor gate length down to 65 nm. Since 2009, Thorsten is leading the R&D group Industrial Solar Cells at the Institute of Solar Energy Research Hamelin (ISFH). His research work focuses on high efficiency industrial-type PERC silicon solar cells and bifacial PERC+ solar cells. Thorsten authored more than 70 journal and conference publications which have been cited more than 1,000 times. He was awarded with the Enercity Energy Efficiency Prize in 2015 for developing a world-record-efficient PERC solar cell. In 2017, Thorsten received the prize of the German Foundation for Industrial Research for developing the bifacial PERC + solar cell. Thorsten is a member of the Scientific Committees of the EU-PVSEC and SNEC Conferences and of the Editorial Advisory Board of Photovoltaics International.

Dr. Shubham Duttagupta is Head of the Monocrystalline Silicon Wafer Solar Cell Group at the Solar Energy Research Institute of Singapore (SERIS)—Singapore's national laboratory for applied solar energy research. Dr. Duttagupta has 10 years of R&D experience with crystalline silicon materials and solar cells. His Ph.D. research focused on the development of industrial multifunctional thin-films required for surface passivation and optics in high-efficiency crystalline silicon wafer solar cells. Since 2014,

Dr. Duttagupta is managing a large R&D group focused on the development of >24% crystalline silicon solar cells for mass production, enabling technology transfer and commercialization of high-efficiency processes. His group has recently developed a highly promising apparatus and method for single-sided deposition of *in-situ* ultrathin oxide and doped low-absorbing layers offering very high deposition rate. In the last ten years of his career in photovoltaics, Dr. Duttagupta has achieved several awards, notably the "Institute of Microelectronics Prize" by the National University of Singapore (NUS), runner-up for the prestigious "SolarWorld Junior Einstein Award," and most recently the "Young Scientist Award" at the PVSEC 2016. He is an active reviewer of several journals and currently serving as the Chairman of Crystalline Silicon Advanced Technology and Material Forum organized by PVQAT TG13 in China.

Roger Görtzen was born in Heerlen, the Netherlands, on August 25, 1977. He received his B.Sc. degree from HTS Utrecht in 2001. From 2001 to 2010, he was with the Dutch research institute TNO in Eindhoven. In the first five years, he started and dedicated himself as a precision mechanical engineer in the field of microsystem technology. Semicon and especially back-end semiconductor were his main field of work. During second 5 years he was the chairman of the works council at TNO and learned more about organization management, finance, and restructuring business. During this same period, he switched his technical background into business development and grew his international business and social skills. In 2007–2008, he obtained his NIMA B Marketing degree and in 2009 he was selected by the TNO to follow the TNO Business School at the Erasmus University in Rotterdam. In the period from 2008 to 2010, his order intake for TNO was in average about 500KEURO/year. The last successful project in TNO was the setup of the Spatial ALD project and the later spin-off called SoLayTec.

In 2010, he was one of the co-founders of SoLayTec, who wrote the business plan, found the right investors, set up the supplier network to work together on a risk/reward basis and started to develop the market for SoLayTec, which resulted in several LOIs and the first product sold before the company was even started. From December 2010 onwards, he was the Marketing and Sales director at SoLayTec.

Roger has published several patents and scientific papers or is the co-author. Furthermore, he writes articles in international magazine of his playing field and is a senior in his area of photovoltaic production equipment and has received several awards with SoLayTec.

Hans Goverde received the M.Sc. degree in Applied Physics from the Eindhoven University of Technology, the Netherlands, in 2012. The title of his M.Sc. project was 'Characterization and optimization of aluminum oxide films as *p*-type Czochralski silicon passivation layer'. In 2016, he obtained the Ph.D. degree (title: Increasing and understanding energy yield of *c*-Si PV modules) from the KU Leuven in collaboration with the IMEC photovoltaics group. Currently, he is working as a researcher for the same group at IMEC.

Ernst Granneman received his degree in Physics from the Technical University Delft, the Netherlands, in 1970. After his military service in the Royal Dutch Navy, he joined the FOM-Institute on Atomic and Molecular Physics in Amsterdam where he received his Ph.D. in 1976 on a topic on multi-photon ionization of alkali-atoms. After that, he became a group leader of the Plasma Physics group at the same institute. In 1980/81, he worked at the Lawrence Livermore lab in Livermore, California, on heating of fusion plasmas.

In 1984, he joined the Dutch Semiconductor Equipment Company ASM as R&D manager of ASM Europe. In that capacity, various metal and dielectric CVD processes were developed.

He was appointed as (part time) professor at the Technical University Delft in 1987, a position he held for 8 years. During that period, he guided students and gave courses on Deposition Technology in the IC industry.

Ernst Granneman worked for ASM during 25 years in various capacities, including CTO in the period 1992–1999 and product manager of the RTP group after that. In his capacity as CTO, he was directly involved with the acquisition of the Finnish ALD company Microchemistry (later renamed in ASM Microchemistry).

In 2009, with his partner Jaap Beijersbergen, he acquired the RTP product line from ASM through a management buy-out. In the years following the MBO, the new company, Levitech BV, developed the Levitrack system, a spatial ALD tool, in which Al_2O_3 films are deposited on c-Si solar cells, resulting in superior passivation of the solar cell surface. Levitech now markets two tools, the Levitor RTP system for the IC industry, and the before-mentioned Levitrack system.

In 2014, he was knighted in the Order of The Netherlands Lion for his role in Dutch Science and for his contributions to the Dutch Equipment Industry.

Ernst Granneman has published more than 100 papers and holds 40 (independent) patents.

After many years as CTO of Levitech, he has stepped down to become an executive scientist at the company, a position he currently still holds.

Ali Hajjiah was born in Kuwait on June 12, 1973. He has been graduated with a B.Sc. in Electrical Engineering from Oklahoma State University, USA, in 1998, and subsequently earned M.Sc. and Ph.D. degrees, in 2003 and 2009, respectively, from Virginia Tech, USA. He is currently at the Kuwait University full time as an assistant professor where he had held key positions in the institute. He was the director of the Semiconductor Laboratory at Kuwait University from 2011 to 2012 and then the director of the Engineering Training and Alumni Center (ETAC) in the College of Engineering and Petroleum from 2015 to 2017. Dr. Hajjiah is currently prime and co-investigator on several research projects funded by the Kuwait Foundation for the Advancement of Science (KFAS). Moreover, he had also joined IMEC Belgium as an Industrial Fellow from January to August 2014 working on enhancing the conversion efficiency of n-type PERT crystalline silicon solar cells. As a visiting scholar, Dr. Hajjiah had joined the Virginia Polytechnic Institute and State University (Virginia Tech), USA, in 2013 and subsequently the Catholic University of Leuven (KU Leuven) in Belgium in 2015 and 2017. He is also a member of several honor societies such as The Golden Key National Honor Society since 1995, the Tau Beta Pi Honor Society—Gamma Chapter since 1997, the Eta Kappa Nu Honor Association—Omega Chapter since 1997 and finally the Phi Sigma Theta National Honor Society since 2002. His research interests include processing and fabrication of Si and III-V semiconductor lasers and thin film solar cells. Currently he is working on efficiency improvement and device physics of Perovskite/Si tandem solar cells.

Associate professor **Bram Hoex** completed both his M.Sc. and Ph.D. degrees at Eindhoven University of Technology in 2003 and 2008, respectively. From 2008 to 2015, he worked at the Solar Energy Research Institute of Singapore (SERIS) at the National University of Singapore (NUS) as a group leader and from 2012 also as the director of the Silicon Materials and Cells Cluster. In 2015, he joined the School of Photovoltaic and Renewable Energy Engineering (SPREE) at UNSW Sydney where he currently serves as Deputy Head of School (Research) and Director International Strategy. During his career, he has raised over A$ 50M in competitive research funding of which A$ 16M as a lead investigator. He has published over 150 journal and conference papers, which have been cited over 4,600 times. His research focuses on the fabrication and characterization of high-efficiency silicon wafer solar cells. His work has received various international recognitions including the 2008 SolarWorld Junior Einstein and the 2016 IEEE PVSC Young Professional awards.

Marc Hofmann received his Diploma degree in Electrical Engineering from the University of Applied Sciences Koblenz, Germany, in 2003, and his Ph.D. degree from the University of Konstanz, Germany, in 2008, for his work on rear-surface conditioning and passivation of crystalline silicon by plasma processes, which he performed with Fraunhofer Institute for Solar Energy Systems, Freiburg, Germany. He is the Head of the Plasma Technology group at Fraunhofer Institute for Solar Energy systems.

Marc Hofmann is a paper reviewer for scientific journals like *SOLMAT, IEEE Journal of Photovoltaics, Thin Solid Films, Journal of Applied Physics*, and *Journal of Materials Science*.

He has published more than 100 scientific papers. His h-index is 19, and he has more than 1,300 citations.

Marc has supervised more than 40 Bachelor, Master, and Ph.D. students creating their thesis.

Machteld Lamers was born in Lienden, the Netherlands, on March 21, 1979. She received her diploma degree in Applied Physics from the Eindhoven University of Technology in Eindhoven, the Netherlands, in 2006. In 2006, she joined as a junior scientist at the Energy Research Centre of the Netherlands (ECN), currently ECN part of TNO, Petten, the Netherlands. She worked within the unit of Solar Energy within the Device Architecture group on the development of innovative solar cells and manufacturing technologies. Her work on Metal-wrap-through cells resulted in 2009 and 2010 in world-record module efficiencies on p-type mc-Si. She received her Ph.D. in Physics from the University of Amsterdam in Amsterdam, the Netherlands, in 2015 on "Application and Analysis of Silicon Nitride Films for Surface Passivation of High Efficiency Silicon Solar Cells". After her Ph.D., she continued as a Medior Scientist to work at ECN, where she widened her field of expertise to innovative module manufacturing technologies and smart integration in applications. In 2018, she has joined as a research fellow at the Delft University of Technology, Delft, the Netherlands, and the Amsterdam Institute for Advanced Metropolitan Solutions (AMS), Amsterdam, the Netherlands, to set up and work on a joined research program on the development of innovative solar energy technologies for application in the urban environment, with the city of Amsterdam as living lab.

Dr. Jozef Poortmans received his degree in Electronics Engineering from the Katholieke Universiteit of Leuven, Belgium, in 1985. He joined the newly build Interuniversity Micro Electronic Centre (IMEC) in Leuven where he worked on laser recrystallization of polysilicon and a-Si for SOI-applications and thin-film transistors. In 1988, he started his Ph.D. study on strained SiGe-layers. Both the deposition and the use of these SiGe-alloys within the base of a heterojunction bipolar transistor were investigated in the frame of this study. He received his Ph.D. degree in June 1993.

Afterwards he joined the photovoltaics group, where he became responsible for the group Advanced Solar Cells. He was involved in initiating IMEC-activities on thin-film crystalline Si solar cells, organic solar cells and III-V solar cells and became Department Director of the PV-department in 2003. In 2008, he started up the Si-PV Industrial Affiliation Program and collected the investment funds to build up the advanced Si-PV R&D-line of IMEC (S-line). As a Program Director PV, he built the industrial partnership active in the S-line. Presently, he is the Scientific Director of the PV and Energy activities of IMEC since 2013. In the same year, he was also appointed IMEC Fellow.

He has been a Board Member of Eurec Agency and is presently member of the Steering Committee of the EU PV Technology Platform. He has also acted as General Chairman of the 21st European Photovoltaic Solar Energy Conference & Exhibition and of the SiliconPV 2012 Conference and has been active in the Scientific Committees of the leading PV-conferences.

Prof. J. Poortmans has authored or co-authored more than 500 papers that have been published in Conference Proceedings and technical journals and he has an h-index of 37. Since 2008, he is part-time professor at the KU Leuven, where he teaches courses on photovoltaics and materials in electrical engineering. In 2013, he became also part-time professor at University Hasselt where he teaches a course on analog electronics. In the same year, he was appointed IMEC Fellow. Since September 2016, he is the Coordinator R&D-strategy of EnergyVille, an institutional partnership between IMEC, VITO, KU Leuven and University Hasselt focused on the themes of Smart Cities and Smart Grids.

Pierre Saint-Cast was born in Lorient, France, in 1982. He received his Master degree in micro- and nanoelectronics from Joseph Fourier University, Grenoble, France, and the Engineering degree from the Polytechnic Institute of Grenoble, both in 2007. In 2012, he received his Ph.D. degree in Physics from the University of Konstanz, Konstanz, Germany. Since 2008, he has been with Fraunhofer Institute for Solar Energy Systems, Freiburg, Germany. Pierre's research interests include the development of passivation layers for solar cell applications, especially plasma-enhanced chemical vapor deposition of Al_2O_3 films, the analytical modeling of the electrical transport in silicon solar cells, and the manufacturing of large area PERC solar cells. His current position is project manager and product manager for PERC solar cells in the group plasma technology at Fraunhofer Institute for Solar Energy Systems. He has published more than 60 papers. His h-index is 13, and he has more than 800 citations.

Hermann Schlemm studied physics at the Technical University at Chemnitz/Germany and received his diploma degree in 1979 and Ph.D. degree in 1983. For more than 35 years, he has been active in his research in the fields of plasma technology and plasma chemistry mostly combined with applications on thin films.

Since 1999, he is with Meyer Burger AG (former Roth & Rau AG), where he is active in photovoltaic technology from the beginning. Currently, he works as a senior principal in basic research for future solar cell technology.

Prof. Jan Schmidt was born in Kiel, Germany, on June 29, 1967.

He received his diploma degree in physics from Clausthal Technical University, Germany, in 1993 and his Ph.D. in physics from the University of Hanover, Germany, in 1998. From 1998 to 2000, he worked as a Humboldt Fellow at the Australian National University, Canberra, Australia, and from 2000 to 2001 as a postdoctoral researcher at the Institute of Physical Electronics, University of Stuttgart, Germany. Since 2001, he is leader of the Photovoltaics Materials Research group at the Institute for Solar Energy Research Hamelin (ISFH), Germany, and since 2010 he is heading the Photovoltaics Department at ISFH. In parallel, Dr. Schmidt is also a full professor in Photovoltaics Materials Research at the Institute of Solid-State Physics, Leibniz University Hanover. Dr. Schmidt's research interests include the development of novel solar cell concepts and fabrication processes, with focus on surface passivation and selective contacts, the development of novel characterization techniques for silicon materials and solar cells, and the analysis and manipulation of defects in mono- and multicrystalline silicon materials. He has authored and co-authored more than 150 peer-reviewed journal papers and more than 110 conference papers in the fields of photovoltaics and semiconductor physics. His research articles have been cited more than 11,000 times with an h-index of 55. He is an inventor of the patent family on dielectric-layer passivation stacks, which are applied today in industrial PERC productions lines worldwide. Dr. Schmidt is a scientific committee member for several conferences (e.g., EUPVSEC, SiliconPV/nPV, CSSC) and an editor of the *IEEE Journal of Photovoltaics*.

Bart Vermang received his M.Sc. degree in physics from the University of Ghent in Belgium, and the Ph.D. degree in electrical engineering from the University of Leuven in Belgium. He performed his M.Sc. final research project at the Norwegian University of Science and Technology (NTNU), where his interest—and career—in renewable energy originated. First, he obtained a Ph.D. for research in silicon Photovoltaics (PV) in IMEC Belgium, followed by two postdoctoral fellowships to work on thin film PV. A Marie Skłodowska-Curie fellowship to move to the University of Uppsala in Sweden, and a fellowship from the Flemish Research Foundation (FWO) to return to IMEC. Recently, Bart obtained a prestigious starting grant of the European Research Council (ERC), and became professor at the University of Hasselt in Belgium. His current research is focused on renewable energy and associated topics, and is performed within the Institute for Materials Research (IMO-IMOMEC) and EnergyVille in Belgium.

Arthur Weeber was born in 1959 in Velsen, the Netherlands. In 1983, he finished his Master of Science in Physics at the University of Amsterdam in the field of solid-state physics. In 1987, he completed his Masters in Chemical Engineering at the same university with specialization anaerobic waste water treatments. After receiving his Ph.D. in 1988 in Physics (topic amorphous alloys), also from the University of Amsterdam, he joined the Energy research Centre of the Netherlands ECN as a solid-state physicist with research topic positron annihilation. Since 1992, he has been performing R&D on crystalline silicon PV technology and has coordinated large national and internal projects with both academic partners and industry ones. After several years of being a senior manager at the ECN, he has become senior scientist at the Solar Energy department in 2015. Since 2015, he has been combining his work at ECN, now ECN part of TNO, with a full professorship position (0.2 fte) at Delft University of Technology in the Photovoltaic Materials and Devices group (Department of Electrical Sustainable Energy, faculty of Electrical Engineering Mathematics). He also is lecturer of Solar Energy within the EUREC Master Renewable Energy at the Hanze University of Applied Sciences.

Arthur Weeber is one of the topic organizers in the field of crystalline silicon solar cells at the EUPVSEC, session chairman at the same conference, member of the executive committee of the SiliconPV (including being one of the chairmen), one of the organizers of the nPV workshop (also being one of the chairmen), and paper reviewer of several journals and international conferences. He is also the author/co-author of over 200 papers and conference contributions.

Preface

In this book, the success story of a solar cell concept is followed that has been proposed in the late 1980s by the Australian scientists of Prof. Martin Green's team at the University of New South Wales. The PERC, passivated emitter and rear cell concept, needed almost 20 years to develop from a lab solar cell to a mass production product that is set to become the main stream cell technology in 2018 with forecasted 75 GW of installed capacity. Since the first lab cells in the 1980s and 1990s and mass production of the 2010s, the silicon solar cell technology has been changing massively. Scientists and engineers all over the world have participated in the development of the PERC from lab concept to industrial product. The key was the understanding of the dominant loss mechanism in aluminum BSF silicon solar cells. Prof. Stefan Glunz from FhG-ISE in Germany explained the successive suppressing of resistive, optical and recombination losses with the leaky bucket analogy. The bucket can in theory be full at the beginning, but leaks through holes with different diameters as depicted in Figure P.1.

Fixing holes will help to maintain the water level in the bucket, with that level symbolizing the overall efficiency of the silicon solar cell. However, this will increase leakage through the remaining holes due to higher water pressure. In the solar cell, due to the complex nature of loss mechanism interaction, the other losses will increase when one "hole" has been closed. Thus, it is important to fix the holes with the largest diameter first, so leakage due to the remaining, smaller holes remain under control or of less relevance.

With the dielectric surface passivation of the front and the rear side of the silicon solar cell, the surface recombination losses are reduced significantly, with a strong impact on the increase in cell efficiency. This important milestone in silicon solar cell development will be the main focus of this book.

As material of choice for the rear passivation, silicon oxide and silicon oxynitride were used in the early days of PERC development. This passivation scheme was the first to enter mass production by cell makers like SolarWorld. In the meantime, aluminum oxide has taken over as the dominating material for rear passivation. This material is deposited by plasma-enhanced chemical vapor deposition (PECVD) equipment as well as in atomic layer deposition (ALD) machines.

The book is build up from discussing fundamental material studies to the description of mass production equipment.

Market situation and market prediction are important instruments to motivate the development of novel cell concepts. The PERC concept in the market evolution

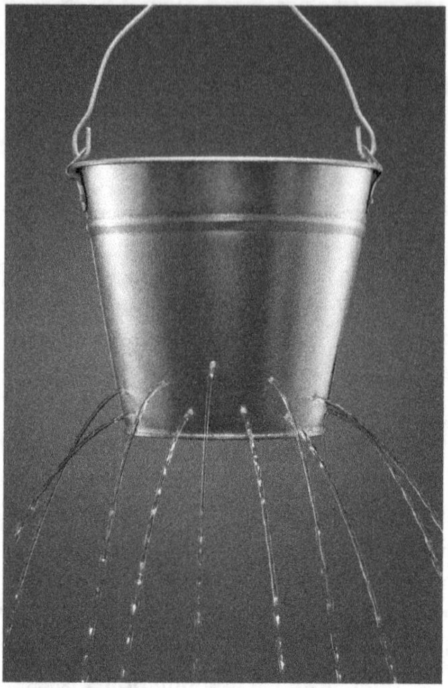

Figure P.1 Leaky bucket analogy symbolizing the different current density loss mechanisms in a silicon solar cell (© Photographerlondon<https://www.dreamstime.com/photographerlondon_info> | Dreamstime.com<https://www.dreamstime.com/>)

from Watt peak (Power) to kilowatt hour (Energy) is described by Prof. Jozef Poortmans and Prof. Ali Hajjiah in Chapter 1.

Prof. Armin Aberle illuminates the history of silicon solar cell passivation from the first attempts until the development toward the industrial product in Chapter 2.

Prof. Bram Hoex explains the physical mechanism of the surface passivation nature of the AlO_x material in Chapter 3.

Prof. Bart Vermang and Dr. Hans Goverde investigates the material properties of the ALD AlO_x passivation layer focusing on the complex relation between charge (Q_f) and interface trap density (D_{it}) in Chapter 4.

Dr. Pierre Saint-Cast and Dr. Marc Hofmann explore the properties of PECVD-AlO_x and discuss different deposition methods in Chapter 5.

Prof. Kristopher Davis and Dr. Lachlan Black describe the properties of the AlO_x layer deposited by APCVD in Chapters 6 and 7, respectively.

Dr. Thorsten Dullweber and Prof. Jan Schmidt describe the implementation of the PECVD AlO_x passivation layer into a p-type PERC concept in Chapter 8.

Dr. Emanuele Cornagliotti investigates the implementation of ALD AlO_x passivation layers in n-type PERT cell concepts in Chapter 9.

Dr. Shubham Duttagupta and Prof. Armin Aberle discuss the application of AlO_x-containing passivation layer stacks for the application in industrial silicon solar cells in Chapter 10.

Dr. Arthur Weeber and Dr. Machteld Lamers provide the study of the front surface passivation and anti-reflection coating material silicon nitride and its implementation in silicon solar cells in Chapter 11.

The important factor for the mass production capability of the PERC concept was the development of high throughput process equipment. This effort is taken into account by the participation of the equipment suppliers in Chapters 11, 12, and 13.

Dr. Hermann Schlemm provides the understanding of microwave plasma sources and their application in high throughput Meyer Burger (aka Roth&Rau until August 31, 2015) PECVD reactors for AlO_x and AlO_x/SiN_x stacks in Chapter 12.

Dr. Ernst Granneman and Dr. Wil van Velzen explain the concept of the spatial ALD Levitech tool and give a detailed analysis of the wafer movement. A clear route to high throughput is depicted in Chapter 13.

Dr. Roger Görtzen describes the development of the SoLayTec spatial ALD tool from a first lab reactor until a high throughput industrial equipment in Chapter 14.

For me, the only thing left is to thank all the authors and wish You, The Reader, a fruitful study of this book.

Yours truly,
Dr. Joachim John

List of acronyms

Abbreviations

AFM	Atomic force microscopy
Al-BSF	Aluminum back surface field
ALD	Atomic layer deposition
AlO_x	Aluminum oxide
amu	Atomic mass unit
APCVD	Atmospheric pressure chemical vapor deposition
ARC	Antireflective coating
BSF	Back surface field
C-AFM	Conductive atomic force microscopy
c-Si	Crystalline silicon
CoO	Cost of ownership
CVD	Chemical vapor deposition
Cz	Czochralski
D_{it}	Interface trap density
EDX	Energy-dispersive X-ray spectroscopy
EQE	External quantum efficiency
FSF	Front surface field
FSRV	Front surface recombination velocity
FZ	Float zone
i-PERC	Industrial passivated emitter and rear cell
IPA	Iso-propanol alcohol
IQE	Internal quantum Efficiency
IR	Infrared
ITRPV	International Technology Roadmap for Photovoltaics
kWp	kilo watt peak
kWh	kilo watt hour
M/S	Metal/Semiconductor
mc-Si	Multi-crystalline silicon
MIS	Metal insulator semiconductor

OED	Oxidation-enhanced diffusion
PCD	Photo-conductance decay
PECVD	Plasma-enhanced chemical vapor deposition
PERC	Passivated emitter and rear cell
PERL	Passivated emitter and rear locally diffused
PERT	Passivated emitter and rear totally diffused
PhD	Doctor of Philosophy
PL	Photoluminescence
PSG	Phosphoric silicate glass
PV	Photovoltaic
PVA	Poly vinyl alcohol
PVD	Physical vapor deposition
Q_f	Oxide-fixed charge
QSSPC	Quasi steady-state photo conductance
RSRV	Rear surface recombination velocity
RTA	Rapid thermal anneal
RTP	Rapid thermal processing
SDE	Saw damage etch
SDR	Saw damage removal
SE	Selective emitter
SIMS	Secondary ion mass spectroscopy
SiO_x	Silicon oxide
SiN_x	Silicon nitride
SPM	Sulfuric peroxide mixture
SRH	Shockley–Reed–Hall
SRV	Surface recombination velocity
STD	Standard deviation
TEM	Transmission electron microscopy
TLM	Transfer length method
UV	Ultra violet
Wh	Watt hour
XPS	X-ray photo-electron spectroscopy

Symbols

χs	Electron affinity
Ω/□	Ohms per square
φ	Implanted dose
φ_m	Metal work function

ρ_c	Specific contact resistivity
$\sigma_{n/p}$	Capture cross section of the defect
τ	Minority carrier lifetime
τ_{eff}	effective minority carrier lifetime
$G(t)$	Generation rate
v_{th}	Thermal velocity
r_2	Linear regression coefficient
C_n	Auger coefficient in n-type material
D_0	Diffusion coefficient
E_a	Activation energy
E_c	Conduction band energy level
E_t	Trap energy level
E_v	Valence band energy level
Eff	Solar cell efficiency (%)
Eta	Solar cell efficiency (%)
FF	Fill factor (%)
J_0	Recombination current density
J_{0n+}	Emitter recombination current density
J_{sc}	Current density
L	Wafer length
L_T	Transfer length
N_A	Acceptor doping density
N_B	Background concentration
N_f	Number of fingers
P_{max}	Maximum power
R_{sheet}	Sheet resistance
U	Recombination rate
V_{oc}	Open-circuit voltage
W	Wafer thickness
W_f	Width of fingers
W_p	Watt peak
X_j	Junction depth

Chapter 1

Market position of PERC silicon solar cells

Jozef Poortmans[1,2,3,4] *and Ali Hajjiah*[5]

1.1 Photovoltaics makes a lot of economical sense

Solar energy is a cornerstone of the future global sustainable energy system. This is recognized by the International Energy Agency (IEA) and a wide range of other public organizations, private companies, and NGOs. Within the portfolio of solar energy technology options, photovoltaic (PV) conversion for electricity generation plays a special role. It is highly modular and suitable for many applications, ranging from small stand-alone systems through medium-sized integrated grid-connected systems to large power plants. Finally, recent sharp cost and price reductions have made PV competitive on retail and even wholesale price level in many countries, which is a crucial step toward the creation of self-sustaining markets. A further reduction by a factor of 1.5–2 will make PV-electricity competitive with electricity generated by fossil fuels on wholesale price level over the largest part of the globe. Such a reduction is very likely to be achieved, the main question is how soon and using which technologies. In this chapter, we will mostly refer to assessments and predictions made within the PV professional associations, like Solar Power Europe or International Technology Roadmap for Photovoltaics (ITRPV) since the predictions of the IEA concerning market prediction and price evolution have consistently been overly pessimistic.

Reduction of investment and operational costs on the level of costs/W_p alone, however, is not sufficient for PV to reach large-scale penetration. A number of very important challenges has to be tackled in parallel. The progress of PV has been based on upscaling and innovation of the cell technology, but it is clear that the "one size fits all" approach will not be sufficient in the future anymore. Different PV-markets are developing with specific cell technologies being well suited to serve these markets. PV module manufacturers will therefore come under pressure to differentiate their products according to esthetics, mechanical strength, and the

[1]IMEC, Leuven, Belgium
[2]KU Leuven
[3]University Hasselt
[4]Coordinator R&D-strategy EnergyVille
[5]EE Department, College of Engineering & Petroleum, Kuwait University, Kuwait

confidence with which the initial and long-term performance can be predicted. In this context, passivated emitter rear cell (PERC) solar cells are expected to serve the higher end of the PV-market, penetrating gradually to the whole PV-market ranging from PV power plants to the PV-specialty markets (seamless integration of PV in the built environment, solar-powered cars, etc.).

From this, it is clear that reliability and qualification of PV-modules is to be shaped according to the location in which these PV-modules will be installed. In addition, given the efficiency increase of PERC PV-cells and modules going to 20% and more, sufficient attention is to be paid to temperature coefficient and degradation phenomena. Moreover, the emphasis will also shift from the performance under standard conditions of temperature and irradiation (expressed as W_p) towards energy generation (kWh) under a broad variety of conditions. At lower illumination conditions, PERC solar cells have a superior performance relative to the full-Al back surface field (BSF) technology. The paper will outline how performance and energy yield increase will have to evolve concurrently to make PV a major electricity source and how PERC-technology is an important milestone in achieving this.

1.2 Photovoltaics, a rapidly growing market

1.2.1 A decade of breathtaking market growth and installed base

The deployment of photovoltaics has been characterized by a tremendous growth rate since the beginning of the previous decade. The installed capacity is described in terms of W_p which is the power generated under standard AM1.5 illumination conditions. An average growth rate for the PV-sector of more than 40%/year since then has propulsed the sector from a small sector in the year with a turnover of 500 M$ /year to a major industrial sector with an annual turnover in the range of 100 B$/year. This is illustrated convincingly in Figure 1.1, showing the evolution of sold and installed PV-modules/year since 2010 [1], and this evolution is not expected to step and present predictions foresee yearly installation levels in the range of 250–400 GW_p/year at the end of the next decade, yet another growth phase by a factor of at least 5.

This sustained growth has brought the global installed PV-capacity presently to the level of nearly 400 GW_p, which generates near to 2% of the global electricity demand, although this is already substantially higher in a number of countries like Italy (8%) and Germany. Presently the strongest growth of the PV-sector is witnessed in the United States and Asia (China, Japan), turning into a real global market instead of being a nearly exclusive European market until 5 years ago. Based on the expected growth, the accumulated capacity would reach 5 TW_p but these predictions are increased every year. Hence, it is not improbable that we will be witnessing a total installed PV-capacity of 10 TWp in 2050, supplying up to 1/3 of the total electricity demand.

1.2.2 A decade of breathtaking price reductions

Concurrently with this strong growth of the industry, the price of PV-modules and PV-systems has come down drastically over the last decade reaching system values in the range of 1.0–2.0 $/$W_p$ for the more mature markets (ranging from commodity to

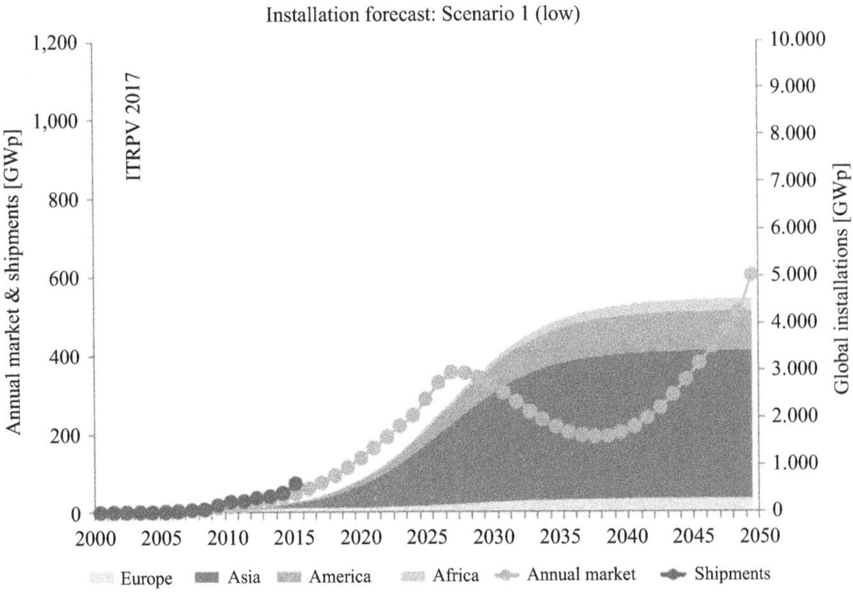

Figure 1.1 Past and future evolution of yearly installed and accumulated PV-capacity (excerpt taken from the ITRPV eighth edition 2017, including a maturity report about the sector)

premium products). The evolution of the average selling price of PV-modules is shown in Figure 1.2. One can see several periods of strong price erosion in the period 2011–2013 and in 2016. The average learning ratio over 30 years is more than 20%, but over the last decade the value is significantly above 30%, a value rarely achieved by any industrial sector over such a period. This price reduction is felt over the full PV value chain (from basic material to module level) as shown in Figure 1.3.

However, the price of a PV-system is determined not only by the price of the PV-module but also by the land area, the mounting structure, the connections, the invertor, the installation, etc. These costs are often referred to as "balance-of-systems costs" (BOS). Whereas in the past, the system price was mainly determined by the PV-module cost, the rapid decline of the latter has led to a situation that at the moment the PV-module costs and BOS-costs are comparable as shown in Figure 1.4 for rooftop PV-systems in Germany (as this is obviously a function of local market conditions).

The effect of this evolution where both price components are becoming of comparable influence is not to be underestimated. The PV-module costs will decline further in accordance with the learning curve characterized by a learning ratio of 20% or higher. BOS-costs decrease at a slower pace. This should however not lead to the erroneous conclusion that further efforts to reduce costs should be only focused on BOS-level. The main lever to reduce BOS-costs is indeed performance. This is easily understood because the more power or energy/unit area is produced by the PV-system, the less installation and mounting costs which scale

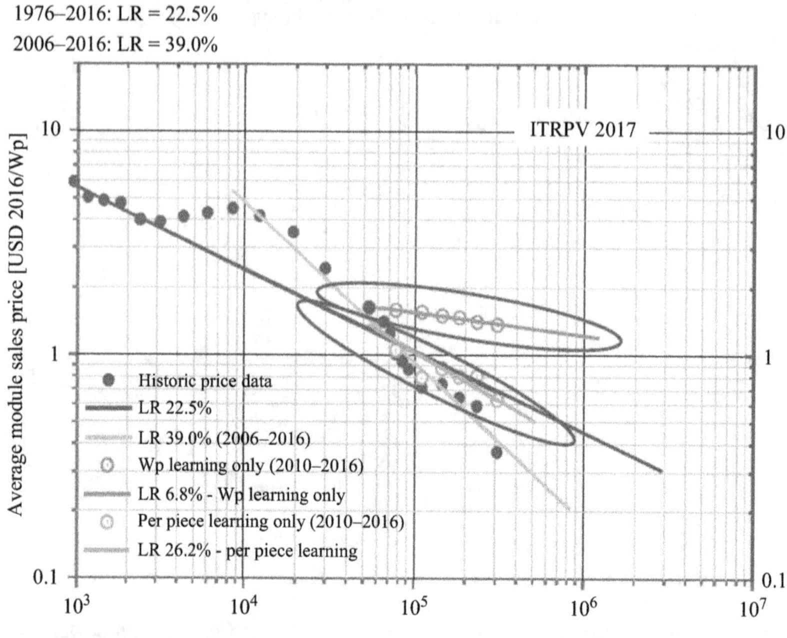

Figure 1.2 Evolution of the average selling price of PV-modules (Taken from ITRPV, eight edition [1])

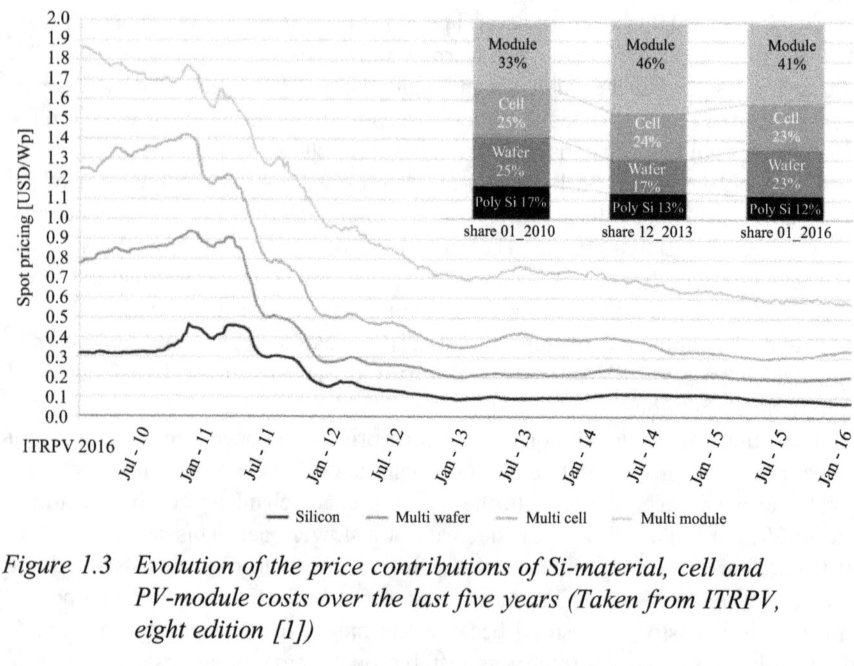

Figure 1.3 Evolution of the price contributions of Si-material, cell and PV-module costs over the last five years (Taken from ITRPV, eight edition [1])

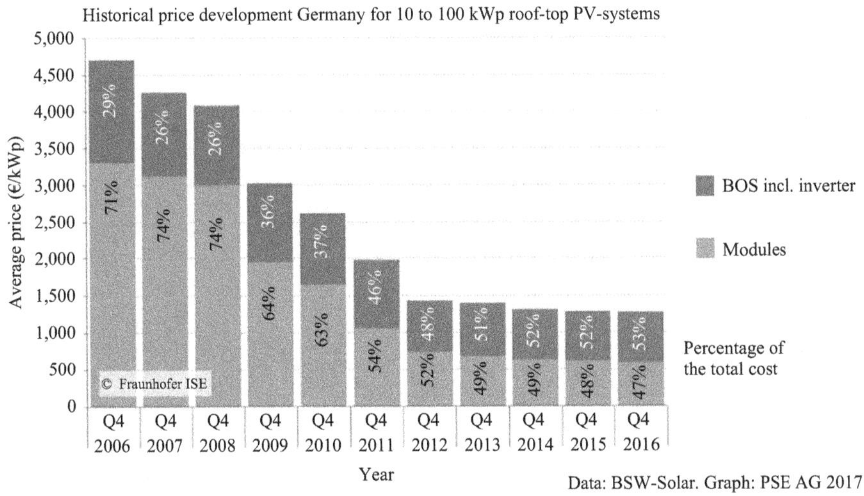

Figure 1.4 Evolution of the PV-system cost subdivided in PV-module cost and BOS-costs (source: PSE, Germany [2])

with area will influence the final investment cost, the sensitivity to which is shown in Figure 1.5. This makes that PV-technologies with a low-energy conversion efficiency, even at low cost level will not survive on the longer term.

The present discussion has up till now strongly focused on the aspects related to the initial investment, but what interests investors in PV power plants mostly is the cost/kWh. In order to compare different energy technologies, one has defined the so-called levelized cost of electricity (LCOE) of each technology. Basically, the LCOE is defined by the following equation:

$$\text{LCOE} = \frac{\sum_{t=1}^{n} \frac{I_t + M_t + F_t}{(1+r)^t}}{\sum_{t=1}^{n} \frac{E_t}{(1+r)^t}} \tag{1.1}$$

where I_t = investment expenditures in year t, M_t = operations and maintenance expenditures in year t, F_t = fuel expenditures in year t, which is zero for photovoltaic electricity, E_t = electricity generation in the year t, r = discount rate and n = investment period considered in years. The energy yield is obviously depending on the exact location of the PV-system, but it is clear that a longer lifetime and high-energy yield will help in reducing the LCOE as one can also deduce from Figure 1.5.

An LCOE comparison between different electricity generation technologies in Germany is shown in Figure 1.6.

As mentioned earlier, making PV a significant contributor to the electricity generation mix it is necessary to reduce the LCOE by an additional factor of 1.5–2. This has to be done by further increasing performance to bring down initial

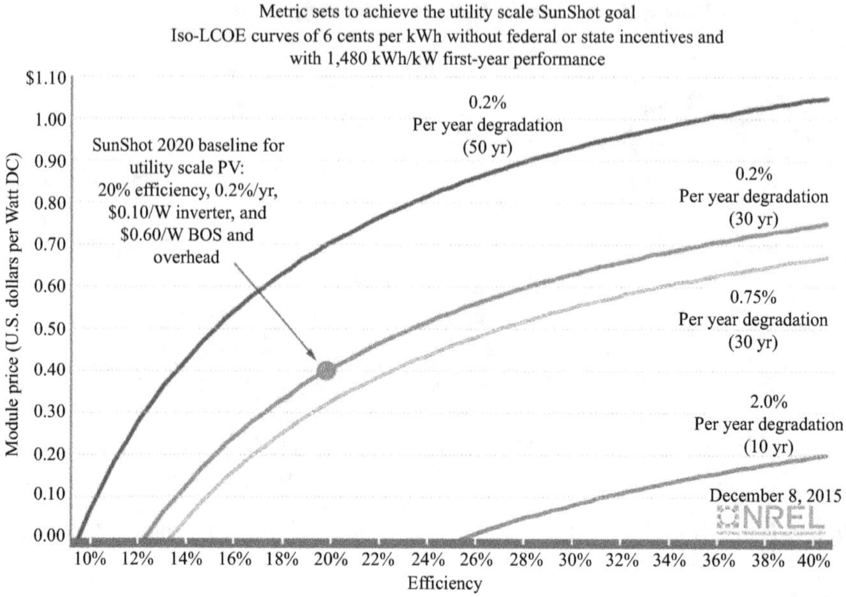

Figure 1.5 Sensitivity of PV-system costs to performance and reliability and lifetime (taken from [3])

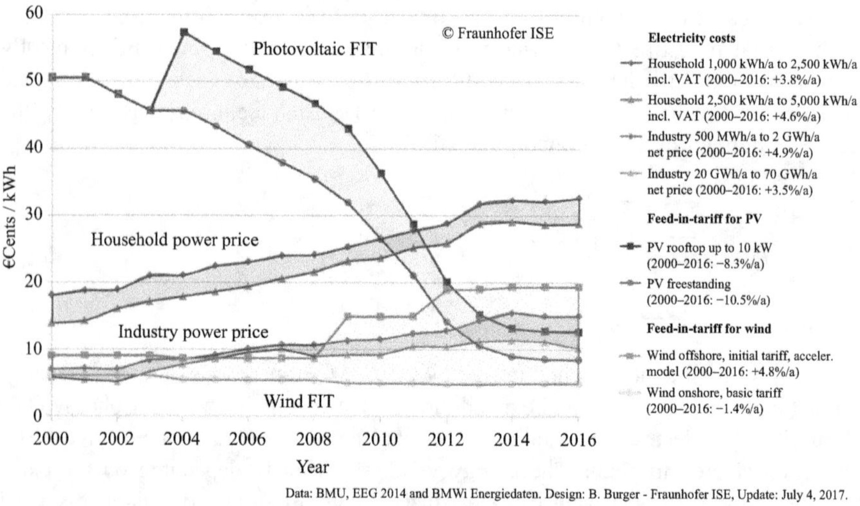

Figure 1.6 LCOE comparison for Germany between different electricity generation technologies (update by Fraunhofer ISE, July 4, 2017, [2])

investment cost as shown in the previous paragraph and increasing lifetime and energy yield to maximize the denominator of the LCOE expression. Nowadays a guarantee for the PV-module that it will deliver at least 80% of its nominal power after 25 years is typically given, but in order to bring the LCOE further down this

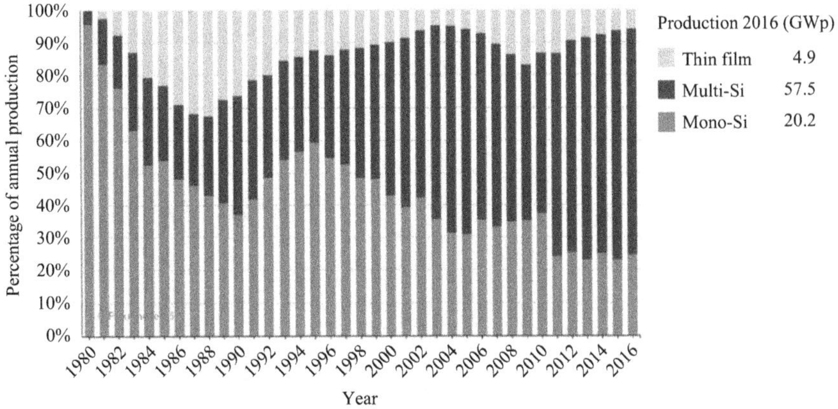

Figure 1.7 Market share taken by different PV-technologies over the last decades

should be extended to 30 years and more emphasizing the role of further reliability studies and improvement.

1.3 PV technologies portfolio

In terms of industrial PV technologies, one can distinguish nowadays between two basic types based on the different light absorbing semiconductor being used: technologies are either based on crystalline Si substrates and thin-film PV technologies. The largest share of the market is taken by crystalline Si solar cells (>95% of the market) and will be discussed in the next paragraph.

1.3.1 Thin-film PV-technologies

The second type of solar cell technology is based on semiconducting absorbers which have in common that, because of their much higher absorption coefficient, the absorbing layer is typically much thinner than in the case of crystalline Si. Typical layer thicknesses are in the range of 1 μm. As a result, these technologies are described as thin-film technologies. There are presently three different types of thin-film technologies. The first is based on hydrogenated amorphous Si layers (sometimes combined with amorphous SiGe-alloys). Industrial a-Si:H solar cells have a much lower efficiency in the range of 6%–7% and this technology might disappear as a result of this low efficiency (see the previous paragraph about the importance of performance). The second type of thin-film PV technology is based on polycrystalline compound solar cells. More specifically these materials are direct bandgap semiconducting materials like CdTe and compounds based on the Cu-In(Ga)-Se(S) compound (CIGS). CdTe has been the most important thin-film PV technology over the last decade, but is being challenged now by the CIGS-technology which is believed to have a higher efficiency potential and does not suffer from the presumed toxicity issues of CdTe. Presently the market share taken by TF-PV technologies is around 5% as shown in Figure 1.7, whereas the market share distribution over the respective thin-film PV-technologies is shown in Figure 1.8.

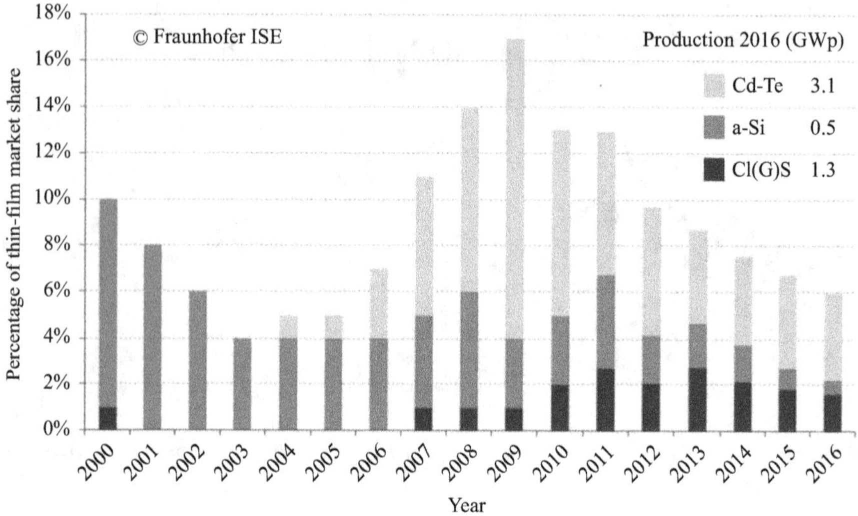

Figure 1.8 Distribution of market share over the different commercially available thin-film PV-technologies

1.3.2 Position of PERC solar cells in the silicon PV-technology market portfolio

The technology, dominating presently the PV-market, is based on crystalline Si and starts from essentially the same starting semiconductor material (polysilicon feedstock) as the microelectronics industry, albeit under the form of mono- and multicrystalline Si substrates which are (semi-)square, much thinner (typically 150 μm) and not polished as compared to their counterparts for microelectronic purposes. The Si-substrates, used in the PV-industry, are in most cases p-type with a resistivity of a few Ω-cm. Most of the cells are being manufactured by an industrial process with a limited number of process steps which are typically: removal of saw damage, P-diffusion to realize an n^+-emitter in the p-Si substrate, parasitic junction removal at the edge, deposition of the anti-reflection coating, screenprinting of Ag-fingers on the frontside and Al at the back followed by firing of the contacts to realize ohmic contacts to the emitter and base. This so-called full Al-BSF basic process results in cells with typical efficiencies of 15%–16.5% for multicrystalline Si solar cells and 17%–19.5% for monocrystalline Si solar cells, but no significant further efficiency improvements are expected for this industrial process as shown in Figure 1.9.

By improvements of the cell architectures and processes to reduce carrier recombination in the volume of the cell, passivation of the surfaces and shadowing losses in the cell and by novel contacting methods industrial cells up to 25% were shown. (Cell results above 25% are reported but these cell technologies are not yet on the market.) The world-record laboratory n-type IBC cell is from Kaneka with

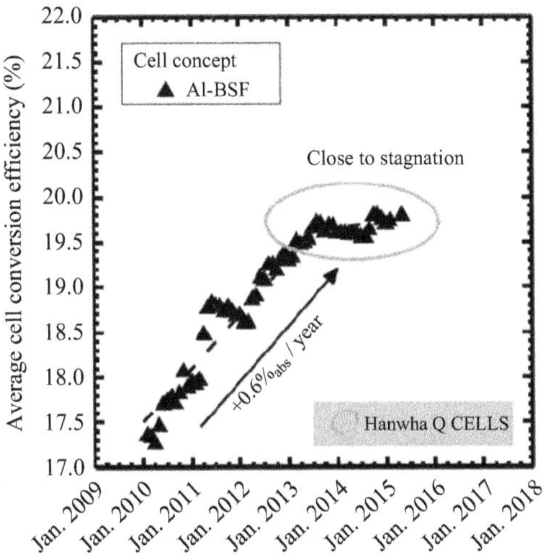

Figure 1.9 Evolution of industrial efficiencies for full Al-BSF cells (taken from [4])

26.6% efficiency, based on heterojunction technology. N-type IBC-cell modules are in the market but only make up a small market share due to their, up to 75%, higher cost. These high-performing cell processes and architectures are presently only manufactured by a limited number of industrial companies (SunPower, Panasonic, Trina Solar, LG, etc.). There is no doubt that these high-efficiency processes based on n-type Si substrates will make their way in the industry in the coming decade as shown in the ITRPV-roadmap [1] for industrial Si solar cell efficiencies.

It is clear from Figure 1.8 that in terms of next-generation technology, the PERC/PERT structure takes a prominent position in these roads to higher performance. The ITRPV cell technology roadmap, shown in Figure 1.8, reflects the consolidated expectations of the PV-industry in 2017. From Figure 1.10, it is clear that the PERC cell structure is expected to play an important role in further increasing the performance of cells in p-type Si substrates to the level of 22%–23% for large-area solar cells. Table 1.1 shows an overview of best large-area cell results based on the PERC/PERT cell architecture reported. The ultimate performance might be even higher when the possibility of very low surface recombination velocities enabled by Al_2O_3 is fully exploited and light-induced degradation by the presence of B-O-complexes in p-type Cz-Si could be fully avoided. This foresight puts a lot of pressure on similar cell architectures like n-PERT solar cells as the cost of the n-type Si substrate is typically 10%–20% higher as compared to p-type Si substrates.

These expected industrial cell perspectives, as outlined in the ITRPV-roadmap 2017, can be compared to Table 1.1 which shows the best efficiencies reported today for the different Si PERC/PERT technologies.

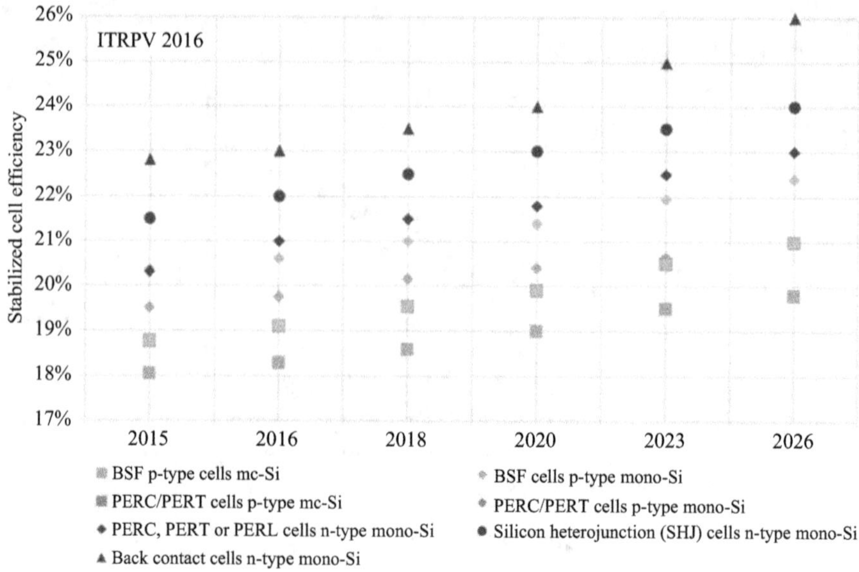

Figure 1.10 Efficiency roadmap for industrial crystalline Si solar cells as outlined in the ITRPV-roadmap (from www.ITRPV.net [1])

Table 1.1 Overview of best PERC/PERT large-area solar cell results (mono = monofacial, bi = bifacial)

Institute/ company	Cell type	Mat.	Size (cm^2)	J_{sc} (mA/ cm^2)	V_{oc} (mV)	FF (%)	Eta (%)	Ref.
Imec	nPERT (mono)	Cz	239	40.7	698	81.3	23.1	Aleman [5]
Imec	nPERT (Bi)	Cz	239	40.5	694	81.1	22.8	Russell [6]
Trina Solar	pPERC (mono)	Cz	243	?	?	?	22.6	Deng [7]
Trina Solar	nPERT (mono)	Cz	243	40.6	685	80.8	22.4	Duan [8]
Solarworld	pPERC (mono)	Cz	(large)	?	?	?	22.0	Press release 2016 [9]
ISFH	nPERT (bi)	Cz	(large)	40.3	673	80.3	21.8	Kiefer [10]
Trina Solar	nPERT (bi)	Cz	243	39.8	673	81.5	21.7	Chen [11]
REC	nPERT (bi)	Cz	(large)	40.4	664	79.8	21.4	Hsu [12]
Hanwha/ Crystal Solar	pPERC (mono)	Cz	239	40.1	671	79.6	21.4	Mertens 15 [13]
Gintech	pPERC (mono)	Cz	(large)	?	?	?	21.4	Press release 2016 [14]

1.3.3 Position of PERC solar cells on the PV-market

The PERC market capacity is currently more than 25 GWp and growing quickly towards 50 GWp of total capacity. Its share of the total PV-market is expected to increase to more than 50% of the total PV-market (which in terms of annual market

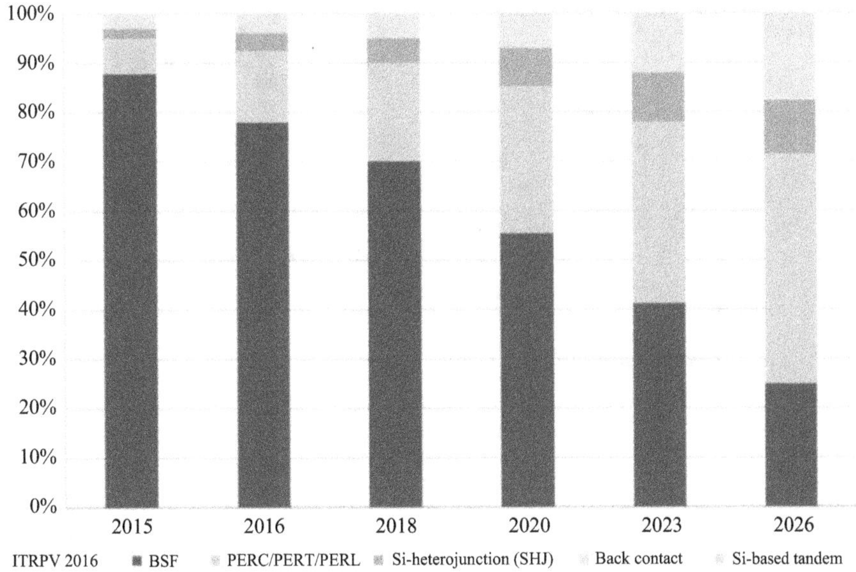

Figure 1.11 Expected technology portfolio evolution for crystalline Si solar cell technologies (taken from [1])

size is to grow to the range of a few 100 GW_p/year as expected in the ITRPV-2017 documents) as can be seen from Figure 1.11.

1.3.4 Energy yield increase by bifaciality and lower T-coefficients

Besides the performance increase enabled by PERC solar cells under standard conditions, one also has to include effects on yearly energy yield by the lower temperature coefficient of these cells and the better performance at lower illumination levels. A prominent example of efforts for increasing the energy yield is realized by the bifacial approach where the energy yield of the PERC-module can be increased by 10%–25%, depending on the albedo with a further concurrent decrease of the LCOE as a consequence. However, there are significant differences between the bifaciality factors of p-type PERC cells, which is typically in the range of 70%, and n-type PERT cells, where bifaciality is in the range of 80%–90% (up to >95% on highly symmetrical cell concepts enabled by co-plating [5]). Most bifacial modules are based on glass-glass modules, shown in Figure 1.12 which has the additional advantage of having very low degradation rates in the range of 0.1–0.2%/year. There is a noticeable exception which is Jolywood, which offers a module based on a transparent backsheet. In the slipstream of the industrial implementation of PERC, the bifacial variant, with partial aluminum metallization at the rearside, is emerging and is adopted by players such as JA Solar, Longi, Solarworld, TRINA and NSP. In the next decade, the market share taken by bifacial PV-modules is expected to increase significantly as shown in Figure 1.13. One of the attractive points of

Figure 1.12 Photograph of bifacial glass-glass module (www.greencoastsolarsystems.com)

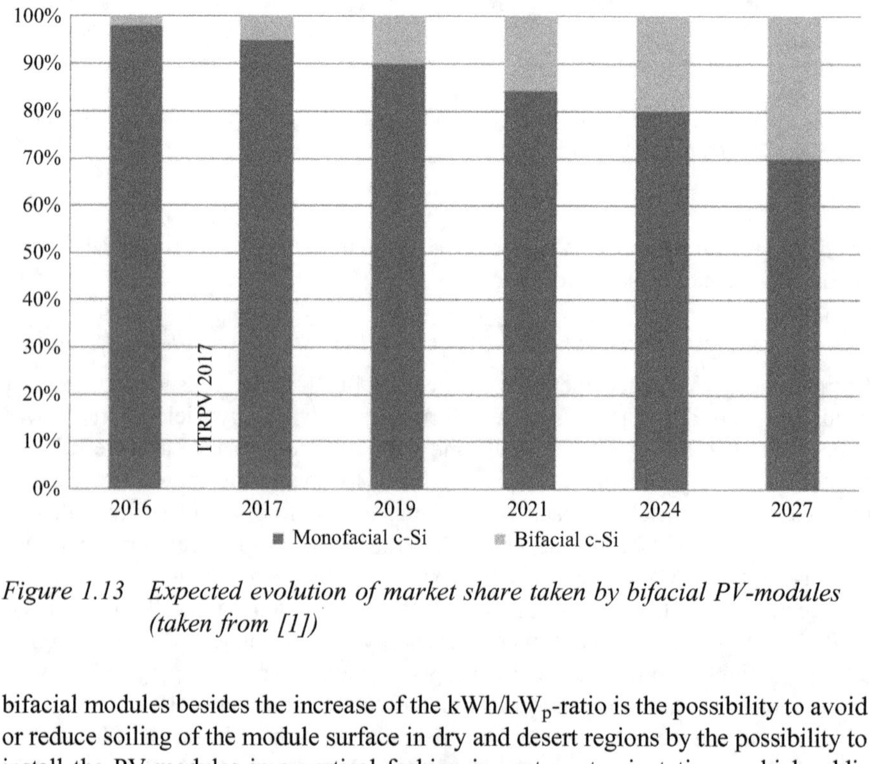

Figure 1.13 Expected evolution of market share taken by bifacial PV-modules (taken from [1])

bifacial modules besides the increase of the kWh/kW$_p$-ratio is the possibility to avoid or reduce soiling of the module surface in dry and desert regions by the possibility to install the PV-modules in a vertical fashion in east-west orientation—which additionally results in a flatter energy generation profile over the day which can simplify grid management.

There is also interest in differentiating products according to their performance and long-term reliability. As PV markets expand into locations with diverse climatic stresses, the challenge of being able to accurately predict long-term performance will

Table 1.2 Overview of T-coefficients of different Si solar cell technologies

Si cell technology	T-coefficient
Full Al-BSF	−0.41 to −0.45
p-type PERC cells	−0.37
n-type PERT cells	−0.34
n-type PERT (selective emitter)	−0.33
n-type heterojunction solar cells	−0.3
CdTe	−0.29
CIGS	−0.47
a-Si:H	−0.24

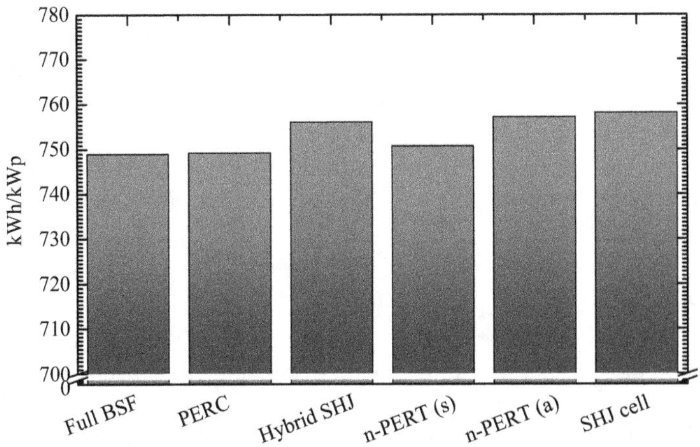

Figure 1.14 Annual energy yield comparison between different Si solar cell technologies for a location in Northern Germany [16]

increase. It becomes more and more clear that the reliability of PV-systems is strongly influenced by the local climatological conditions. In order to take these aspects into account, different climate typologies have been introduced: continental, desert, hot and humid, maritime, Mediterranean, polar and steppe and each of these climate topologies has a different influence on PV-module reliability [15]. Crucial parameters are then the T-coefficient of the power output under standard conditions and the behavior at lower illumination levels and in both aspects PERC solar cells enable a significant improvement relative to traditional solar cells with a full Al-BSF. In Table 1.2, one can find an overview of the experimentally extracted T-coefficients for the different Si solar cell technologies and in comparison those of thin-film PV technologies.

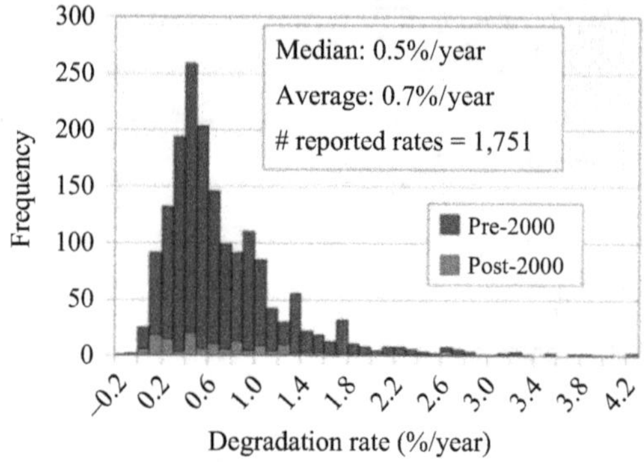

Figure 1.15 Distribution plot for degradation rate of crystalline Si PV-modules (taken from ref. [3])

Using the T-coefficients, we used high accuracy to estimate difference in energy yield over a year with the meteorological and irradiation conditions prevailing at a German location (Oldenburg) and measured with high time granularity. The resulting energy yield is shown in Figure 1.14. It is clear that especially the n-PERT case with selective emitter is particularly promising in the sense that it is significantly better than the full Al-BSF and nearly as attractive as the heterojunction cell technology.

1.3.5 Stability

It is striking that, despite the nearly 1,000-fold growth of the PV-industry since the beginning of the 1980s, crystalline Si PV-technology has been dominating the PV-market with a market share which is still as high as 95% nowadays. This dominance is not expected to significantly change over the next decade. This dominance in terms of market share is strongly related to the stability of PV-modules based on crystalline Si. As shown in a broad study by the National Renewable Energy Laboratory (NREL) [17], degradation of thin-film PV modules is still larger as compared to crystalline Si-PV-modules as shown in Figure 1.15 and Figure 1.16.

Most of these PV-modules in the NREL-study are however still based on full-Al BSF cells. In this context, it is important to keep the well-documented stability of crystalline Si solar cells for the upcoming PERC/PERT-technologies. It has been observed for p-PERC solar cells that these are subject to light-induced degradation. The physical cause behind this phenomenon is a decrease of the minority carrier lifetime induced by the excess carrier generation when light is absorbed [18]. Under the influence of excess minority carriers and their recombination, B-O complexes are being formed. A determining factor in the case of monocrystalline Si substrates is the presence of B and O in the wafer [16,19] as shown in Figure 1.17.

Whereas for monocrystalline Si substrates the phenomenon is relatively well understood, this is definitely not the case for multicrystalline Si solar cells.

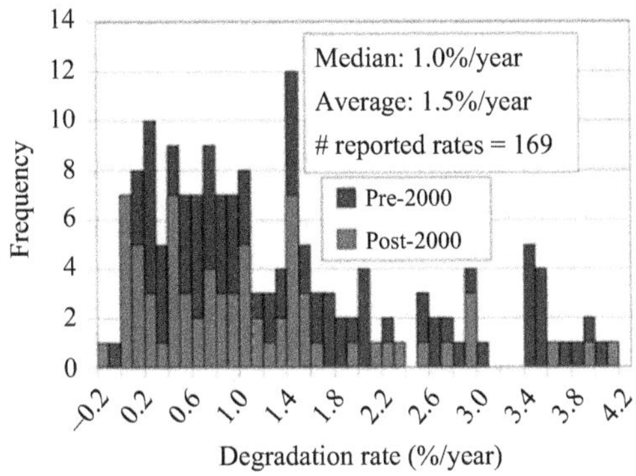

Figure 1.16 Distribution plot for degradation rate of thin-film PV-modules (taken from ref. [3])

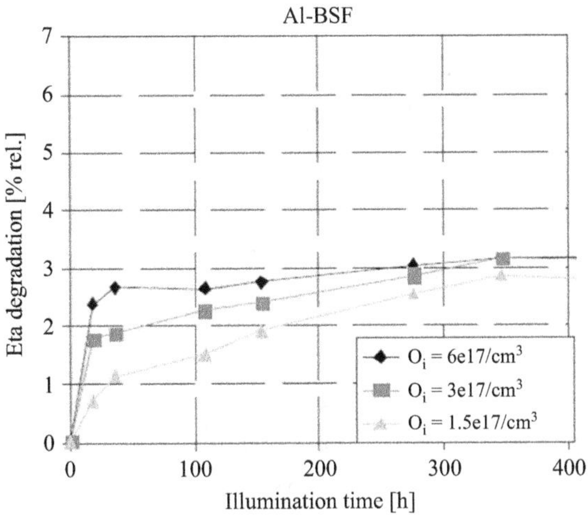

Figure 1.17 Effect of [O]-concentration on performance degradation (taken from [17])

For multicrystalline Si PERC solar cells, unexpectedly strong light-induced-degradation is found when illuminated at temperatures above 50 °C. This seemed to be independent from the choice of material and is not related to boron–oxygen complex formation, neither by iron–boron pair dissociation processes.

For monocrystalline Si solar cells, this effect can be mitigated by using Si-materials with $[O_i] < 3 \times 10^{17}$ cm^{-3} or by using n-type Si substrates. An alternative solution for p-type Si substrates consists in applying a light-induced regeneration treatment for p-Si [20]. It consists of a treatment at 140–150 °C under intense light (>1 sun). Also the presence of hydrogen during the treatment seems to play an

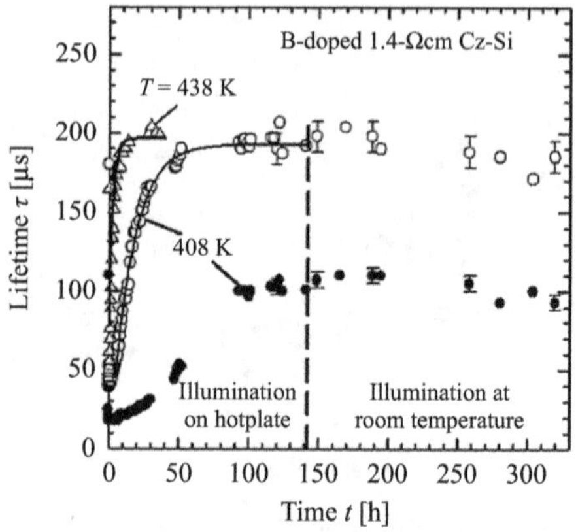

Figure 1.18 Effect of illumination on minority carrier lifetime (taken from [18])

important role [21]. After the treatment, the minority carrier lifetime is recovered and the material remains stable afterwards as can be seen in figure 1.18.

But it is clear that this additional treatment will increase the production cost of the solar cells whereas the long-term stability after the treatment is still the subject of discussion and experimental investigation. The effect of the treatments is influenced by a wide range of device-related parameters and the carrier density distribution within the silicon wafer. Also the thermal history of the wafer plays an important role as well as substrate hydrogenation. All these factors complicate the final conclusions to be drawn from the presently running investigations.

1.3.6 Major players on the market for PERC-based PV-modules

The market makes a basic distinction between mono- and multicrystalline Si PV-modules. The multicrystalline PV-modules are presently dominating the market, but the monocrystalline share is expected to grow in the next 5 years as to reach parity at the end of this period. This is based on two evolutions: the first is related to the adoption by monocrystalline Si wafer manufacturers of diamond-coated wire sawing with lower costs as compared to slurry-based sawing. The second is based on the obvious trend toward higher efficiencies where the potential of mono-crystalline Si is higher than multicrystalline Si. The observed efficiency gain for PERC-like monocrystalline Si solar cells is presently in the range of 0.8%–1%, whereas the gain is more modest in the range of 0.4%–0.8% for multicrystalline Si solar cells. Looking at the commercial solar module level, the highest efficiencies using traditional monocrystalline Si technology is around 18% (see, e.g., the solar modules of Longi). With PERC-technology this goes up to values slightly below 19% (see e.g. the solar modules offered by Trina Solar). In Figure 1.19, the top solar modules for different commercial manufacturers are shown and again it is clear that the PERC-technology is strongly enabling.

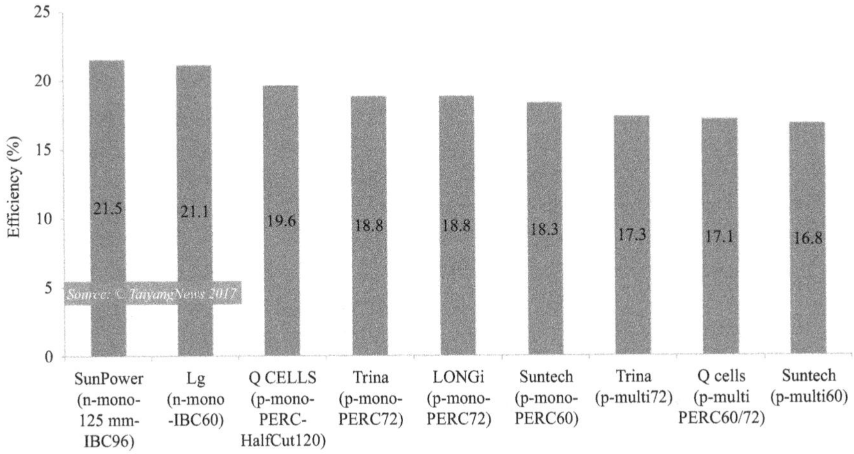

Figure 1.19 Top solar module efficiencies of leading manufacturers

References

[1] International Technology Roadmap for Photovoltaics, Eight edition, September 2017. The document can be found at http://itrpv.net/Reports

[2] Photovoltaics report, prepared by Fraunhofer Institute for Solar Energy with support of PSE AG, July 4, 2017. The report can be found at https://www.ise.fraunhofer.de/content/dam/ise/de/documents/publications/studies/Photovoltaics-Report.pdf

[3] R. Jones-Albertus, D. Feldman, R. Fu, *et al.*, "Technology advances needed for photovoltaics to achieve widespread grid price parity", Progress in Photovoltaics: Research and Applications, Volume 24, Issue 9, 2016, 1272–1283.

[4] F. Fertig, A. Lantzsch, A. Mohr, *et al.*, "Mass production of p-type Cz-Silicon solar cells", Energy Procedia, Volume 124, 2017, 338–345.

[5] F. Duerinckx, M. Aleman, and C. Jia, "2-side contacted n-type cells: The route from nPERT towards poly-Si passivating contacts", nPV-workshop, Lausanne, March 2018.

[6] R. Russell, L. Tous, E. Cornagliotti, *et al.*, "Simultaneous fabrication of n & p contacts for bifacial cells by a novel co-plating process", Proceedings of the EU-PVSEC, 2017, Amsterdam, September 25–29, pp. 212–217, 2017.

[7] W. Deng, F. Ye, R. Liu, *et al.*, "22.61% Efficient fully screen printed PERC solar cell", 44th IEEE Photovoltaic Specialists Conference, Washington, June 25–30, 2017.

[8] W. Duan, S. Yuan, Y. Sheng, *et al.*, "A route towards high efficiency n-type PERT solar cells", Proceedings of the EU-PVSEC, Munich, June 20–24, 2016, pp. 399–402, 2016.

[9] SolarWorld, Press Release, Bonn, May 18, 2016, "SolarWorld presents the Bisun System: bifacial solar cell technology combined with optimized

racking and 300 watt high performance module at Intersolar Europe", 2016.

[10] F. Kiefer, J. Kruegener, F. Heinemeyer, *et al.*, "Bifacial fully screen-printed n-PERT solar cells with BF_2 and B-implanted emitters", Solar Energy Materials and Solar Cells, 157, p. 326, 2016.

[11] Y. Chen, S. Yuan, Y. Cui, *et al.*, "Opportunity and challenge: >21% large-area n-type PERT bifacial solar cells in research and production", presentation can be found at http://bifipv-workshop.com/fileadmin/images/bifi/miyazaki/presentations/, 2016.

[12] A. Hsu, "315W 60-Cell bifacial module using n-type cells", presentation can be found at http://bifipvworkshop.com/fileadmin/images/bifi/miyazaki/presentations/, 2016.

[13] V. Mertens, S. Bordihn, A. Mohr, *et al.*, "21.4 % Efficient fully screen printed n-type solar cell on epitaxially grown silicon wafers with built-in boron rear side emitter", Proceedings of the EU-PVSEC, 2015, Hamburg, September 14–18, pp. 1000–1002, 2015.

[14] Gintech, Press Release, January 6, 2016, "Gintech unveils new solar cells technology, Records 21.44% maximal efficiency-new record for solar tech at a wide scale", 2016.

[15] D. C. Jordan, S. R. Kurtz, K. VanSant, and J. Newmiller, "Compendium of photovoltaic degradation rates", Progress in Photovoltaics: Research and Applications, Volume 24, Issue 7, 2016, 978–989.

[16] H. Goverde, D.G. Anagnostos, J. Govaerts, *et al.*, "Accurately simulating PV energy production: Exploring the impact of module build-up", Proceedings of the EU-PVSEC, 2017, Amsterdam, September 25–29, pp. 1643–1646, 2017.

[17] D. C. Jordan and S. R. Kurtz, "Photovoltaic degradation rates—An analytical review", Progress in Photovoltaics: Research and Applications, Volume 21, 2013, 12–29.

[18] J. Schmidt, A. Cuevas, S. Rhein, and S. W. Glunz, "Impact of light-induced recombination centres on the current–voltage characteristic of Czochralski silicon solar cells", Progress in Photovoltaics, Volume 9, Issue 4, 2001, 249–255.

[19] K. Ramspeck, S. Zimmermann, H. Nagel, *et al.* "Light induced degradation of rear passivated mc-Si solar cells", Proceedings of the 27th EU-PVSEC, Frankfurt, September 2012, 861–865 or S. Dubois, F. Tanay, J. Veirman, *et al.*, "The Bolid Project: Suppression of the boron-oxygen related light-induced degradation: Main Results", Proceedings of the 27th EU-PVSEC, Frankfurt, September 2012, pp. 749–754.

[20] B. Lim, K. Bothe, and J. Schmidt, "Deactivation of the boron–oxygen recombination center in silicon by illumination at elevated temperature", Physica Status Solidi (RRL) 2, No. 3, 93–95 (2008)/DOI 10.1002/pssr.200802009

[21] H. Li, J. Xu, K. Chen, *et al.*, "Comparison of industrial solutions to light induced degradation of high efficiency Cz PERC cells", Proceedings of the EU-PVSEC, 2017, Amsterdam, September 25–29, pp. 983–985, 2017.

Chapter 2

Introduction to surface passivation of industrial crystalline silicon solar cells

Armin G. Aberle[1]

In the last 200 years, the global energy system has primarily relied on fossil fuels. Unfortunately, fossil fuels are polluting the environment and the air which we breathe. Owing to atmospheric pollution, the Earth's temperature is rising, deserts are expanding, fresh water supplies are dwindling and sea levels are rising. In parallel to these alarming trends, the global population increases by about 80 million every year – approximately the size of Germany – and with it the global energy demand keeps going up. To maintain liveable conditions for future generations, mankind urgently needs a new energy system that is based on clean and renewable energies.

Fortunately, there is hope from a technological perspective. Rapid progress in the past 40 years with the costs of wind power and solar power (see Figure 2.1, [1]) provides opportunities for massive changes on a global scale. Since the 1980s, the globally deployed generation capacity of solar photovoltaic (PV) systems has grown exponentially to reach close to 300 GW_p (Gigawatt-peak) in 2016. It will exceed the global installed capacity of nuclear power plants in 2018, will exceed 1,000 GW_p in 2023 and will reach tens of thousands of GW_p in 2050. A similar growth scenario is possible for wind power. Grid parity has already been reached for wind and solar power in many countries, and every year their costs are being reduced further. In very sunny places such as the United Arab Emirates, PV electricity can now be generated at costs of less than 3 US cents per kWh [2].

Of course, despite the excellent progress, the transition to renewables will not be easy. As we all know, the intensity of sunlight varies during the day due to clouds, and wind speeds are fluctuating. There are also periods where there is no wind and no sunshine. Nevertheless, by clever monitoring, weather forecasting, power demand management, energy efficiency, energy storage and flexible gas turbine power plants, the transition to smart energy grids and a cleaner and smarter energy world will be possible.

The main technology responsible for the phenomenal progress with the costs of solar PV electricity has been solar cells (and PV modules) based on crystalline

[1]Solar Energy Research Institute of Singapore, National University of Singapore, 7 Engineering Drive 1, Singapore

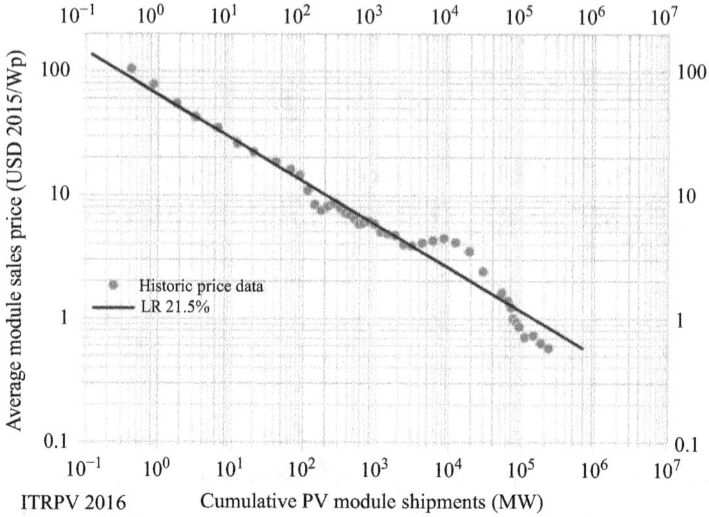

Figure 2.1 Average PV module sales price (in $\$/W_p$, symbols) vs. cumulative PV module shipments ('learning curve'), from 1976 to December 2015. As shown by the solid line, the learning rate has been about 21.5% over this 39-year period. The learning rate is the relative reduction in PV module sales price for every doubling of the cumulative PV module shipments. From Ref. [1]

silicon (Si) wafers. The solar cells with the largest market share (>80%) are made from p-type Si wafers (either multicrystalline or monocrystalline) and are called 'Al-BSF cells' due to a p^+ doped back surface field (BSF) layer [3] at their rear surface formed by screen printing of an aluminium paste and subsequent 'firing' in a belt furnace at high temperature (>750 °C) for a few seconds to form a molten Si–Al alloy. A schematic cross section of an Al-BSF solar cell is shown in Figure 2.2(a). The p^+ BSF layer serves three purposes: (i) to provide a low-resistance ohmic contact between the rear metal and the c-Si wafer, (ii) to suppress rear surface recombination losses of minority carriers and (iii) to getter impurities from the bulk of the cell, boosting the cell's effective minority carrier lifetime. The fact that the most successful solar cell ever produced (>250 GW_p) has been given this name by the PV community is testimony of the enormous importance of surface passivation for silicon solar cells. The effective surface recombination velocity of an Al-BSF structure (i.e., a pp^+/Al stack) on a p-type Si wafer is typically of the order of 10^3 cm/s [4], and thus at least 10 times lower than that of a p/Al structure.

The main reasons for the massive cost reductions of Al-BSF silicon solar cells and modules over the past 40 years have been economies of scale, improved and cheaper raw materials (poly-Si feedstock, silicon wafers, encapsulants, backsheets, etc.), a cheaper wafering method (wire sawing instead of blade sawing), the development of screen-printable silver metal pastes that are able to contact phosphorus-diffused n^+ emitters with high sheet resistance (>80 Ω/square) and thus

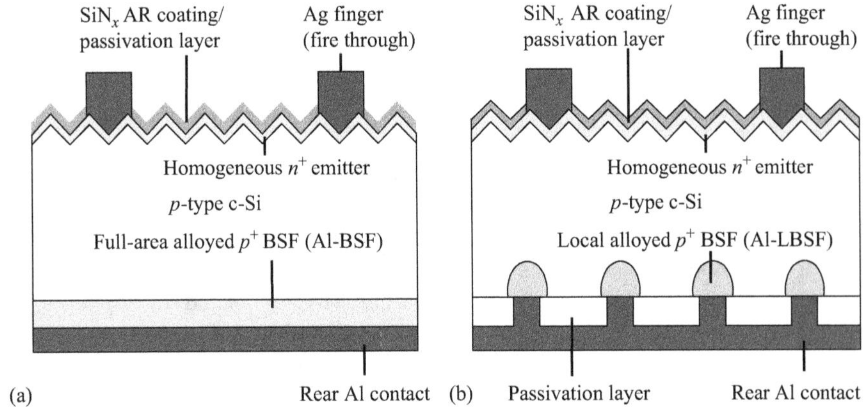

Figure 2.2 Schematic cross sections of (a) an industrial Al-BSF silicon solar cell and (b) an industrial monofacial Al-LBSF or PERC silicon solar cell. The n^+ emitter is realised via a phosphorus (P) diffusion process. The contacts on the front and rear surfaces are formed by screen printing and firing of metal pastes

better blue response and, last but not least, improved front surface (and bulk) passivation via hydrogen-rich amorphous silicon nitride films (a-SiN$_x$:H or, in short, SiN$_x$) deposited by plasma-enhanced chemical vapour deposition (PECVD). Prior to the PV industry's shift to plasma silicon nitride in the 1990s [4–9], thermally grown SiO_2 or atmospheric pressure chemical vapour deposition (APCVD) TiO_2 passivation layers were used, both of which were a compromise in either antireflection (AR) performance (SiO_2) or passivation performance (TiO_2). In the case of TiO_2, the screen-printed contacts could be applied either before or after the AR coating using fire-through pastes [10,11]. PECVD SiN$_x$ simultaneously provides excellent passivation of phosphorus diffused emitters and excellent AR performance (due to its refractive index in the 2.0–2.2 range at a wavelength of 600 nm). Moreover, with the advent of fire-through pastes for SiN$_x$ [6], the presence of hydrogen in the SiN$_x$ film enables silicon bulk passivation by effusion during co-firing, which is especially beneficial for multi-Si cells [7,12]. The implementation of plasma SiN$_x$ as a 'surface and bulk passivating AR coating' [4] not only provided a general efficiency boost for silicon wafer solar cells but also significantly narrowed the efficiency gap between multi-Si and mono-Si solar cells and thereby provided increased viability for the cheaper multi-Si wafers. As a result of all these innovations, the average sales price of PV modules made from Al-BSF Si solar cells had fallen to 0.58 \$/W$_\text{p}$ in 2015 (see Figure 2.1), with module efficiency typically around 16%.

In research labs, it has long been known that dielectric films provide far superior electronic passivation to alloyed aluminium. For example, interdigitated back contact (IBC) silicon solar cells [13,14] rely upon dielectric passivation of most of both surfaces. It thus makes sense to integrate the benefits of dielectric surface passivation into the rear surface of an alloyed BSF solar cell. However, an important inhibiting factor in the 1980s in removing the alloyed aluminium layer was the poor cleanliness of process furnaces in most laboratories. The alloyed Al layer is an excellent sink for impurities, allowing moderate minority carrier lifetimes to be achieved even with dirty furnaces.

It seems that, in 1988, researchers at Fraunhofer ISE in Germany were the first to publish results on this solar cell structure [15]. They used an H-pattern type alloyed rear Al electrode and coined the name 'local back surface field cell' (LBSF cell). The exposed rear surface of their bifacial solar cell was passivated using a high-temperature oxidation process which simultaneously created the alloyed local BSF. The modest open-circuit voltage achieved (630–650 mV [16]) is indicative of the difficulty experienced at the time in maintaining furnace cleanliness.

Research in the late 1980s by Blakers *et al.* in Martin Green's group at the University of New South Wales (UNSW) demonstrated 22.8% efficiency in 1989 for the so-called Passivated Emitter and Rear Cell (PERC) [17,18]. Importantly, high furnace cleanliness was maintained through the inclusion of a small amount of trichloroethane (a source of chlorine gas) in the furnace ambient, as shown by the high open-circuit voltage obtained (705 mV). The monofacial PERC cell combined rear oxide passivation with an evaporated Al layer, which provided both low surface recombination and high rear reflectance. Metal/silicon contact recombination at the rear was kept low by a matrix of point-like openings in the oxide (occupying ~1% of the rear surface) that were formed prior to Al evaporation. Both non-alloyed and alloyed Al back contacts were investigated, with non-alloyed contacts giving higher solar cell efficiency [19]. The non-alloyed contacts were formed by a 15-minute thermal anneal ('sintering') in forming gas at a temperature of about 400 °C [18], which also acted as an anneal step to reduce the recombination rate at the silicon-oxide interface. Boron diffusions into point-openings in the rear surface oxide were also utilised in some devices, giving open-circuit voltages of up to 675 mV [19].

In 1996, Aberle *et al.* at the German research institute ISFH replaced the high-temperature grown thermal oxide by low-temperature plasma-deposited silicon nitride, which had been shown to provide very good surface passivation on both p and n-type c-Si surfaces [20,21]. Good efficiencies of over 19% were realised for front and rear SiN_x passivated bifacial Al-LBSF cells with evaporated electrodes [21].

Although lab-scale Al-LBSF and PERC solar cells were already made in the 1980s, the concept only gained industrial attention around 2008. The main reason was the emergence of improved screen-printable Al pastes that can produce a good-quality local BSF without destroying the passivation quality of the rear dielectric film (which depends to a significant degree on the field-effect passivation mechanism provided by fixed electrical charges in the dielectric [22,23]). Thus, using laser ablation for local opening of the dielectric film, large-area monofacial Al-LBSF cells or PERC cells could be realised by means of a screen-printed and fired Al paste, see Figure 2.2(b). By 2011, several institutions had achieved pilot-scale screen-printed versions of such cells with efficiencies of over 19%, for example SERIS [24], Schott Solar [25], Q-Cells [26], ISFH [27] and Suntech [28].

The best results were obtained using a PECVD AlO_x/SiN_x passivation stack at the rear. The first industrial-scale Al-LBSF/PERC solar cell was probably Centrotherm's Centaurus cell, which in 2012 achieved above 19.5% efficiency using an SiO_2/SiN_x passivation stack at the rear [29,30]. In the following years, numerous solar cell manufacturers established this solar cell technology in their factories, resulting in a global manufacturing capacity of about 12 GW_p at the end of 2016 [31]. If this trend continues in the next few years, the Al-LBSF/PERC cell will replace the Al-BSF cell as the PV industry's 'standard' solar cell. The author estimates that, by 2020, PV modules based on multi-Si Al-LBSF/PERC cells will sell for less than 0.3 $\$/W_p$, with module efficiency of around 19%. In the PV industry, this solar cell structure is now generally referred to as PERC solar cell [1].

In parallel to multi-Si PERC solar cells, advanced monocrystalline silicon (mono-Si) solar cell architectures and new types of PV modules (for example, using half-cut or bifacial solar cells) will be industrially fabricated in increasing volumes in the coming years, enabling more than 21% module efficiency at module prices of below 0.35 $\$/W_p$ in 2020. Examples for these solar cells are p-mono PERC cells and n-mono phosphorus BSF cells (now often termed PERT cells, for Passivated Emitter, Rear Totally diffused [32]). Even higher module efficiencies of more than 23% will be realised using heterojunction cells [33] and IBC cells [13,14]. All these cell architectures require advanced or novel surface passivation techniques, for example, heterojunction stacks or tunnel oxide passivated contact schemes [34,35].

As described above, surface passivation technologies have played a crucial role in the development of industrial silicon solar cells over the past 40 years, and they will continue to do so in the future. The various chapters of this book describe a variety of surface passivation techniques that will help to take industrial silicon solar cells to even greater heights, ensuring ongoing reductions of the levelized cost of electricity (LCOE) from PV systems and their global deployment on a massive scale.

References

[1] International Technology Roadmap for Photovoltaic (ITRPV), Seventh Edition 2016, http://www.itrpv.net/Reports/Downloads (10 Nov 2016).

[2] "Cheapest Solar on Record Offered as Abu Dhabi Expands Renewables", Bloomberg, http://www.bloomberg.com/news/articles/2016–09–19/cheapest-solar-on-record-said-to-be-offered-for-abu-dhabi (20 Sep 2016).

[3] Mandelkorn J. and Lamneck J.H., Simplified fabrication of back surface electric field silicon cells and novel characteristic of such cells, Proc. 9th IEEE Photovoltaic Specialists Conference, New York, p. 66 (1972).

[4] Aberle A.G., Surface passivation of crystalline silicon solar cells – A review, Progress in Photovoltaics, 8, 473–487 (2000).

[5] Jäger K., Roth P., Schum B., Bende G., Hoffmann W., and Hezel R., Thin crystalline silicon 100 mm × 100 mm MIS-inversion layer solar cells with bifacial sensitivity, Proc. 11th European Photovoltaic Solar Energy Conference, Montreux, Switzerland, pp. 168–171 (1992).

[6] Electrical contacts and method of manufacturing same, International Patent Application PCT/US91/06445, Mobil Solar Energy Corporation, Billerica, USA (2 Apr 1992).
[7] Ruby D.S., Wilbanks W.L., Fleddermann C.B., and Hanoka J.I., The effect of hydrogen-plasma and PECVD-nitride deposition on bulk and surface passivation in string-ribbon silicon solar cells, Proc. 13th European Photovoltaic Solar Energy Conference, Nice (Stephens, Bedford, 1995), p. 1412 (1995).
[8] Aberle A.G., Lauinger T., and Hezel R., Remote PECVD silicon nitride – A key technology for the crystalline silicon PV industry of the 21st century?, Proc. 14th European Photovoltaic Solar Energy Conference, Barcelona, pp. 684–689 (1997).
[9] Aberle A.G., Overview on SiN surface passivation of crystalline silicon solar cells, Solar Energy Materials and Solar Cells, 65, 239–248 (2001).
[10] Cheek G.C., Mertens R.P., Van Overstraeten R., and Frisson L., Thick-film metallization for solar cell applications, IEEE Transactions on Electron Devices, 31, 602–609 (1984).
[11] Mitchell K.W., Aldrich D.L., Walle J.R., *et al.*, Characterization of commercial single crystal Si cell technology, Proc. 10th European Photovoltaic Solar Energy Conference, Lisbon, Portugal, p. 310 (1991).
[12] Winderbaum S., Leo A.J., Shea S.P., Koval T.D., and Kumar B., Characterisation of industrial-scale remote PECVD SiN depositions, Proc. 3rd World Conference on Photovoltaic Energy Conversion, Osaka, Japan, pp. 1128–1130 (2003).
[13] Sinton R.A., Kwark Y., Gan J.Y., and Swanson R.M., 27.5-percent silicon concentrator solar cells, IEEE Electron Device Letters, 7, 567–569 (1986).
[14] Verlinden P., Vande Wiele F., Stehelin G., and David J.P., An interdigitated back contact solar cell with high efficiency under concentrated sunlight, Proc. 7th European Photovoltaic Solar Energy Conference, Sevilla, Spain, pp. 885–889 (Oct 1986).
[15] Knobloch J., Aberle A.G., Warta W., and Voss B., Dependence of surface recombination velocities at silicon solar cell surfaces on incident light intensity, Proc. 8th European Photovoltaic Solar Energy Conference, Florence, Italy, pp. 1165–1170 (May 1988).
[16] Aberle A.G., Investigations on surface passivation of high-efficiency silicon solar cells, PhD thesis (in German), University of Freiburg, Germany (Sep 1991).
[17] Blakers A.W., Wang A., Milne A.M., Zhao J., Dai X. and Green M.A., 22.6% efficient silicon solar cells, Conf. Record 4th International Photovoltaic Science and Engineering Conference, Sydney, p. 801 (Feb 1989).
[18] Blakers A.W., Wang A., Milne A.M., Zhao J., and Green M.A., 22.8% efficient silicon solar cell, Applied Physics Letters, 55, 1363 (1989).
[19] Blakers A.W., Australian National University, Canberra, private communication, November 2016.

[20] Lauinger T., Schmidt J., Aberle A.G., and Hezel R., Record low surface recombination velocities on 1 Ohmcm p-silicon using remote plasma silicon nitride passivation, Applied Physics Letters, 68, pp. 1232–1234 (1996).

[21] Hübner A., Aberle A.G., and Hezel R., Novel cost-effective bifacial silicon solar cells with 19.4% front and 18.1% rear efficiency, Applied Physics Letters, 70, pp. 1008–1010 (1997).

[22] Aberle A.G., Glunz S., and Warta W., Impact of illumination level and oxide parameters on Shockley-Read-Hall recombination at the Si-SiO_2 interface, Journal of Applied Physics, 71, 4422–4431 (1992).

[23] Aberle A.G., Glunz S., and Warta W., Field effect passivation of high efficiency silicon solar cells, Solar Energy Materials and Solar Cells, 29, 175–182 (1993).

[24] Boreland M.B. and Aberle A.G., Pilot line processing of silicon wafer solar cells using industry-scale equipment, Proc. 26th European Photovoltaic Solar Energy Conference, Hamburg, Germany, pp. 1996–2000 (Sep 2011).

[25] CleanTechnica, "Screen-Printed Solar Cell Sets New Efficiency Record – 20.2%", https://cleantechnica.com/2011/08/25/screen-printed-solar-cell-sets-new-efficiency-record-%E2%80%93–20–2/ (25 Aug 2011).

[26] Engelhart P., Manger D., Klöter B., *et al.*, QANTUM – Q-Cells next generation high-power silicon cell & module concept, Proc. 26th European Photovoltaic Solar Energy Conference, Hamburg, Germany, pp. 821–826 (Sep 2011).

[27] Institut für Solarenergieforschung ISFH, "20.1%-efficient screen printed silicon solar cell with novel rear passivation layer", press release, Hamelin, Germany, http://www.isfh.de/institut_solarforschung/pressemitteilungen.php?_l=1 (2 Mar 2012).

[28] Suntech Power Holdings, "Suntech sets world record 20.3% efficiency for Pluto cell technology", press release, http://am.suntech-power.com/, http://www.prnewswire.com/news-releases/suntech-sets-world-record-203-efficiency-for-pluto-cell-technology-142301885.html (12 Mar 2012).

[29] Münzer K.A., Schöne J., Teppe A., *et al.*, Rear side passivated and locally contacted solar cells with laser diffused selective emitter, Energy Procedia, 15, 1–9 (2012).

[30] Centrotherm Photovoltaics AG, "Centrotherm Photovoltaics' Centaurus technology for high efficiency solar cells successfully transferred into industrial-scale production", press release, www.centrotherm-pv.com, http://www.presseportal.de/pm/68053/2253243/eans-news-centrothermphotovoltaics-centaurus-technology-for-high-efficiency-solar-cells (15 May 2012).

[31] "IHS Markit: PERC and N-type technologies to greatly reduce market share of traditional solar cells by 2020", Solar Server Global Solar Industry website, http://www.solarserver.com/solar-magazine/solar-news/current/2016/kw35/ihs-markit-perc-and-n-type-technologies-to-greatly-reduce-market-share-of-traditional-solar-cells-by-2020.html (1 Sep 2016).

[32] Zhao J., Wang A., and Green M.A., High-efficiency PERL and PERT silicon solar cells on FZ and MCZ substrates, Solar Energy Materials and Solar Cells, 65, 429–435 (2001).
[33] Taguchi M., Kawamoto K., Tsuge S., *et al.*, HIT^{TM} Cells – High-efficiency crystalline Si cells with novel structure, Progress in Photovoltaics, 8, 503–513 (2000).
[34] Blakers A.W. and Green M.A., 678-mV open-circuit voltage silicon solar cells, Applied Physics Letters, 39, 483 (1981).
[35] Feldmann F., Bivour M., Reichel C., Steinkemper H., Hermle M., and Glunz S.W., Tunnel oxide passivated contacts as an alternative to partial rear contacts, Solar Energy Materials and Solar Cells, 131, 46–50 (2014).

Chapter 3
Material properties of AlO_x for silicon surface passivation

Bram Hoex[1]

3.1 Introduction

Aluminium oxide (Al_2O_3) has in the recent years emerged as the *de facto* standard film to passivated lightly and highly doped *p*-type c-Si surfaces such as used in *p*-type PERC (passivated emitter and rear contact) and diffused *n*-type solar cell. The potential of Al_2O_3 was already recognised in the late 1980s by Hezel and Jaeger [1]. However, the results were not yet outstanding at that time, and these results went mostly unnoticed. It was only in 2006 when the work by Agostinelli *et al.* [2] and Hoex *et al.* [3] ignited a significant interest from the photovoltaic field. This significant interest was mainly due to the excellent quantitative results demonstrated on dedicated test samples [3–7] and solar cell devices [8,9]. Subsequently, hundreds of scientific papers have been published about Al_2O_3 c-Si surface passivation. In this chapter, I will predominantly focus on the surface passivation mechanism of Al_2O_3 and how it is related to its material properties. The chapter starts off with a short description of the fundamentals of surface recombination stressing the importance of the electronic properties of the c-Si/Al_2O_3 interface with respect to defect density and fixed charge. Next, an overview is presented of the electronic properties for Al_2O_3 layers synthesised on c-Si substrates by various techniques. Finally, these electronic properties are related to the atomic-scale material properties for Al_2O_3 films with a particular focus on the, typically unintentional, presence of an interfacial silicon oxide film.

3.2 Surface recombination fundamentals

The recombination at a crystalline silicon surface can be described by the Shockley–Read–Hall (SRH) mechanism [10,11]. The recombination rate U_s at the surface for a single defect can be expressed as

$$U_s = \frac{\left(n_s p_s - n_i^2\right) v_{th} N_{it}}{\frac{n_s + n_1}{\sigma_p} + \frac{p_s + p_1}{\sigma_n}}, \tag{3.1}$$

[1]School of Photovoltaic and Renewable Energy Engineering, University of New South Wales, Australia

where n_s and p_s are the electron and hole concentration at the surface, n_i is the intrinsic carrier density, v_{th} is the thermal velocity, N_{it} is the interface defect density, n_1 and p_2 are statistical factors for electrons and holes, and σ_n and σ_p are the electron and hole capture cross-section. Typically, the product of N_i and the capture cross-section, the so-called surface recombination velocity *S*, is used to quantify the recombination activity of an interface. In reality, there is a continuum of defects at the interface (3.1) should be replaced with an integral over the whole bandgap energy using $D_i(E)$ instead of N_{it} as is done in the extended SRH formalism [12] and this will be discussed in detail later in this chapter. From the numerator of (3.1), we can learn that the recombination rate depends on the deviation from thermal equilibrium (expressed by $n_s n_p - n_i^2$) and the defect density N_{it} or D_{it}. Hence, a reduction of the interface defect density, typically referred to as *chemical passivation*, is an effective way to reduce the recombination rate. From the denominator of (3.1), we can learn that the recombination rate is highest when $p_s/n_s \sim \sigma_n/\sigma_p$ and is reduced in all other cases. As a σ_n and σ_p are determined by the nature of the defect, and thus not easily changeable, the easiest way to decrease the surface recombination rate is to reduce either the electron or hole concentration at the surface. In most cases, the recombination rate is ruled by the *minority carrier* concentration at the surface. This can be done by either static electric fields or dopant profiles and is typically referred to as *field-effect passivation* or more recently *charge carrier population control.* The effect of charge carrier control is illustrated using EDNA 2 simulations [13] in Figure 3.1 for moderately doped *n*- and *p*-type c-Si. Let us first consider the case of *n*-type silicon. The surface recombination rate J_{0_s} for a negative fixed charge density of 10^{10} cm^{-3} is entirely ruled by the interface defect density and is equal to the recombination rate at the *p*-type surface which was simulated using identical defect parameters. When the negative fixed charge density is increased from 10^{10} cm^{-3} to 10^{11} cm^{-3}, the hole concentration at the surface, which is the minority charge carrier in the bulk of the *n*-type silicon wafer, is increased and this results in an *increase* in the recombination rate at the surface which peaks at a negative fixed charge density of $\sim 4 \times 10^{11}$ cm^{-3}. At this negative fixed charge density, the electron and hole concentration are equal and consequently, the recombination rate is maximised. When the negative charge density is further increased, the electron becomes the *minority carrier* at the surface, and its concentration is further reduced when increasing the negative fixed charge density. For very high negative fixed charge densities of $>10^{12}$ cm^{-3}, the recombination rates for the *n*-type and *p*-type surface are roughly equal again. It can also be shown that the recombination rate in this region inversely scales with the square of the fixed charge density [14], making charge carrier control a very effective way to reduce surface recombination. The recombination behaviour for the *p*-type c-Si wafer is more straightforward as electrons are already the minority carrier and their concentration is further reduced by the negative fixed charge density. Figure 3.1 thus also clearly shows that it is imperative to have the correct polarity of fixed charge for surface passivation as in some cases it can significantly *increase* the surface recombination rate as, for example, is explained in detail in a recent paper by To *et al.* [15] in the case of heavily doped *p*-type c-Si passivated by the positive charge dielectric silicon nitride.

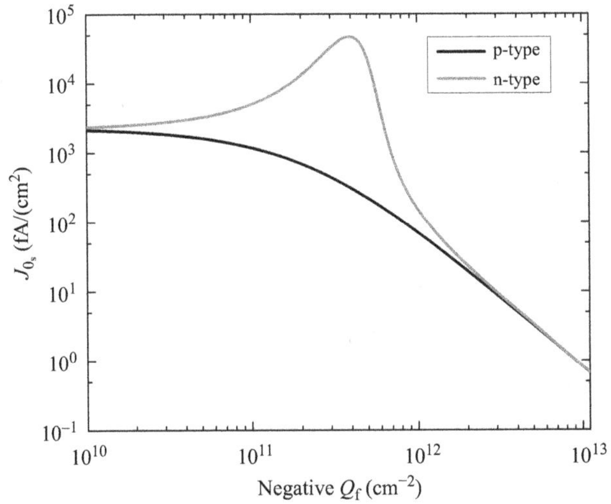

Figure 3.1 Simulated surface recombination rate J_{0_s} as a function of the negative fixed charge density Q_f at the surface for a moderately doped p-type and n-type silicon wafer. The dopant concentration was 1×10^{17} cm^{-3} for both wafers and for simplicity the capture cross-sections for holes and electrons were assumed to be equal resulting in a surface recombination velocity of 10^4 cm^{-2} for both holes and electrons. These simulations were done using the EDNA 2 software hosted at PV Lighthouse [13]

3.3 Electronic properties of the c-Si/Al_2O_3 interface

From the discussion in Section 3.2, it is clear that the D_{it} and Q_f values which are obtained after surface passivation films are deposited on c-Si are key for its surface passivation. In Table 3.1, I have summarised the electronic properties (D_{it} at mid-gap and Q_f) for Al_2O_3 synthesised by a wide variety of methods. In almost all cases a fixed negative charge density is reported, particularly after a post-deposition anneal. A fixed negative charge density reduces the electron concentration at the surface and consequently, Al_2O_3 is particularly effective in passivating *p*-type c-Si. It can also be seen that most deposition techniques can be optimised in a way that very low interface defect densities in the order of 10^{11} cm^{-2} can be achieved. Combined with a negative fixed charge density in excess of 10^{12} cm^{-2}, this provides an excellent level of surface passivation on lightly and highly doped *p*-type surfaces.

In order to fully model the recombination characteristics at the c-Si/Al_2O_3 interface, more detailed information is required than given in Table 3.2. The most rigorous work in this respect has been done by Black and McIntosh [27]. They were able to determine the energy-dependent defect density at the c-Si/Al_2O_3 interface for Al_2O_3 films grown by atmospheric chemical vapour deposition and the results

Table 3.1 Bulk and interface electronic properties for Al_2O_3 films deposited on c-Si as reported in the literature. All films were deposited on H-terminated c-Si unless indicated otherwise. Q_f was extracted from the flat band voltage shift determined from either conventional or contactless capacitance-voltage (CV) capacitance-voltage measurements of metal-oxide-semiconductor (MOS) structures. The interface defect density was extracted from CV and/or conductance measurements of MOS structures. Most samples were analysed after a post-deposition anneal; however, some samples were also measured as-deposited indicated with "a–d" in the first two columns. This table was adapted from Hoex et al. [16]

Negative Q_f (10^{12} cm^{-2})	D_{it} at mid-gap (10^{11} eV^{-1} cm^{-2})	Deposition method	Deposition temperature (°C)	Post-deposition thermal treatment	Ref.
0.2–0.5	–	Pyrolysis of $Al(C_3H_8O)_3$ on c–Si with 23–100 nm thermal SiO_2	425	30 min FGA[a] at 425 °C	[17]
0.2–0.5	–	CVD from $AlBr_3$ + NO on c-Si with 23–100 nm thermal SiO_2	910	30 min FGA[a] at 425 °C	[17]
−0.5 (a–d) 3.2	15 (a–d) 0.8	Pyrolysis of $Al(C_3H_8O)_3$ on c-Si with 1.5 nm native SiO_2	360	15 min FGA[a] at 510 °C	[1]
10–11 (a–d)	–	Pyrolysis of $AlCl_3$ in CO_2/H_2 environment[b]	900	None	[18]
–	0.3	Oxidation of evaporated Al	<400	800 °C in N_2	[19]
7	1–3	ALD from $AlCl_3$ and H_2O[c]	370	None	[20]
5–10 (a–d)	2–5[c] (a–d)	Reactive target sputtering in Ar/O_2	380	FGA[a]	[21]
–	2–10 (a–d)	ALD from $AlCl_3$ and H_2O	300–800	None	[22]
3–9	–	Molecular beam epitaxy from Al and N_2O on c-Si with an interfacial layer formed from Al and chemical SiO_2.[d]	750	None or anneal at 600 °C in O_2, N_2, H_2 or air	[23]
Up to 10	–	ALD from $Al(CH_3)_3$ and H_2O[b]	<300	Yes, not specified	[2]
7	–	Remote plasma-enhanced CVD from metal organic[e] and O_2 on c-Si[f]	300	30 s anneal at 800 °C and 30 min FGA[a] at 425 °C	[24]
1–3 (a–d) 6	100 (a–d) 1	Plasma ALD using $Al(CH_3)_3$ and an O_2 plasma	200	30 min anneal at 400 °C	[25]
0.13 (a–d) 2.4	3 (a–d) 1	Thermal ALD using $Al(CH_3)_3$ and H_2O	200	30 min anneal at 400 °C	[25]
5.3 (a–d) 3.4	100 (a–d) 1	Thermal ALD using $Al(CH_3)_3$ and O_3	200	30 min anneal at 400 °C	[25]
1–2	1	PECVD using $Al(CH_3)_3$ and N_2O	350	Industrial firing at 800 °C	[26]

[a]Forming gas anneal, typically 5%–10% H_2 in N_2.
[b]The pre-treatment of the c-Si wafer was not specified.
[c]The defect density could be lowered by the addition of Zr or Si in the Al_2O_3 to values in the mid 10^{10} eV^{-1} cm^{-2} range.
[d]Polycrystalline Al_2O_3 was grown in this study due to epitaxy process.
[e]$(CH_2)_3Al_2(C_4H_9O)_3$.
[f]Films were grown at H-terminated c-Si and c-Si with 1.0 nm or 1.6 nm thermal SiO_2 with no significant impact on the amount of negative Q_f.

Table 3.2 Properties of the defect distribution at the c-Si/Al_2O_3 interface. The defect distribution was determined by a combination of capacitance-voltage and conductance measurements by Black and McIntosh [27] and two defects were used to describe the mid-gap interface defect distribution. Defect A is the dominant defect when describing the interface recombination activity of p-type c-Si wafers. σ_n and σ_p are the capture cross-section for electrons and holes, respectively

Parameters (units)	Defect A	Defect B
σ_n (cm^{-2})	1.8×10^{-14}	1×10^{-16}
σ_p (cm^{-2})	1.3×10^{-16}	2×10^{-14}
$E_t - E_v$ (eV)	0.51	0.64
Standard deviation (eV)	0.09	0.07
$D_{it,peak}/D_{it,mid\text{-}gap}$	0.8	0.13

are summarised in Table 3.2. They showed that these interface defect parameters in combination with the interface defect density were able to describe the interface recombination behaviour. In addition, they showed that these parameters could also be used to reproduce measurements on lifetime samples passivated by Al_2O_3 grown using other deposition techniques such as ALD and PECVD, in particular when electrons are the minority charge carrier as is the case for *p*-type c-Si.

These defect parameters have also been shown to be applicable for highly *p*-type doped surfaces. The effective surface recombination rate at a highly doped surface can be described as [28]:

$$S_{n_0} n_A \frac{2kT\varepsilon_{si}}{Q_f^2}, \tag{3.2}$$

where S_{n_0} is the fundamental surface recombination rate for electrons, k is the Boltzmann constant, T is the temperature in, ε_0 is the permittivity of silicon and N_A is the doping concentration at the interface. From (3.2), we can learn that the effective recombination rate at a doped surface scales with the dopant density. Consequently, it appears that highly doped surfaces are more difficult to passivate compared to lightly doped surfaces and this is often incorrectly attributed to a higher defect density at the interface between a dielectric and doped c-Si. In the case of a boron-diffused surface, there is another effect that is further complicating the analysis. As boron diffusions are typically significantly deeper compared to phosphorus diffusions, this results in significantly non-uniform surface concentrations at the surface when non-planar surfaces such as pyramid surfaces are used. Ma *et al.* were able to accurately simulate the position-dependent boron surface concentration using multidimensional process simulations in Sentaurus TCAD [29]. Subsequently, they showed that identical surface recombination parameters could be used for planar and pyramid textured surfaces as shown in Figure 3.2. When using a more traditional approach which only takes an increased surface area

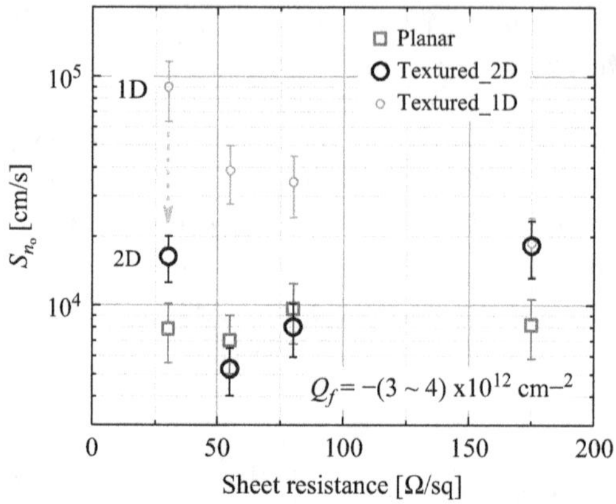

Figure 3.2 Simulated S_{n_0} values of planar and pyramidally textured samples as a function of the sheet resistance of the boron emitters on planar samples. For the textured samples, the actual S_{n_0} values were comparable to the results of the planar samples when a rigorous 2D simulation was used (Textured_2D); When the standard 1D simulation was used for textured samples (Textured_1D), the effective S_{n_0} values were found to be much higher (up to ten times) than the actual S_{n_0} values for the planar samples, while only a factor of 1.73 is expected based on the surface area ratio between a textured and planar surface. Reprinted from Ref. [29], with the permission of AIP Publishing

into account, the surface recombination at a pyramid textured surface was calculated to up to 10 times higher, particularly for more heavily diffused surfaces. Hence, these results confirm that the interface properties of the c-Si/Al_2O_3 do not depend on the surface morphology or surface dopant concentration, at least for the ranges which are most relevant for silicon wafer-based solar cells.

3.4 Atomic composition of the c-Si/Al_2O_3 interface and its relation to its electronic properties

In Figure 3.3(a), a transmission electron microscopy (TEM) image is shown of an Al_2O_3 film on c-Si grown by plasma-assisted ALD [30]. It can clearly be seen that the film is not uniform and the presence of an interfacial silicon oxide layer is evident from the electron energy loss spectrum shown in Figure 3.3(b). The presence of an interfacial silicon oxide layer is also apparent from Fourier transform infrared measurements shown in Figure 3.5, in particular after annealing, for an Al_2O_3 film grown by plasma-assisted ALD. It is well known that silicon oxide affords the lowest defect density in combination with c-Si due to the excellent lattice match, while Al_2O_3 would result in a relatively defective interface [31]. The original quality of the silicon oxide depends strongly on the synthesis method used

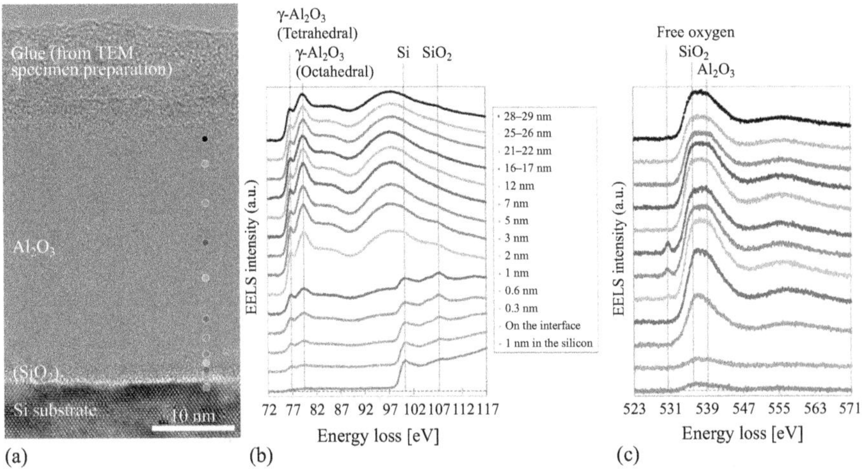

Figure 3.3 (a) Cross-sectional transmission electron microscopy (TEM) image of a c-Si/Al_2O_3 specimen. The Al_2O_3 film was grown by plasma-assisted ALD and received a post-deposition anneal. In this TEM image, the positions of the electron energy loss spectroscopy (EELS) measurements are indicated by coloured dots. (b) Low-energy EELS spectra showing the Al- and Si-$L_{2,3}$ edges. (c) High-energy EELS spectra showing the oxygen K-edge. Reprinted from Ref. [30], Reproduced with permission from Wiley Publishing

for the Al_2O_3. This is apparent from Table 3.1 where a significant deviation is found in the interface defect density after film synthesis for the various techniques and particularly a poor interface quality is, for example, reported for ALD films grown using an oxygen plasma as an oxidant. The formation of this interfacial oxide layer can, for example, be detected using *in situ* spectroscopic ellipsometry. From Figure 3.4, it can be seen that the initial growth during ALD can strongly depend on the precursors used [32]. It is well known that the oxidation strength of an oxygen plasma is significantly higher than that of water. This is clearly apparent in the first 5 ALD cycles where an ~1 nm film is grown in the case of plasma-assisted ALD and only an ~0.3 nm film in the case of thermal ALD using DMAI as the Al-precursor. It was found that the plasma-assisted process oxidises the c-Si substrate resulting in an 0.5 nm silicon oxide layer, while the relatively high reactivity of DMAI with the hydrogen-terminated Si surface results in a significantly thinner silicon oxide film. Fortunately, the quality of the silicon oxide film can be significantly improved by a post-deposition anneal where the film restructures as shown in Figure 3.5 and potentially some silicon dangling bonds can be passivated by atomic hydrogen emanating from the Al_2O_3 film [16]. The silicon oxide film is typically formed unintentionally. However, various groups are now using, for example, chemical methods to grow an intentional silicon oxide layer

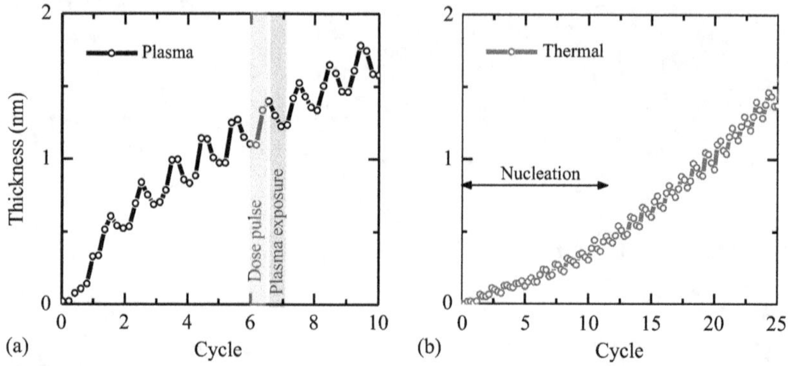

Figure 3.4 *The thickness of (a) plasma-assisted and (b) thermal atomic layer deposited Al_2O_3 on c-Si as measured by in-situ spectroscopic ellipsometry. The Al-precursor in this case was DMAI while an oxygen plasma and water were used as oxygen precursor for the plasma-assisted and thermal ALD process, respectively. This figure was originally published by Lee et al. [32], Reproduced with permission*

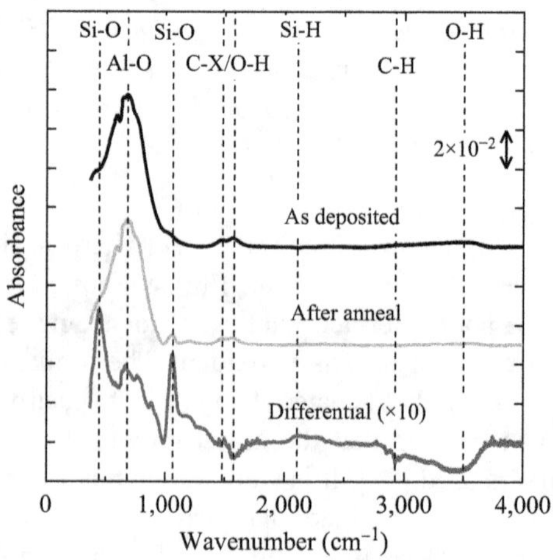

Figure 3.5 *Infrared absorption spectra of a 30 nm thick Al_2O_3 film which was synthesised by plasma-assisted ALD on both sides of a high resistivity c-Si substrate. Data is presented in the as-deposited and annealed state. Absorption peaks related to Al–O, Si–O, C–H, O–H and Si–H bonds can clearly be distinguished in the spectra as indicated. For clarity also the differential absorption spectrum is shown indicating the changes in the Al_2O_3 resulting from the post-deposition anneal. Reprinted from Ref. [16], with the permission of AIP Publishing*

prior to Al_2O_3 synthesis to improve process robustness [33] and it was found to be mandatory when using DMAI as the Al-precursor [32].

The silicon oxide layer also plays a crucial role in the origin of the negative fixed charge in the Al_2O_3 film. The negative charge has been attributed to defects in the Al_2O_3 film where the most likely defects are oxygen interstitials and aluminium vacancies [34,35]. An alternative way to look at the Al_2O_3 material structure was proposed by Johnson *et al.* [24]. They described bulk Al_2O_3 as a mixture of tetrahedrally and octahedrally coordinated Al, where the tetrahedral configuration has a net negative charge and the octahedral configuration has a net positive charge. In Figure 3.3, it is shown that both configurations can be detected in an Al_2O_3 film grown on c-Si. However, interestingly the EELS peak associated with tetrahedral Al seems to be more prominent close to the interface compared to the bulk of the film, indicating an excess of tetrahedral Al at near the interface, which is consistent with the fact that the negative fixed charge density is located near the interface [30]. As silicon oxide has a tetrahedral configuration of the Si atoms, it appears that it 'forces' the first Al_2O_3 atomic layers to be tetrahedrally coordinated as well before going to steady-state conditions where the tetrahedral and octahedral configuration are in equilibrium. A similar observation was already made earlier by Kimoto *et al.* for Al_2O_3 films studied for application as high-dielectric for complementary metal oxide semiconductor applications [36]. It has been shown by Dingemans *et al.* that the silicon oxide thickness can even be used to control the fixed charge density of the SiO_2/Al_2O_3 stack [37]. They showed that the negative fixed charge density in the Al_2O_3 film decreases when the silicon oxide thickness is increased to 5 nm and even charges polarity to positive fixed charge for thicker silicon oxide films.

3.5 Long-term stability of the silicon surface passivation by Al_2O_3

As the typical warranty period for a photovoltaic module is 25 years, it is of key importance that the thermal and ultraviolet stability of surface passivation films is investigated thoroughly before they are applied in high volume manufacturing. The stability of Al_2O_3 under one-sun light exposure was investigated in detail by Liao *et al.* [38] and the main result is shown in Figure 3.6. In this study, a typical symmetrically passivated wafer was light soaked for a long time and it was found that the effective surface recombination velocity decreased as function of time during the light soaking. The improvement in surface passivation was attributed to an increase in the negative fixed charge density by the injection of electrons from the valence band into empty defect states in the Al_2O_3 film. After removing the sample from the light soaking chamber, the level of surface passivation slowly decreased again to its initial level before light soaking. This behaviour could be modelled by a two-stage trapping model, which captures electrons during light soaking and release the captured electrons again when stored in the dark. The trapping was, however, found to be significantly faster than the de-trapping

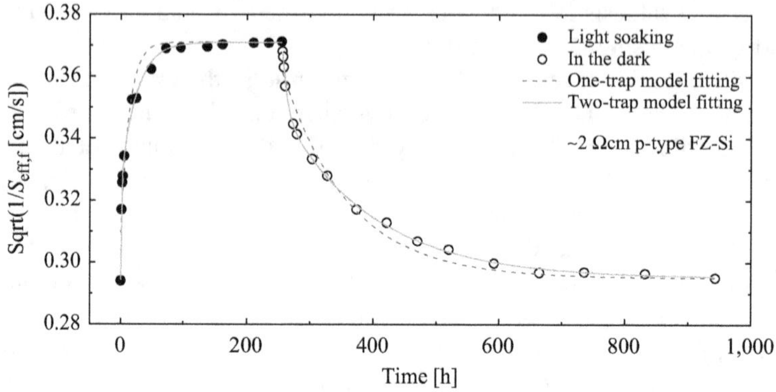

Figure 3.6 Time-dependent S_{eff} of a p-type c-Si wafer symmetrically passivated by a 30 nm Al_2O_3 film. The sample was light soaked under one Sun for 255 h and subsequently stored in the dark. As can be seen, the measured data can be reproduced well by assuming a two-trap model. A one-trap model does not fully capture the trapping and de-trapping fully. Reprinted from Ref. [38], with the permission of AIP Publishing

process. A more fundamental study was conducted by Gielis *et al.* [39]. They used second harmonic generation to study the strength of the electric field in c-Si resulting from mirroring the charge in the Al_2O_3 film. They were able to measure significant electrical fields in all studied samples, even for samples which did not show a significant level of surface passivation. This clearly indicated that charge carrier control alone is not sufficient for Al_2O_3 and that chemical passivation is also pivotal. For well-passivated samples they observed that their measurement was changing the electric field strength in the c-Si substrate and this effect strongly depended on the wavelength of the probe beam. By measuring this effect as a function of the wavelength of the probe beam, they were able to demonstrate that this effect was most significant when the photon energy of probe beam was half of the difference between the silicon valence band and the electron affinity of Al_2O_3 film (3.1 eV in this case).

3.6 Conclusion

In this chapter, I went into detail into the crystalline silicon surface passivation mechanism of Al_2O_3. I showed that both chemical passivation and charge carrier control by a high fixed negative charge density explain the excellent surface passivation provided by Al_2O_3 and this is particularly suitable for the passivation of *p*-type c-Si surface as the electrons are the minority carriers in this case. The desired electronic properties can be achieved by a wide range of synthesis methods and seem to be related to the intrinsic material properties of the system as shown

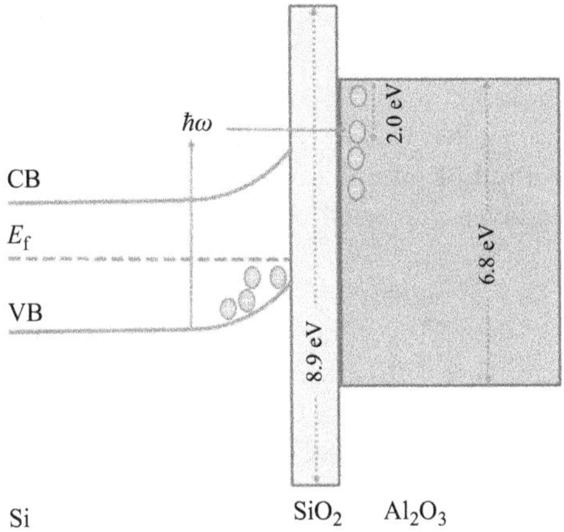

Figure 3.7 Schematic illustration of the band diagram of the c-Si/SiO_2/Al_2O_3 system (CB denoted conduction band, VB valence band and E_f Fermi level). The Al_2O_3 film has negatively charged defects resulting from either oxygen interstitials or aluminium vacancies, which are resulting from the tethrahedral match with SiO_2. The negative charge in the Al_2O_3 film is mirrored by holes in c-Si resulting in field-effect passivation. The SiO_2 film ensures a relatively low interface defect density aided by hydrogenation from the Al_2O_3 film during a post-deposition anneal

schematically in Figure 3.7. The good interface quality of the system is ensured by the presence of a thin interfacial silicon oxide layer, which is either resulting from the synthesis process or is intentionally grown prior to Al_2O_3 deposition. Typically, a post-deposition anneal is required to improve the quality of the silicon oxide film, and in addition this can provide hydrogenation to improve the interface defect density further. The negative fixed charge density in the Al_2O_3 film is due to either oxygen interstitials or aluminium vacancies, which are resulting from the match of the first atomic layers of the Al_2O_3 with the interfacial silicon oxide film.

References

[1] Hezel R and Jaeger K. 'Low-temperature surface passivation of silicon for solar-cells'. Journal of the Electrochemical Society. 1989;**136**:518.

[2] Agostinelli G, Delabie A, Vitanov P, *et al.* 'Very low surface recombination velocities on p-type silicon wafers passivated with a dielectric with fixed negative charge'. Solar Energy Materials and Solar Cells. 2006;**90**:3438.

[3] Hoex B, Heil SBS, Langereis E, van de Sanden MCM, and Kessels WMM. 'Ultralow surface recombination of c-Si substrates passivated by plasma-assisted atomic layer deposited Al_2O_3'. Applied Physics Letters. 2006;**89**: 042112.

[4] Hoex B, Schmidt J, Bock R, Altermatt PP, van de Sanden MCM, and Kessels WMM. 'Excellent passivation of highly doped p-type Si surfaces by the negative-charge-dielectric Al_2O_3'. Applied Physics Letters. 2007;**91**:112107.
[5] Hoex B, van de Sanden MCM, Schmidt J, Brendel R, and Kessels WMM. 'Surface passivation of phosphorus-diffused n(+)-type emitters by plasma-assisted atomic-layer deposited Al_2O_3'. Physica Status Solidi-R. 2012;**6**: 201105445.
[6] Dingemans G, van de Sanden MCM, and Kessels WMM. 'Plasma-enhanced chemical vapor deposition of aluminum oxide using ultrashort precursor injection pulses'. Plasma Processes and Polymers. 2012;**9**:761–71.
[7] Saint-Cast P, Kania D, Hofmann M, Benick J, Rentsch J, and Preu R. 'Very low surface recombination velocity on p-type c-Si by high-rate plasma-deposited aluminum oxide'. Applied Physics Letters. 2009;**95**:151502.
[8] Benick J, Hoex B, van de Sanden MCM, Kessels WMM, Schultz O, and Glunz SW. 'High efficiency n-type Si solar cells on Al_2O_3-passivated boron emitters'. Applied Physics Letters. 2008;**92**:253504.
[9] Schmidt J, Merkle A, Brendel R, Hoex B, van de Sanden MCM, and Kessels WMM. 'Surface passivation of high-efficiency silicon solar cells by atomic-layer-deposited Al_2O_3'. Progress in Photovoltaics. 2008;**16**:461–6.
[10] Hall R. 'Electron-Hole Recombination in Germanium'. A Physical Review. 1952;**87**:387.
[11] Shockley W and Read W. 'Statistics of the recombinations of holes and electrons'. Physical Review. 1952;**87**:835–42.
[12] Girisch RBM, Mertens RP, and Dekeersmaecker RF. 'Determination of Si-SiO_2 interface recombination parameters using a gate-controlled point-junction diode under illumination'. IEEE Transactions on Electron Devices. 1988;**35**:203.
[13] McIntosh KR and Abbott MD. *EDNA 2*, Available from: https://www2.pvlighthouse.com.au/calculators/EDNA%202/EDNA%202.aspx [Accessed on 12 May 2018]
[14] Kuhlmann B, Aberle AG, and Hezel R, editors. 'Two-dimensional numerical optimisation study of inversion layer emitters of silicon solar cells'. *Proceedings of the 13th European PVSEC*; 1995; Nice, France: Stephens Bradford.
[15] To A, Ma F-J, and Hoex B. 'Improved understanding of the recombination rate at inverted p+ silicon surfaces'. Japanese Journal of Applied Physics. 2017;**56**:08MB5.
[16] Hoex B, Gielis JJH, de Sanden M, and Kessels WMM. 'On the c-Si surface passivation mechanism by the negative-charge-dielectric Al_2O_3'. Journal of Applied Physics. 2008;**104**:113703.
[17] Aboaf JA, Kerr DR, and Bassous E. 'Charge in SiO//2-Al//2O//3 double layers on silicon'. Journal of the Electrochemical Society. 1973;**120**: 1103–6.
[18] Mehta DA, Butler SR, and Feigl FJ. 'Electronic charge trapping in chemical vapor-deposited thin films of Al_2O_3 on silicon'. Journal of Applied Physics. 1972;**43**:4631–8.

[19] Chin A, Wu YH, Chen SB, Liao CC, and Chen WJ, editors. 'High quality La_2O_3 and Al_2O_3 gate dielectrics with equivalent oxide thicknesses 5–10 A'. *Proceedings of the Symposium on VLSI Technology*; 2000; Honolulu: IEEE, Piscataway, NJ, USA.

[20] No SY, Eom D, Hwang CS, and Kim HJ. 'Property changes of aluminum oxide thin films deposited by atomic layer deposition under photon radiation'. Journal of the Electrochemical Society. 2006;**153**:F87–F93.

[21] Manchanda L, Morris MD, Green ML, *et al.* 'Multi-component high-K gate dielectrics for the silicon industry'. Microelectronic Engineering. 2001;**59**: 351–9.

[22] Duenas S, Castan H, Garcia H, *et al.* 'Influence of single and double deposition temperatures on the interface quality of atomic layer deposited Al_2O_3 dielectric thin films on silicon'. Journal of Applied Physics. 2006; **99**:54902.

[23] Shahjahan M, Okada T, Sawada K, and Ishida M. 'Effect of annealing on physical and electrical properties of ultrathin crystalline gamma-Al_2O_3 high-k dielectric deposited on Si substrates'. Japanese Journal of Applied Physics Part 1-Regular Papers Short Notes & Review Papers. 2004;**43**: 5404–8.

[24] Johnson RS, Lucovsky G, and Baumvol I. 'Physical and electrical properties of noncrystalline Al_2O_3 prepared by remote plasma enhanced chemical vapor deposition'. Journal of Vacuum Science & Technology a-Vacuum Surfaces and Films. 2001;**19**:1353.

[25] Dingemans G, Terlinden NM, Pierreux D, Profijt HB, van de Sanden MCM, and Kessels WMM. 'Influence of the oxidant on the chemical and field-effect passivation of Si by ALD Al_2O_3'. Electrochemical and Solid State Letters. 2011;**14**:H1–H4.

[26] Duttagupta S, Ma FJ, Lin SF, Mueller T, Aberle AG, and Hoex B. 'Progress in surface passivation of heavily doped n-type and p-type silicon by plasma-deposited AlO_x/SiN_x dielectric stacks'. IEEE Journal of Photovoltaics. 2013;**3**:1163–9.

[27] Black LE and McIntosh KR. 'Modeling recombination at the Si-Al_2O_3 interface'. IEEE Journal of Photovoltaics. 2013;**3**(3):936–43.

[28] McIntosh KR and Black LE. 'On effective surface recombination parameters'. Journal of Applied Physics. 2014;**116**:014503.

[29] Ma FJ, Duttagupta S, Shetty KD, *et al.* 'Two-dimensional numerical simulation of boron diffusion for pyramidally textured silicon'. Journal of Applied Physics. 2014;**116**(18).

[30] Hoex B, Bosman M, Nandakumar N, and Kessels WMM. 'Silicon surface passivation by aluminium oxide studied with electron energy loss spectroscopy'. Physica Status Solidi-Rapid Research Letters. 2013;**7**(11):937–41.

[31] Lucovsky G and Phillips JC. 'Limitations for aggressively scaled CMOS Si devices due to bond coordination constraints and reduced band offset energies at Si-high-k dielectric interfaces'. Applied Surface Science. 2000;**166**:497–503.

[32] Chang-Yeh L, Shaozhou W, Xin C, *et al.* 'Improving the silicon surface passivation by aluminum oxide grown using a non-pyrophoric aluminum precursor'. Physica Status Solidi (RRL) – Rapid Research Letters. 2018:1800156.

[33] Bordihn S, Engelhart P, Mertens V, *et al.* 'High surface passivation quality and thermal stability of ALD Al($_2$)O($_3$) on wet chemical grown ultra-thin SiO($_2$) on silicon'. Energy Procedia. 2011;**8**:654–9.

[34] Matsunaga K, Tanaka T, Yamamoto T, and Ikuhara Y. 'First-principles calculations of intrinsic defects in Al_2O_3'. Physical Review B. 2003; **68**(21): 85110.

[35] Liu YY, Stradins P, Deng HX, Luo JW, and Wei SH. 'Suppress carrier recombination by introducing defects: The case of Si solar cell'. Applied Physics Letters. 2016;**108**(2):022101.

[36] Kimoto K, Matsui Y, Nabatame T, *et al.* 'Coordination and interface analysis of atomic-layer-deposition Al_2O_3 on Si(001) using energy-loss near-edge structures'. Applied Physics Letters. 2003;**83**:4306–8.

[37] Dingemans G, Terlinden NM, Verheijen MA, van de Sanden MCM, and Kessels WMM. 'Controlling the fixed charge and passivation properties of Si(100)/Al_2O_3 interfaces using ultrathin SiO_2 interlayers synthesized by atomic layer deposition'. Journal of Applied Physics. 2011;**110**:93715.

[38] Liao BC, Stangl R, Mueller T, Lin F, Bhatia CS, and Hoex B. 'The effect of light soaking on crystalline silicon surface passivation by atomic layer deposited Al_2O_3'. Journal of Applied Physics. 2013;**113**:6.

[39] Gielis JJH, Hoex B, van de Sanden MCM, and Kessels WMM. 'Negative charge and charging dynamics in Al_2O_3 films on Si characterized by second-harmonic generation'. Journal of Applied Physics. 2008;**104**:73701.

Chapter 4

Material properties of Al_2O_3 grown on Si: interface trap density (D_{it}) and fixed charge density (Q_f)

Bart Vermang[1,2,3] *and Hans Goverde*[3]

Chapter 4 gives a summary composed from [1], and consistent with more general overview papers [2,3]. First, Section 4.1 describes recombination at Si surfaces theoretically to clarify two key passivation approaches: chemical and field effect passivation. Then, Section 4.2 quantifies the chemical and field effect passivation of Al_2O_3, i.e., its interface trap density and fixed charge density, respectively. Finally, Section 4.3 provides a larger picture by comparing Al_2O_3 to other Si surface layers, and showing that Al_2O_3 surface passivation can also be applied to passivate thin film photovoltaics.

4.1 Theoretical preface

A Si substrate surface can be represented by a large density of defect levels within the band-gap, whose recombination dynamics are described by the Shockley–Read–Hall (SRH) formalism. The surfaces or interfaces of a silicon substrate represent large numbers of partially bonded silicon atoms, which give rise to many dangling bonds. Such a severe discontinuity of the crystalline structure can be represented by a large density of defect levels within the band-gap near the semiconductor surface. These discrete energy levels greatly facilitate recombination through a two-step process whereby a free electron from the conduction band first relaxes to the defect level, and then relax to the valence band where it annihilates a hole. Shockley, Read, and Hall were the first to formulate the rate of surface recombination (U_s). This U_s for a single defect level is given in the equation below, as a function of the concentrations of electrons and holes at the surface (n_s and p_s), the intrinsic carrier concentration (n_i), the thermal velocity of the electrons (v_{th}), the interface defect density (N_{it}), the statistical factors n_1 and p_1, and the electron and hole capture cross sections (σ_n and σ_p). The energy dependence of all these

[1]University of Hasselt, Partner in Solliance and EnergyVille, Belgium
[2]IMEC division IMOMEC, Belgium
[3]IMEC, Partner in Solliance and EnergyVille, Belgium

parameters is, for the sake of generalization, neglected by assuming a single defect at mid-gap. In reality, U_S should be expressed by the extended SRH formalism with an integral over the bandgap energies while replacing N_{it} by D_{it} [4,5]:

$$U_s = \frac{\left(n_s p_s - n_i^2\right) v_{th} N_{it}}{\frac{n_s + n_1}{\sigma_p} + \frac{p_s + p_1}{\sigma_n}} \tag{4.1}$$

This SRH formalism can be used to explain the two key methods to reduce recombination at Si surfaces, i.e., chemical and field effect passivation. Recombination at the surface of crystalline silicon can be suppressed by one of the following two actions, or a combination of both: first, by chemical bonding, which can be regarded as the surface passivation "par excellence" and second, by modifying the relative concentrations of electrons and holes. The SRH formalism shows that U_s can be decreased by a reduction in N_{it} (or D_{it}), which is referred to as chemical passivation. By growing an appropriate dielectric layer such as silicon oxide, many of the dangling silicon bonds are passivated with oxygen or hydrogen atoms and the N_{it} (or D_{it}) is reduced. Another way to reduce the recombination is by a significant reduction in the density of one type of charge carrier at the surface (n_s or p_s) by an electric field. This is called field-effect passivation. It can be achieved by doping the semiconductor surface to repel the minority carries (such as in an emitter region or a back surface field). Alternatively, fixed charges in an overlying dielectric layer can be used to either repel the minority carriers (for a p-type wafer one would use negative charges to repel the free electrons), or in the extreme case invert the surface (large amounts of fixed positive charges would invert the surface of a p-type silicon wafer) [2,3].

4.2 Al_2O_3 material properties

In this section, the chemical and field effect passivation of atomic layer deposition (ALD) of Al_2O_3, because of its unique features, are quantified. Today, high efficiencies have been realized by the implementation of ultrathin Al_2O_3 films in laboratory-type and industrial solar cells. It can be deposited by thermal, plasma-enhanced (PE) or plasma-assisted (PA) ALD, plasma-enhanced chemical vapor deposition (PECVD), atmospheric pressure (AP) CVD, or reactive sputtering. For all these deposition techniques, very low effective surface recombination velocities (or very high minority carrier lifetimes) have been reported [1]. In this chapter, the focus is on thermal ALD, mainly because of its excellent uniformity and growth control, and its compatibility with high-throughput manufacturing. Figure 4.1(a) shows the effective lifetime of ALD-Al_2O_3 passivated p-type and n-type float zone (FZ) Si as a function of injection level. Note that both samples also had an optimized annealing in forming gas prior to the measurement. It is clear that the Al_2O_3 film leads to high effective minority carrier lifetimes for both p-type and n-type FZ Si. In following paragraphs, it is shown that the ALD Al_2O_3 Si surface passivation mechanism is based on chemical passivation—a low density of interface defects—and field-effect passivation—caused by a high density of fixed negative charges

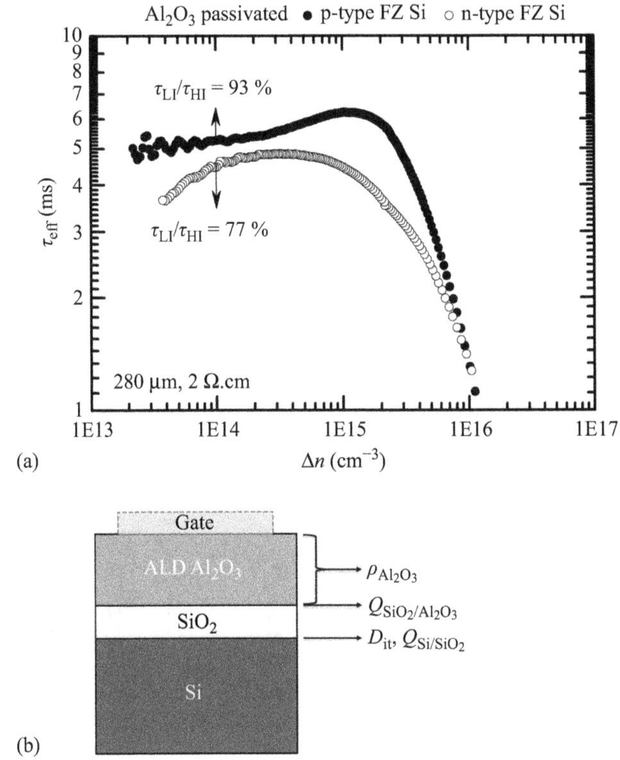

Figure 4.1 Injection level-dependent effective lifetime measured by quasi-steady-state photoconductance for 2 Ω.cm p-type and n-type FZ Si passivated by ALD Al_2O_3. The samples are 280 µm thick, and underwent an optimized annealing in forming gas prior to measurement. The ratio between effective lifetime at low and high injection (LI and HI) is also quantified, where $\Delta n_{LI} = 4 \times 10^{13}$ cm^{-3} and $\Delta n_{HI} = 2 \times 10^{14}$ cm^{-3}. (b) Schematic cross section of ALD-Al_2O_3 chemical and field effect passivation, or interface trap density (D_{it}) and fixed charge density (Q_f), respectively. All charge locations used in the extraction of Q_f are indicated. The charge at the gate-electrode/ALD-Al_2O_3 interface is assumed to be included in the work function

within the Al_2O_3 layer. Figure 4.1(b) shows a schematic model of the ALD-Al_2O_3 chemical and field effect passivation.

Table 4.1 quantifies the chemical Si surface passivation of ALD-Al_2O_3, as obtained by a combination of a high-quality SiO_2 interfacial layer and the extra hydrogenation of remaining dangling bonds. Figure 4.2(a) gives the density of interface states versus energy with respect to the valence band maximum (as calculated from deep-level transient spectroscopy (DLTS) measurements [6]) for an as-deposited Al_2O_3 passivated Si sample, after a forming gas anneal (FGA) at

Table 4.1 Calculated density of interface states. A schematic model can be found in Figure 4.2

$D_{it}(Si/SiO_2(/Al_2O_3))$ ($\times 10^{12}$ cm^{-2} eV^{-1})
≤2.4

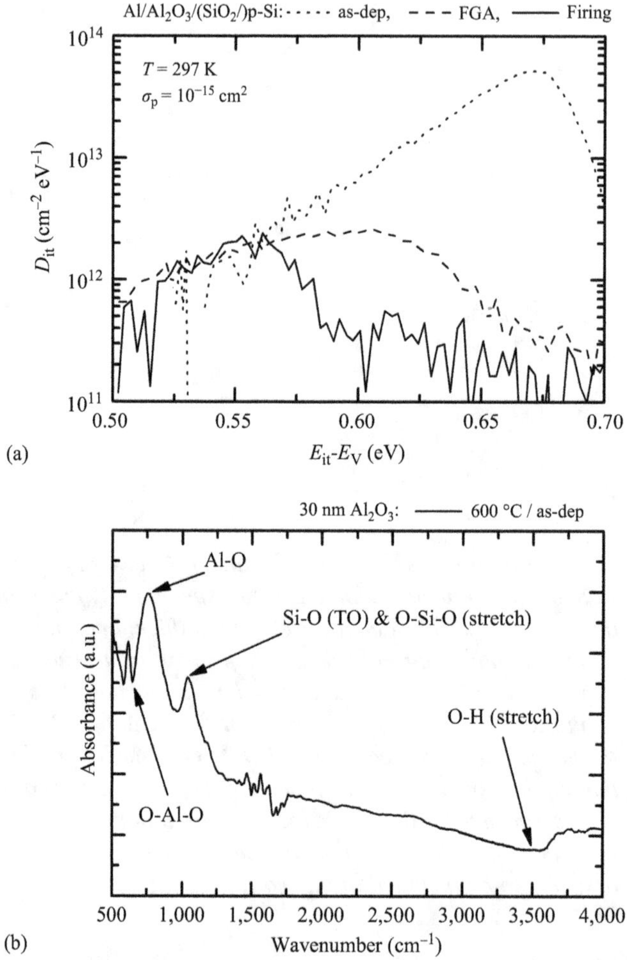

Figure 4.2 (a) Density of interface states versus energy with respect to the valence band maximum for an as-deposited ALD Al_2O_3 passivated p-type Si sample, after a forming gas anneal at 350 °C and after a co-firing step. (b) Differential Fourier transform infrared spectroscopy spectrum of an ALD Al_2O_3 (30 nm) passivated p-type Si sample annealed at 600 °C and as-deposited. Also the absorption band assignments are indicated

Table 4.2 Calculated areal and volumetric charge carrier densities. A schematic model can be found in Figure 4.2

Q_{Si/SiO_2} ($\times10^{10}$ cm^{-2})	Q_{SiO_2/Al_2O_3} ($\times10^{11}$ cm^{-2})	$\rho_{Al_2O_3}$ ($\times10^{16}$ cm^{-3})
6.6 ± 0.3	−3.9 ± 1.3	−3.2 ± 2.1

350 °C and after a high-temperature co-firing step. It shows that the FGA and firing lowers the D_{it} at the Si interface and at the same time moves the distribution towards lower activation energies. Figure 4.2(b) shows the differential Fourier transform infrared spectroscopy spectrum of an Al_2O_3 passivated Si sample annealed at high-temperature and as-deposited [7]. This spectrum shows: (i) broad absorption features between 450 and 1,000 cm^{-1} that are characteristic for aluminum oxide, (ii) an absorption peak at 1,060 cm^{-1} indicating that a SiO_2 interface layer is formed between the bulk Al_2O_3 and its Si surface, and (iii) a wide feature at 3,500 cm^{-1}: broken O–H bonds and a release of hydrogen. The passivation properties of the interfacial SiO_2 are very similar to a thin thermally grown SiO_2, as confirmed by electron spin resonance (ESR) showing the presence of about 10^{12} cm^{-2} dangling bond centers [6]. Consequently, it can be concluded that the lower D_{it} and the shift to lower activation energies seen in Figure 4.2(a) are a combination of a high-quality SiO_2 interfacial layer, and additional hydrogen passivation of residual dangling bonds. The passivating quality of such an optimally annealed Al_2O_3 film has been shown in Figure 4.1(a), where effective minority carrier lifetimes in the order of several milliseconds are measured.

Table 4.2 quantifies the field effect Si surface passivation of ALD-Al_2O_3, as obtained by a negative charge density in the Al_2O_3 layer. A slant-etched capacitance–voltage (C–V) measurement technique has been used to quantify the field effect passivation [8]. It provides statistically more reliable data and even allows a decoupling of the location of the charge within the passivation stack Si/SiO_2(interfacial)/Al_2O_3. Estimation of the fixed charge components is based on a methodology where both the interfacial SiO_2 and Al_2O_3 thicknesses are varied, and combined with the one-dimensional Gauss Law. This equation is given below, with Q_{Si/SiO_2} the charge density at the Si/SiO_2 interface; ε_{SiO_2} the relative permittivity of SiO_2; ϕ_{ms} the work function difference between the metal gate and the semiconductor material; $\rho_{Al_2O_3}$ the bulk charge density within the Al_2O_3 layer; $EOT_{Al_2O_3}$ the equivalent oxide thickness of this Al_2O_3 layer; Q_{SiO_2/Al_2O_3} the charge density at the SiO_2/Al_2O_3 interface; and the flat-band voltage (V_{fb}) and the total equivalent oxide thickness (EOT). Hence, employing a slant-etching technique – where the thermally grown SiO_2 interlayer thickness is continuously scaled across a wafer – an estimation of Q_{Si/SiO_2} can be obtained by plotting the V_{fb} as a function of EOT. See Figure 4.3(a), where a straight-line fit is obtained for the V_{fb}-EOT plot. In a similar way, V_{fb}-EOT plots for slant-etched wafers deposited with various thicknesses of Al_2O_3 films can be linearly fitted. The intercepts of these linear fits with the V_{fb}-axes can be expressed as a function of $EOT_{Al_2O_3}$, as is shown in

Figure 4.3(b). The second degree polynomial fit of this plot leads to an estimation of $\rho_{Al_2O_3}$ and Q_{SiO_2/Al_2O_3}. The results indicate positive fixed charge at the interface Si/SiO_2, negative charge at the interface SiO_2/Al_2O_3 and a non-negligible charge density in the Al_2O_3 bulk, as is summarized in Table 4.2. Hence, the overall charge density for Al_2O_3 passivation layers is negative, which causes field effect passivation. This fixed negative charge is consistent with a model for the local atomic bonding of amorphous Al_2O_3 that has two different bonding environments for the

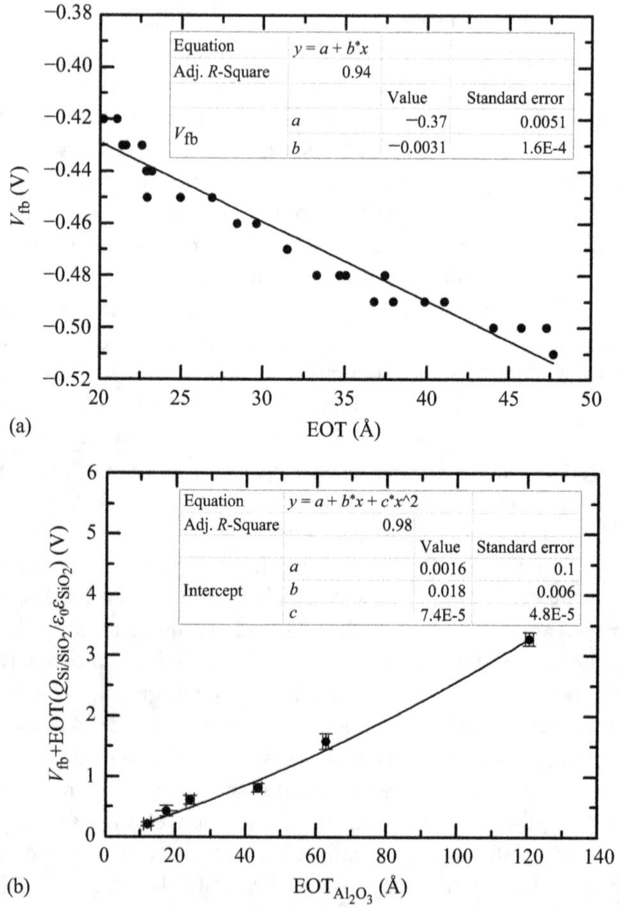

Figure 4.3 Flat-band voltage versus equivalent oxide thickness plot for slant-etched SiO_2 wafers. Data are from three identical SiO_2 wafers that have been slant-etched in HF to reach a nominal range of 1 to 6 nm of SiO_2. (b) $V_{fb}+EOT(Q_{Si/SiO_2}/\varepsilon_0\varepsilon_{SiO_2})$ as a function of $EOT_{Al_2O_3}$ as explained in the text. The fit is a second-degree polynomial indicating the presence of bulk charge in the ALD Al_2O_3 layer, as can be seen from the equation in the text

Al atoms: (i) a tetrahedrally coordinated Al site that has a net negative charge, and (ii) an octahedrally coordinated site in which the Al has a valence charge of 3+. The tetrahedrally coordinated Al is dominant at the SiO_2/Al_2O_3 interface, attributed to the fact that Si in the interfacial SiO_2 film also has a tetrahedral coordination [9]:

$$V_{fb} = \mathrm{EOT}\left[-\frac{Q_{Si/SiO_2}}{\varepsilon_0\varepsilon_{SiO_2}}\right] + \left\{\varphi_{ms} - \frac{Q_{SiO_2/Al_2O_3}\mathrm{EOT}_{Al_2O_3}}{\varepsilon_0\varepsilon_{SiO_2}} - \frac{\rho_{Al_2O_3}(\mathrm{EOT}_{Al_2O_3})^2}{2\varepsilon_0\varepsilon_{SiO_2}}\right\} \tag{4.2}$$

4.3 Epilog

Al_2O_3 surface passivation combines field-effect passivation by negative fixed charges and a low interface defect density, which is a very unique combination as compared to other typical Si surface passivation layers and ideal to passivate p-type Si surfaces. Figure 4.4 gives an overview of typical industrial passivating coatings on silicon in terms of fixed charge density and interface trap density, which shows that the Al_2O_3 passivation properties are indeed unique. The Al_2O_3 layer contains a negative charge density, which will help to suppress p-type Si surface recombination due to accumulation conditions and thus increased majority carrier density. However, this is not ideal to passivate n-type Si, where the surface will become inverted. In the case of an inverted surface, the electron and hole density will become equal a distance away from the interface, which will enhance recombination in the subsurface when bulk defects are present. A phenomenon that will become more effective at low injection. This is also seen in Figure 4.1(a), where the ratio between effective lifetime at low and high injection is shown for both graphs. This ratio is the highest (and very close to 1) in case of p-type Si passivation by Al_2O_3 [2,3].

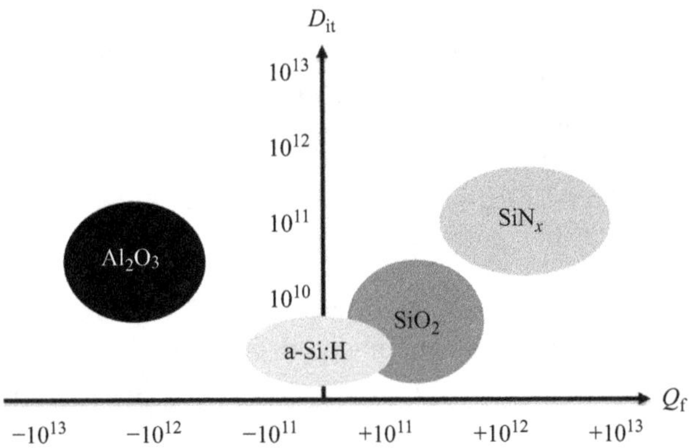

Figure 4.4 Overview of industrial passivating coatings on silicon in terms of fixed charge density and interface trap density (based on [3])

Latest results indicate that Al_2O_3 grown on thin film photovoltaic (PV) materials has similar unique material properties, where it has been successfully integrated as a surface passivation layer. The electronic properties of the ALD Al_2O_3 grown on Cu(In,Ga)Se_2 (CIGS) thin film PV materials have been experimentally extracted and numerically simulated. That study confirms the excellent CIGS passivation quality for optimized (annealed) ALD Al_2O_3 films, due to a negative fixed charge-induced field-effect passivation combined with chemical passivation, as described in [10]. Such an Al_2O_3 surface passivation layer has been used to fabricate rear-passivated CIGS solar cells. Using very thin CIGS absorber layers, these cells show a clear increase in all cell characteristics compared with corresponding unpassivated reference cells. The most important advances are a clear increase in open circuit voltage due to improved rear surface passivation, a clear increase in short circuit current due to enhanced rear internal reflection, and a clear increase in fill factor due to more ideal cell behavior [11].

Acknowledgments

This work has received funding from the European Research Council (ERC) under the European Union's Horizon 2020 research and innovation program (grant agreement no. 715027). In addition, B. Vermang also acknowledges the financial support of the Flemish Research Foundation FWO (mandate 12O4215N).

References

[1] B. Vermang, "Aluminum oxide as negatively charged surface passivation for industrial crystalline silicon solar cells", Leuven, Belgium: Leuven University Press, ISBN 978-94-6018-531-1, 2012.
[2] G. Dingemans and W.M.M. Kessels, "Status and prospects of Al_2O_3-based surface passivation schemes for silicon solar cells", Journal of Vacuum Science and Technology A 30, 040802, 2012.
[3] A. Cuevas, T. Allen, J. Bullock, Y. Wan, D. Yan, and X. Zhang, "Skin care for healthy silicon solar cells", Proceedings of the 42nd IEEE PVSC (New-Orleans), pp. 1–6, 2015.
[4] W. Shockley and W.T. Read, "Statistics of the recombinations of holes and electrons", Physical Review, 87, 835–842, 1952.
[5] R.N. Hall, "Electron-hole recombination in germanium", Physical Review, 87, 387, 1952.
[6] E. Simoen, A. Rothschild, B. Vermang, J. Poortmans, and R. Mertens, "Impact of forming gas annealing and firing on the Al_2O_3/p-Si interface state spectrum", Electrochemical and Solid-State Letters, 14, H362–H364, 2011.
[7] H. Goverde, B. Vermang, T. Bearda, *et al.*, "Al_2O_3 surface passivation characterized on hydrophobic and hydrophilic Si by combination of QSSPC, CV and FTIR", Energy Procedia, 27, 355–360, 2012.

[8] A. Rothschild, B. Vermang, X. Loozen, B.J. O'Sullivan, J. John, and J. Poortmans, "ALD-Al_2O_3 passivation for solar cells: Charge investigation", Proceedings of the 25th EU PVSEC (Valencia, Spain), pp. 1382–1385, 2010.
[9] B. Hoex, M. Bosman, N. Nandakumar, and W.M.M. Kessels. "Silicon surface passivation by aluminium oxide studied with electron energy loss spectroscopy", Physica Status Solidi RRL, 7(11), 937–941, 2013.
[10] R. Kotipalli, B. Vermang, J. Joel, R. Rajkumar, M. Edoff, and D. Flandre, "Investigating the electronic properties of Al_2O_3/Cu(In,Ga)Se_2 interface", AIP Advances, 5(10), 107101, 2016.
[11] B. Vermang, J.T. Wätjen, C. Frisk, *et al.*, "Introduction of Si PERC rear contacting design to boost efficiency of Cu(In,Ga)Se_2 solar cells," IEEE Journal of Photovoltaics, 4(6), 1644–1649, 2014.

Chapter 5
PECVD-AlO_x

Pierre Saint-Cast[1] and Marc Hofmann[1]

5.1 Introduction

5.1.1 Atomic layer deposition and earlier work

The first reported use of AlO_x for solar cell (SC) passivation dates back to the 1980s [1]. The AlO_x layers were deposited on a thin thermally grown SiO_2 layer, using atmospheric pressure chemical vapor deposition (APCVD). In this study [1], the surface recombination velocity (SRV) is reported to be approximately 200 cm s^{-1}.

After this, Si-SC specialists forgot about AlO_x for more than ten years, before it surfaced back in 2004 in the work of Agostinelli *et al.* [2,3] with the aim of applying AlO_x layers for the rear surface passivation of *p*-type solar cells. For the deposition of the layer, Agostinelli *et al.* used spin-on sol-gel [3] and atomic layer deposition (ALD) [3].

Shortly after this work, a series of excellent results published by Hoex *et al.* has attracted the attention of the photovoltaic (PV) community. The plasma-assisted ALD (PA-ALD) AlO_x layers from Hoex *et al.* proved their quality in two breakthrough results:

- Efficiencies above 20% [4], on *p*-type rear passivated SC
- Efficiencies above 23% [5], on *n*-type SC, where AlO_x was used for the passivation of the emitter.

However, classical ALD techniques are limited by a very low deposition rate, which forbid its application on the industrial production level. The scientists needed to search for alternatives.

5.1.2 Alternatives to classical ALD

The development of alternatives consisted in reaching high passivation quality using deposition techniques with a much higher throughput potential than classical ALD. The first promising result was obtained by Miyajima *et al.* [6] on parallel plate plasma-enhanced chemical vapor deposition (PECVD) in 2008. This result was followed by linear antenna microwave (MW) PECVD [7] and radiofrequency sputtering [8]

[1]Plasma Technology, Division Photovoltaics, Fraunhofer Institute for Solar Energy Systems ISE, Germany

(RF sputtering) in 2009. Then, in 2010 and in 2011, respectively, Spatial ALD [9], APCVD [10], and inductively-coupled plasma (ICP) PECVD [11] were proven to be valuable alternatives. Within the profusion of processes recently developed, PECVD offers very high deposition rates without significantly compromising the AlO_x layer quality. In addition, PECVD presents the advantage to be already present within most of the SC production lines for the fabrication of SiN_x anti-reflective coating (ARC).

5.1.3 PECVD deposition techniques

PECVD offers a very large variety of deposition techniques apart from linear antenna MW PECVD. As mentioned before, AlO_x layers were also developed on parallel plate PECVD system [6]. More recently excellent passivation qualities were reported on inductively coupled plasma sources (ICP) [11]. This deposition technique allows very high deposition rates. Dingemans *et al.* [12] used the flexibility of PECVD technique in a creative way, by pulsing the precursor gas into the chamber using ultrafast valves. This technique allows the control of the density and the quality of the layer by varying the duration of the gas pulse.

The main advantage of PECVD techniques is its cost-to-quality relation. Despite ALD techniques often offer a better conformity and control of the deposition, PECVD can reach high production throughput without sacrificing the electrical quality, which makes PECVD the actual favorite deposition technique for AlO_x in solar cell production.

5.2 Plasma and layer characteristics of MW-PECVD

5.2.1 Plasma characteristics

This part focus on an AlO_x deposition processes developed using an inline industrial-type PECVD system (Roth&Rau SiNA or Maia). These systems are currently in extensive use in the crystalline Si PV industry, typically applied to deposit a-SiN_x:H ARC layers on commercial mono- and multi-crystalline silicon solar cells. Industrial inline systems of this type reach production throughputs of more than 3,000 wafers per hour, and are not a limiting step in production.

The plasma is created by 2.45 GHz MW pulses, which are introduced to the reaction chamber via a linear antenna. Two MW generators on each side provide the power to the linear antenna. During the process, the chamber pressure is approximately 0.15 mbar and its temperature is around 300 °C. An oxidant gas (N_2O, O_2, CO_2, etc.), trimethylaluminum (TMA, $Al(CH_3)_3$), and argon (Ar) are used as reactant gases. Before the deposition the wafer are preheated, then they are transported on a trail under the glowing plasma (see Figure 5.1). The plasma is continuously glowing and the aluminum oxide layer is deposited at a constant rate. The thickness of the layer is controlled by the velocity of the trail under the plasma source.

5.2.2 Homogeneity of the process

The PECVD system was originally used for the deposition of SiN_x and needed to be adapted to allow for a high-quality deposition of Al_2O_3 layers. The reaction of

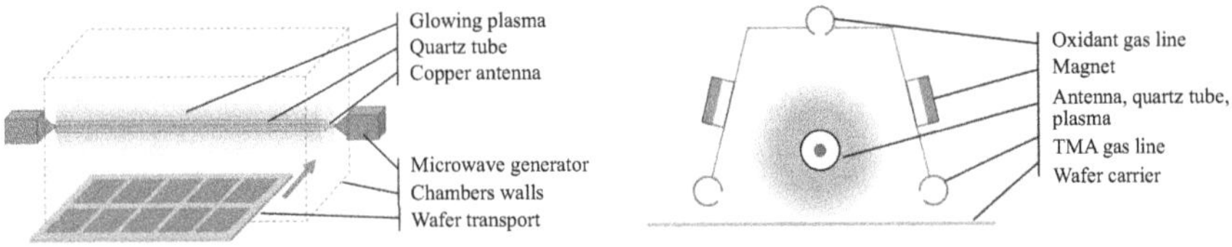

Figure 5.1 Schematic of the MW-PECVD setup with the plasma source on the left, a cross section showing the gas injection on the right

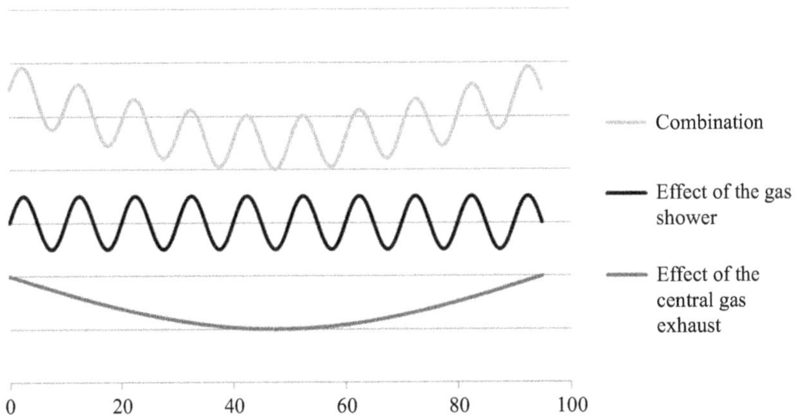

Figure 5.2 Schematic of the thickness of the layer over the trail for different configurations of the exhaust and the gas shower

TMAl with the oxidant gas is very local, close to the TMAl gas shower. The local deposition can be explained by the fact that TMAl does not need the plasma ignition to be absorbed on the silicon surface. The deposition rate depends also on the homogeneity of the pressure in the chamber. The use of a central pump exhaust might lead to inhomogeneity in the deposited film. These problems can be solved using a smaller pitch between the gas shower injectors and using an exhaust distribution. The effect of the gas distribution on the layer thickness homogeneity is shown in a schematic Figure 5.2.

The MW is introduced into the deposition chamber via a linear antenna. Two MW sources are connected to this antenna, one on each side of the antenna. The power in the MW antenna decays with the distance to the MW source. Therefore, if the power level is insufficient the power distribution can exhibit inhomogeneities (see Figure 5.3(a)).

The peak power of the MW pulse and the mean power of the MW also greatly influence the homogeneity of the deposited layer. The MW power decreases alone the linear antenna. As the system presents a MW generator on each side, the power decrease of one side is compensated by the other side (as explained in Figure 5.3(b)).

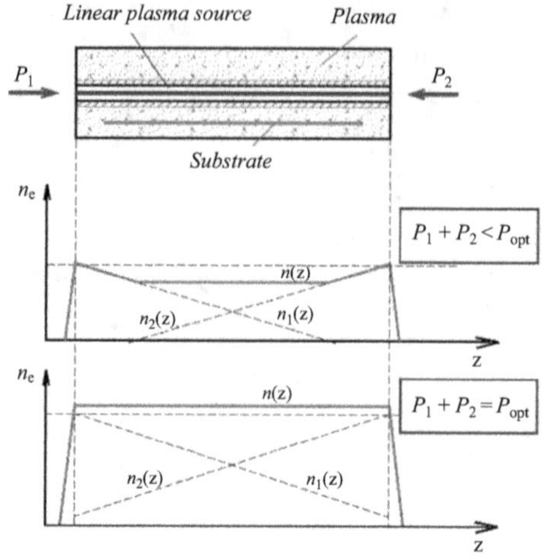

Figure 5.3 (a) Explicative sketch of inhomogeneity due to linear decrease of the plasma power [13]

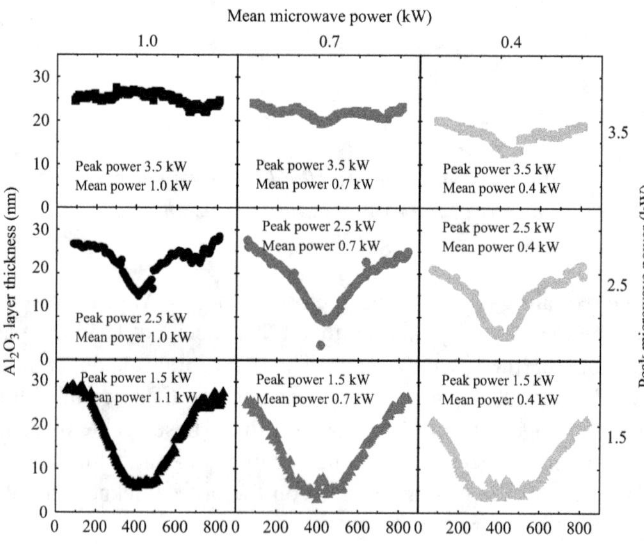

Figure 5.3 (b) Aluminum oxide layer thickness as a function of the position on the trail

If the power is not sufficient, then the decreases are too quick and the middle of the antenna receives less power. In Figure 5.3(b), the Al_2O_3 layer thickness over the trail is plotted for three mean power and three peak power settings. It can be observed that the homogeneity of the thickness is mainly affected by the peak power; however, the

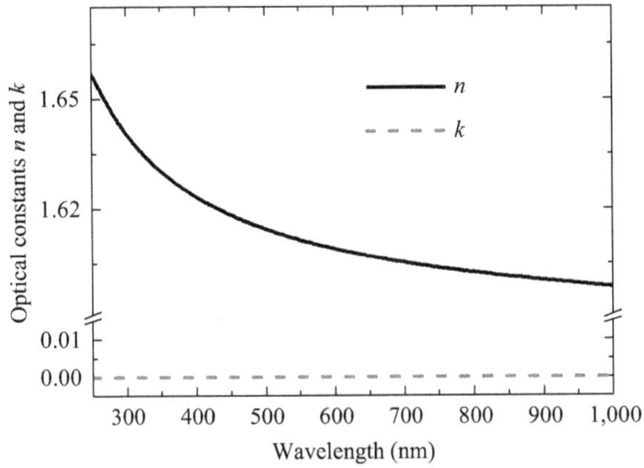

Figure 5.4 Refractive index n and extinction coefficient k of PECVD aluminum oxide as a function of the incident photon wavelength, measured using ellipsometry

mean power is slightly influencing the homogeneity, and it has a noticeable influence on the deposition rate.

5.2.3 Layer characteristics

5.2.3.1 Potential application of aluminum oxide

The first application of aluminum oxide layer for c-Si photovoltaics is the passivation of lowly doped silicon, corresponding to the rear surface of a *p*-type passivated emitter rear cell (PERC). The second application of this layer is the passivation of boron-doped emitters for *n*- type solar cells. Both applications require different layer properties:

- For PERC application, the layer should passivate lowly doped *p*-type silicon, the passivation quality should also withstand high temperature treatments like firing. In order to increase the internal reflection, the layer refractive index should be as low as possible and the thickness as high as possible. In order to thicken the layer, another layer (e.g., silicon nitride) might be used.
- For *n*-type application, the aluminum oxide layer is combined with silicon nitride to form an anti-reflective stack. The light absorption in the layer needs to be as low as possible. As the reflective index of aluminum is lower than the one of SiN_x, its thickness should be as low as possible in order to decrease the reflection. For this application, the passivation of silicon surface highly doped with boron is needed.

5.2.3.2 Optical characteristics and chemical composition

The refractive index and the extinction coefficient of typical PECVD AlO_x layers are shown Figure 5.4. In the relevant wavelength for solar cells (250 to 1,000 nm), the refractive index is between 1.60 and 1.65, while the photon absorption was below the detection limit. This is comparable to PA-ALD Al_2O_3 values[85] reported to be between 1.60 and 1.65. These deposited films are amorphous and have a significantly lower refractive index as crystalline aluminum oxide like sapphire (1.77).

These optical properties are suitable for front or back surface application on solar cells. Benick *et al.* [5] showed that thin aluminum oxide layers can be used in combination with silicon nitride for ARC layers, while Schmidt *et al.* [4] showed that aluminum oxide exhibits similar behavior to SiO_2 for internal back reflection at the rear of Si solar cells.

The composition of the layer has been determined by X-ray photoelectron spectroscopy. The layer is found to be quasi-stoichiometric with 60% oxygen (O), 36% aluminum (Al), and 4% carbon (C), where nitrogen (N) and silicon (Si) were not detected. The small carbon content is attributed to the methyl groups of TMA. As the studied aluminum oxide layer is proven to be very close to stoichiometric Al_2O_3, "Al_2O_3" can be also used for the PECVD aluminum oxide.

5.2.3.3 Passivation of lowly doped p-type silicon surface

The quality of the surface passivation is usually evaluated on the so-called lifetime samples. These are wafers passivated on both top and bottom surfaces with the same passivation layer, forming a symmetrical sample. The highest material and surface preparation qualities are required in order to avoid limiting the charge carrier lifetime by other factors than the surface recombination (especially bulk recombination). The effective charge carrier lifetime (τ_{eff}) is measured preferentially using a method which is sensitive to the injection level. The recombination rate (U) in the sample can be described as

$$U = W\frac{\Delta n}{\tau_{eff}} = 2 \cdot S \cdot \Delta n + U_{bulk}, \tag{5.1}$$

where W is the wafer thickness, Δn the minority carrier density, and S the SRV. A maximum value for the SRV (S_{max}) is obtained by neglecting the recombination rate (U_{bulk}) in the bulk, then

$$S_{max} = \frac{W}{2 \cdot \tau_{eff}}. \tag{5.2}$$

As the goal is to minimize the SRV, S_{max} is a pessimistic evaluation of it. A more realistic value can be obtained by calculating U_{bulk} using parametric value for the radiative and Auger recombination [14].

The carrier directly after deposition (as-deposited lifetimes) is the range of 1–100 μs, which is partly several orders of magnitude below the limit of the material. The passivation needs to be activated, and a forming gas anneal (FGA) can be applied with temperatures ranging 400–450 °C for a duration of about 5–30 min. The passivation might be also activated by a firing process (1–3 s at 800–850 °C). This firing process is typically applied in industry for the sintering of the screen-printed contacts. In Figure 5.5, the effective carrier lifetime depending on the excess carrier

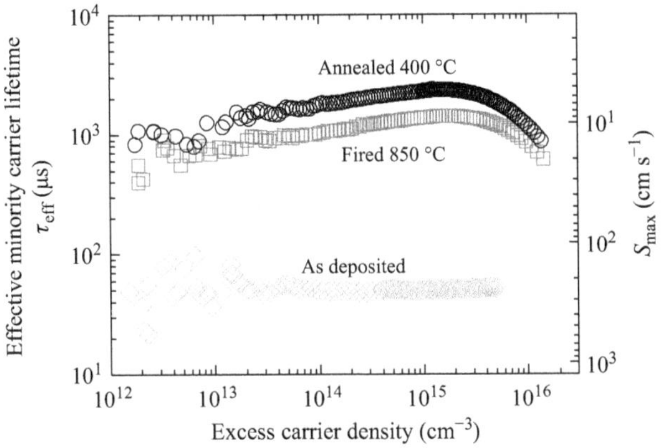

Figure 5.5 Effective minority carrier lifetime τ_{eff}, and the corresponding S_{max} of c Si (FZ (100), p-type, 1 Ω cm, 250 μm thick). The sample are passivated by PECVD AlO_X several post process tempering step are applied

density is presented. For solar cells under standard irradiation (~1 sun), the excess carrier density typically varies from 10^{13} cm^{-3} to 10^{15} cm^{-3}. In this range, PECVD Al_2O_3 shows a relatively stable minority carrier lifetime of more than 1.2 ms for the annealed state corresponding to S_{max} values below 10 cm/s. Unlike a-SiN_x:H and SiO_2 the lifetimes achieved with PECVD Al_2O_3 are not observed to decrease strongly at low injection levels. B. Hoex *et al.* [15] interpreted this phenomenon as an effect of the accumulation layer induced by negative-charge passivation.

A recombination event at the surface requires a minority carrier, a majority carrier, and an available state (defect) at the surface. The surface passivation leads to a decrease of the probability of the recombination. We can define two types of passivation, the chemical passivation and the field effect passivation. The chemical passivation corresponds to the saturation of the dangling bonds which decreases the density of interfacial states (D_{it}). The other passivation is the field effect passivation; it can be obtained by doping the surface or in the case of aluminum oxide by charges in the layer. The field obtained repels the majority carriers from the surface, which decreases the probability to reach the surface states. In the case of aluminum oxide the built-in charges are negative which is well suited for *p*-type surfaces passivation.

High-frequency and quasi-static capacitance-voltage (CV) measurements (method of Castagné and Vapaille [16]) of metal insulator semiconductor (MIS) capacitance is a common characterization technique to extract the charge density (Q_{tot}) and the interface trap density (see Figure 5.6). For PECVD, aluminum oxide values of $Q_{tot} = -2.1 \times 10^{12}$ cm^{-2} and $D_{it} = 1.7 \times 10^{11}$ eV^{-1} cm^{-2} were measured for a 30 nm thick PECVD aluminum oxide layer after annealing. The high density of negative charges induces an accumulation layer on *p*-type silicon, producing a field effect passivation. Low D_{it} associated with high negative surface charge density can

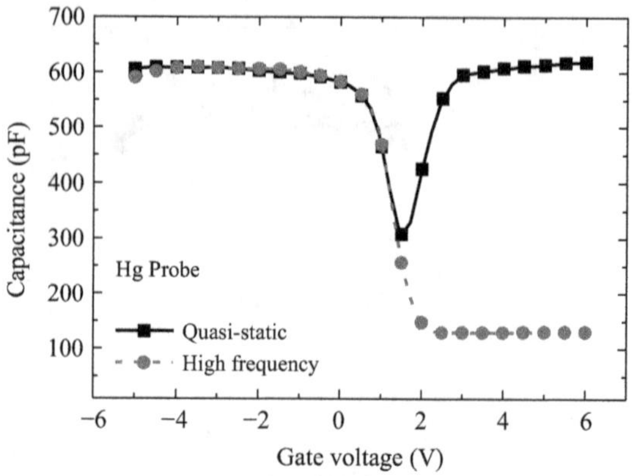

Figure 5.6 High-frequency and quasi-static capacitance-voltage measurements of a 30 nm thin PECVD Al_2O_3 layer deposited on a p-type doped silicon substrate (1 Ωcm). A charge density of $Q_{tot} = -2.1 \times 10^{12}$ cm^{-2} was calculated from these curves

Table 5.1 Summary of aluminum oxide deposition techniques used for passivation of c-Si p-type surfaces. The values reported are after annealing

Deposition technique	Si substrate	S_{eff} (cm s^{-1})	Q_{tot} (cm^{-2})
APCVD [1]	p-type, 2 Ωcm	210	-3×10^{12}
Sol-gel [2]	FZ p-type, 1 Ωcm	100	-1.6×10^{12}
PA-ALD [15]	FZ p-type, 2 Ωcm	~6	-1×10^{13}
PECVD [6] (with CO_2^*)	Cz p-type, 10 Ωcm	~10	-2.1×10^{12}
RF sputter AlO_x [10]	FZ p-type, 0.8 Ωcm	~55	-3×10^{12}
PECVD (with N_2O or O_2^*) [7]	FZ p-type, 1 Ωcm	<10	-2.1×10^{12}

explain the excellent passivation shown by PECVD Al_2O_3 layer on *p*-type silicon. PECVD and ALD are not the only deposition technique allowing the building of negative charge and a high-quality surface passivation. In Table 5.1, a summary of reported SRV and charge density for different deposition process is provided.

5.2.3.4 Passivation of highly doped p-type silicon surface

The quality of the surface passivation on highly doped surface is evaluated on diffused wafers which feature boron-doped emitter on both sides. The samples used are also symmetrical like for the lifetime samples. The contribution for the recombination in the emitter is described by the emitter saturation current density (j_{0_e}) which can be calculated as [17]:

$$\frac{1}{\tau_{\text{eff}}} - \frac{1}{\tau_{\text{Auger}}} = \frac{1}{\tau_{\text{SRH}}} + 2\frac{j_{0_e}(N_D + \Delta n)}{qn_i^2 W}, \tag{5.3}$$

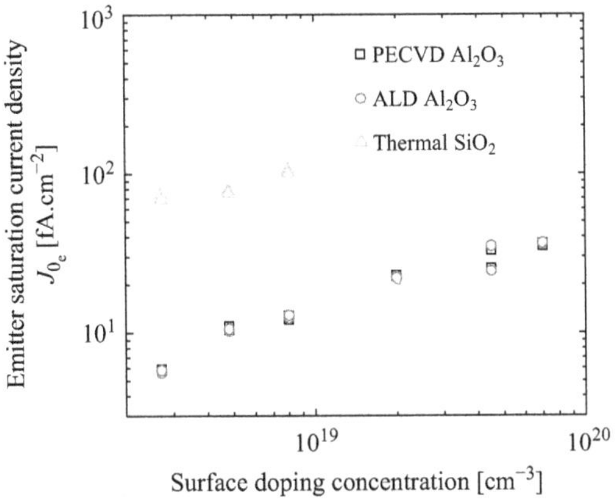

Figure 5.7 Emitter saturation current density (j_{0_e}) as a function of the surface boron doping concentration (N_{surf}). The measurement has been performed at an injection level of 2×10^{16} cm^{-3}, on the planar sample

where τ_{eff} is the effective lifetime measured, τ_{Auger} is the Auger lifetime [14], τ_{SRH} is the trap assisted recombination in the bulk (Shockley–Read–Hall (SRH) recombination [18,19]), n_i is the intrinsic carrier concentration ($n_i = 1 \times 10^{10}$ cm^{-3}) [20], q is the elementary charge, and W is the wafer thickness. In Figure 5.7, the j_{0_e} is plotted as a function of the surface doping concentration (N_{surf}) of each emitter. The results given by the samples passivated with ALD and PECVD Al_2O_3 cannot be clearly distinguished from each other showing that both passivation layers are on very similar level. However, the three lower surface concentrations show much higher j_{0_e} when passivated by thermal silicon dioxide (SiO_2). Therefore, the two passivation layers can be considered as equivalent on boron emitters. The samples passivated by SiO_2 present clearly higher values of j_{0_e}. This shows that silicon dioxide cannot passivate boron emitter as effectively as Al_2O_3.

The emitter saturation current density does not only depend on the recombination at the surface but also on the recombination in the emitter (Auger, SRH, etc.) and on the transport of the carriers thought the emitter or in other words it depends on the field effect generated by the emitter. Therefore, a numerical simulation is necessary to obtain the separation of the recombination in the emitter with the recombination at the surface [21]. The simulation allows calculating the j_{0_e} for an emitter from which the profile is known for a certain SRV. When the SRV is set to zero, the j_{0_e} is minimum, this value is $j_{0_{min}}$. As the accuracy of the measurement and the simulation of j_{0_e} is limited, there is also a minimum surface recombination velocity (SRV_{min}) value which can be determined by this method. In Table 5.2, the values of $j_{0_{min}}$ and SRV_{min} are reported for each emitter and the j_{0_e} measured for the emitter passivated with Al_2O_3. It can be noticed that the j_{0_e} values follow the same tendency as the theoretical $j_{0_{min}}$ values. Additionally for the emitters 2 to 6, the measured j_{0_e} is equal or even below $j_{0_{min}}$. For these emitters, the only conclusion that can be made is that the SRV values can only be equal or lower than the SRV_{min} values.

Table 5.2 Measured emitter saturation current density (j_{0_e}), theoretical minimum of the emitter saturation current ($j_{0_{min}}$), and surface recombination velocity (SRV_{min}) of six emitters. The emitter recombination was simulated

N_{Surf} (cm^{-3})	Measured j_{0_e} ($fA \cdot cm^{-2}$)	Simulated $j_{0_{min}}$ ($fA \cdot cm^{-2}$)	SRV ($cm \cdot s^{-1}$)
Emitter 1	<35	22.4	50 ± 2
Emitter 2	18–26	24.6	150 ± 5
Emitter 3	16–18	21.0	45 ± 2
Emitter 4	8.6–10	13.4	50 ± 2
Emitter 5	6.5–7.0	10.4	35 ± 2
Emitter 6	5.0–5.5	5.6	6 ± 1

For highly doped boron surfaces, the passivation quality obtained by PECVD and ALD aluminum oxide are very similar, and its quality is much higher than using silicon dioxide passivation. The j_{0_e} obtained are so low that they correspond to the simulated lower limit. The emitters are therefore not limited by the passivation quality.

5.3 Plasma and layer characteristics of ICP-PECVD

Inductively coupled plasma PECVD (ICP-PECVD) allows very effective gas dissociation [22] due to the very high plasma density and therefore high ionization of the gas species. Veith *et al.* [11] and Roos *et al.* [23] have shown that highly passivating aluminum oxide layers can be manufactured using this technique by applying a gas mixture of TMAl, Ar, and N_2O. Applying ICP-PECVD aluminum oxide layers as rear passivation for PERC-type solar cells led to efficiency of above 20% [11,23].

Within this chapter, a closer look will be given at a SINGULUS Singular ICP-PECVD tool and process that was applied for the fabrication of thin AlO_x layers. Sketches of the tool can be found in Figures 5.8 and 5.9. Figure 5.9 elucidates the way the wafers are taken through the deposition tool. A carrier with four 156×156 mm^2 wafers enters the vacuum through the load lock and then is moved clockwise to the IR station and onwards to the AlO_x and the three SiN_x stations. Here, the ICP-sources operating at 13.56 MHz are mounted and are process gas-separated by individual vacuum locks.

The deposition experiments and carrier lifetime investigations were carried out using float-zone (FZ) monocrystalline silicon wafers (p-type, boron doped) with a resistivity of 1 . . . 1.5 Ω cm and a thickness of about 250 . . . 300 μm.

Plasma analysis applying spectroscopic plasma monitoring (SPM) technique based on optical emission spectroscopy provides deep time-resolved insights into the dynamics of the plasma species. Therefore, it can be recognized what kinds of species are present and how to control their ratios. In the investigations presented

Figure 5.8 SINGULUS Singular PECVD tool applied in the investigations here [23]

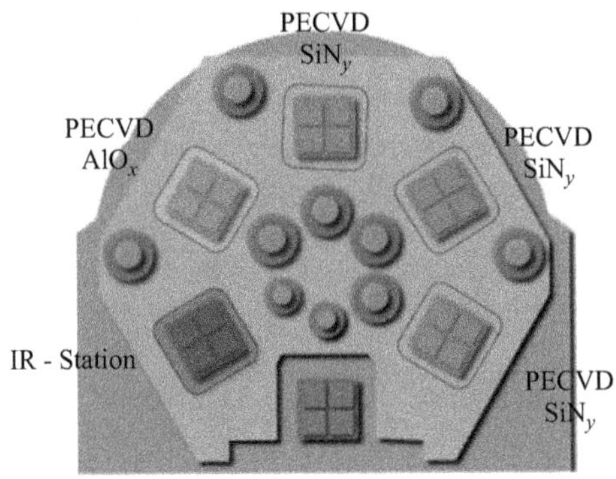

Figure 5.9 Deposition station top view schematic of the singular PECVD tool showing the load lock at the bottom of the image, IR station for wafer heating, and the deposition chambers for AlO_x and SiN_x. Since the SiN_x layers are typically prepared much thicker than the AlO_x layer, three SiN_x stations are used sequentially. For all process stations, the same process time is available due to the synchronized clockwise movement of the four wafers from one process station to the other. See also [23]

here, a PLASUS EMICON 3MC system was applied. The analysis of the emitted spectra was partly performed using the actinometry method where the spectra are normalized to a nonreactive gas which was in our case the emission from the argon gas being kept at constant partial pressure throughout the experiments.

The combination of this *in-situ* characterization method with *ex-situ* analysis of the created layers is of benefit to achieve a correlation of parameters and an optimization. Contactless capacitance-voltage (COCOS, applying a Semilab SDI CFORS tool), spectroscopic ellipsometry (SE, applying a J. A. Woollam M-2000 tool), and carrier lifetime measurements (QSSPC, WCT-120 tool by Sinton Instruments) were performed. Detailed characteristics of the plasma technology were elucidated by particle-in-cell Monte Carlo (PIC-MC) simulations of the

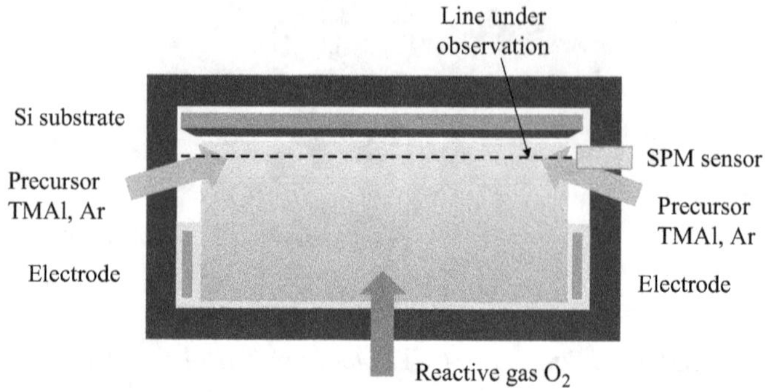

Figure 5.10 Schematic drawing of the setup of the AlO_x deposition chamber. The deposition takes place via the plasma which is located below the substrates. The reactive gas (oxygen containing in the case of AlO_x deposition) is passing the inductively coupled plasma (ICP) and gets dissociated. The precursor gas (trimethylaluminum TMAl) is entering the deposition chamber near the substrates. An SPM sensor is installed to observe the plasma close to the substrate. [25] PIC-MC plasma simulations help to understand the physics within the PECVD process

ICP-PECVD process chamber [24] providing local electron temperature and electron density distributions in the PECVD reactor that can facilitate an efficient deposition process.

Since PERC-type and other solar cell types applying passivated surfaces are becoming increasingly important not only in academia goal also industrially, a thorough understanding of the AlO_x deposition process is of high relevance and allows for stable and highly efficient processes.

One goal of the investigations presented here is the reduction of the AlO_x layer thickness from 20–30 nm to 5 nm or even less still allowing the same excellent surface passivation properties.

5.3.1 Plasma simulations

For the simulation of the AlO_x deposition plasma process, which was performed by M. Siemers and A. Pflug at Fraunhofer Institute for Surface Engineering and Thin Films (IST), the setup of the process chamber was used as sketched in Figure 5.10, which is in accordance with the actual reactor geometry.

Figures 5.11 and 5.12 show simulated electron density and electron energy distributions for a 2D cross section through the reaction chamber applying argon as working gas. Argon is often used for benchmarking and with argon experimental plasma measurements are easily done to compare against the simulations. Additionally, the applied power was about 1/100 of the reality due to limitations in the simulation. However, it is expected that the simulation results still allow qualitatively correct interpretations.

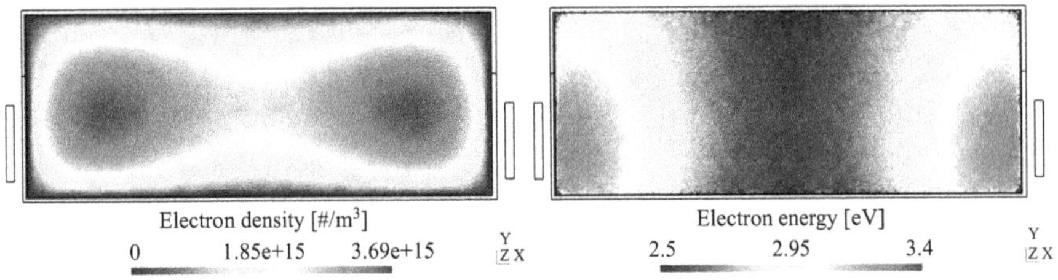

Figure 5.11 To investigate general plasma behavior, intense simulations were carried out using the particle-in-cell Monte Carlo (PIC-MC) method. The left and right parts show the two-dimensional distributions of electron density and electron energy, respectively, in the reactor vessel (horizontal width about 500 mm) for a gas (Ar) pressure of p = 2 Pa [25]

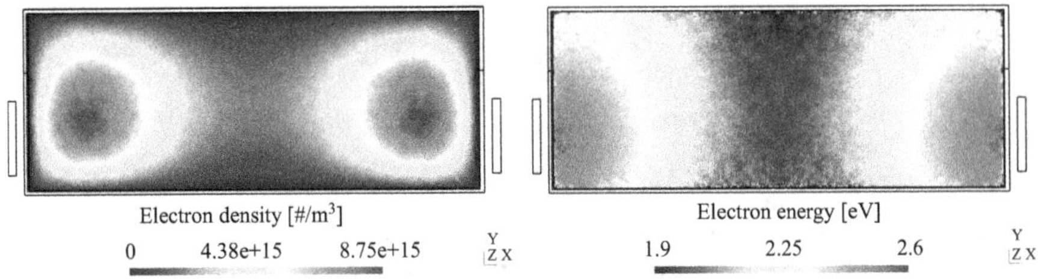

Figure 5.12 PIC-MC simulation results of electron density (left) and electron energy (right) for a process pressure of p = 10 Pa. Please compare with Figure 5.11 [25]

In the conducted simulations, a variation of the process pressure was applied with $p = 2$ Pa, in Figure 5.11 and $p = 10$ Pa, in Figure 5.12. Increasing the gas pressure leads to higher plasma density and lower electron energy. This higher plasma density causes more shielding of the inner plasma volume resulting in a stronger plasma confinement in close vicinity of the electrode. Electron densities up to 3×10^{15} m^{-3} or 9×10^{15} m^{-3} and low maximum electron energies of 3.4 eV or 2.5 eV for process pressures of $p = 2$ Pa or $p = 10$ Pa, respectively, were results of the simulation. Low electron energies are typical for an ICP–plasma generating low voltages and thus avoiding high-energy ions, which could damage the cell surface. The electron densities are relatively low which can be attributed to the low RF power (13.56 MHz) applied in the simulation compared to the real power used in the experimental setup. In the electron density distribution, one can find the "secondary coil" shape of the ICP setup.

5.3.2 In-situ plasma monitoring investigation

SPM based on optical emission spectroscopy was used to observe the plasma created for the deposition of AlO_x layers. A gas mixture of trimethylaluminum (TMAl,

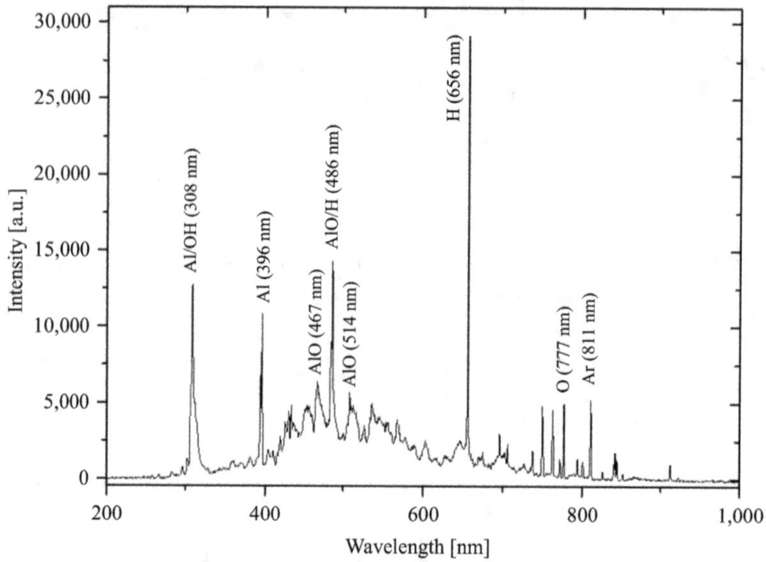

Figure 5.13 An example of a plasma monitoring spectrum [25]

$Al(CH_3)_3$), oxygen (O_2), and argon (Ar) was applied. In the photon spectrum, a lot of optical emission lines could be detected and assigned to various atoms or molecules in the plasma with the main contributions from species like Ar, H, O, Al, and AlO. Since Ar is inert and an Ar emission line is separated from all others, one can use it as base for an actinometric analysis of the data. Actinometry was presented by Coburn and Chen. More information can be found in [26]. Shortly, the Ar signal can be used to normalize the other signals and to allow an easier cross-interpretation over different parameter sets including start-up and shut-down characteristics.

One example of an investigated process can be found in Figures 5.13 and 5.14. Figure 5.13 shows a typical spectrum of the process with the spectral lines of the observed species. In Figure 5.14, the temporal behavior of the species over process time is presented where the top graph shows the data as acquired and the bottom graph's data underwent an actinometric calculation.

In the case presented here with each deposition cycle taking 7.5 s, we find that in the first 1.8 s after switching on the plasma, start-up conditions occur for the formation of the different gas species after the plasma ignition. A peak of H is most prominent. H must be coming from cracked TMAl molecules. Since H is beneficial for interface passivation, enough visible H in the spectra can be interpreted as a first positive sign for the surface passivation properties of the layer. At the same time, the amount of detected O species is relatively slowly increasing to a saturation value. One might wonder why the oxygen signal is so small in the start-up phase. We see that during start-up AlO species are providing an increased emission behavior showing that the formation of these species is one drain for oxygen. Thus,

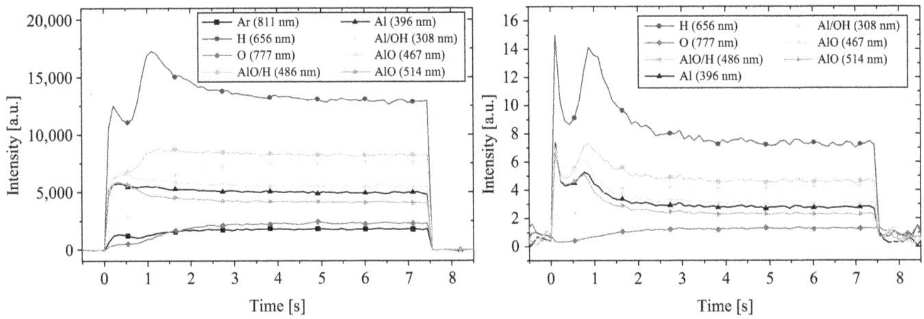

Figure 5.14 An example of gas dynamics within one deposition cycle is shown as acquired with SPM technique. The intensity on the y-axis equals the relative counts of photons at the mentioned spectral lines. The different species can be identified and separated. Aluminum and/or oxygen-containing species create the AlO$_x$ *layer. Hydrogen plays an additional crucial role in the creation of highly passivating layers due to its ability to lower surface defect density. Since Argon is inert, one can use it as reference gas for further analysis based on actinometry. Left: SPM data as acquired. Right: SPM data after actinometry based on Ar [25]*

a very effective decomposition of TMAl in the beginning of the plasma process and/or the presence of a large amount of TMAl seems to occur which provides bonding partners for oxygen in form of AlO, OH, and presumably other species such as CO_x as well. This is leading to stable process conditions after about 1.8 s.

The very steep curves in the lower graph at $t = 0$ s are due to actinometric calculations and should not be over-interpreted.

5.3.3 *Ex-situ investigation of layer and interface properties*

PECVD process parameter variations were carried out with simultaneous plasma monitoring and deposition of passivating AlO_x layers on wafer samples. One of these investigations was focusing on an O_2 flux variation. Here, the O_2 flux was changed from 20% to 100% (relative units) while keeping all other process parameters constant. The process pressure was increased slightly due to the higher total gas flux. Figure 5.15 shows an overview of the lifetime results. A strong increase of lifetime was found for relatively low oxygen fluxes. In Figure 5.16, plasma monitoring spectra of the O line (right) and the H line (left) in dependency of the O_2 gas flux at constant flux of the aluminum precursor in the PECVD process, after actinometric data analysis, are displayed. The oxygen signal ($\lambda = 777$ nm) is increased as expected for the process with higher O_2 gas flux. However, the H signal is decreased with increasing O_2 flux.

We assume that the decreased H signal correlates to the observed lifetime reduction, probably due to a reduced level of H incorporation in the passivating AlO_x layer. Also the AlO signal is increased with increasing O_2 flux (not displayed

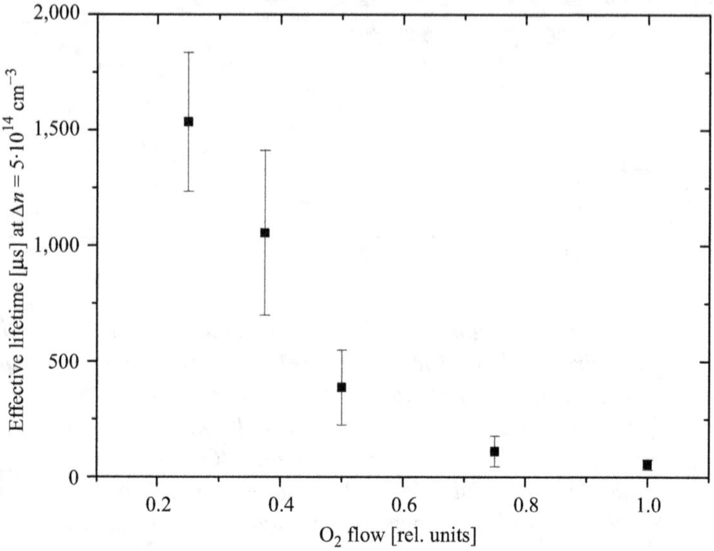

Figure 5.15 Effective carrier lifetime as a function of the O_2 gas flux (given in relative units) [23]

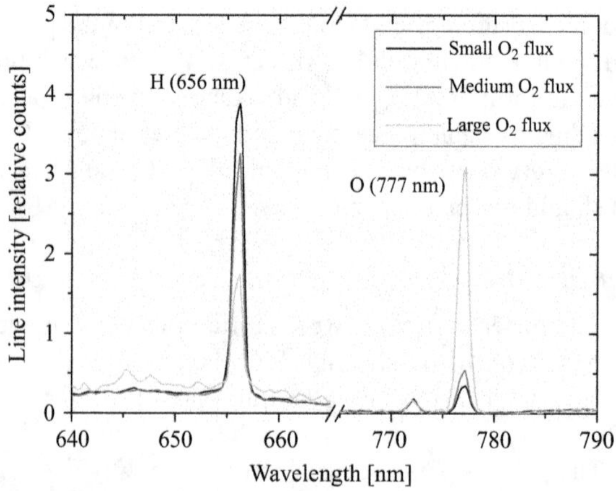

Figure 5.16 Plasma monitoring spectra of the O line (right) and the H line (left) in dependency of the O_2 gas flux at constant flux of the aluminum precursor in the PECVD process, after actinometric data analysis [25]

as graph here) indicating a more prominent presence of AlO compounds already in the gas phase. The influence of these compounds on the AlO_x layer structure, the interface defect density (D_{it}), and the negative charge density (Q_{tot}) is subject to further investigation.

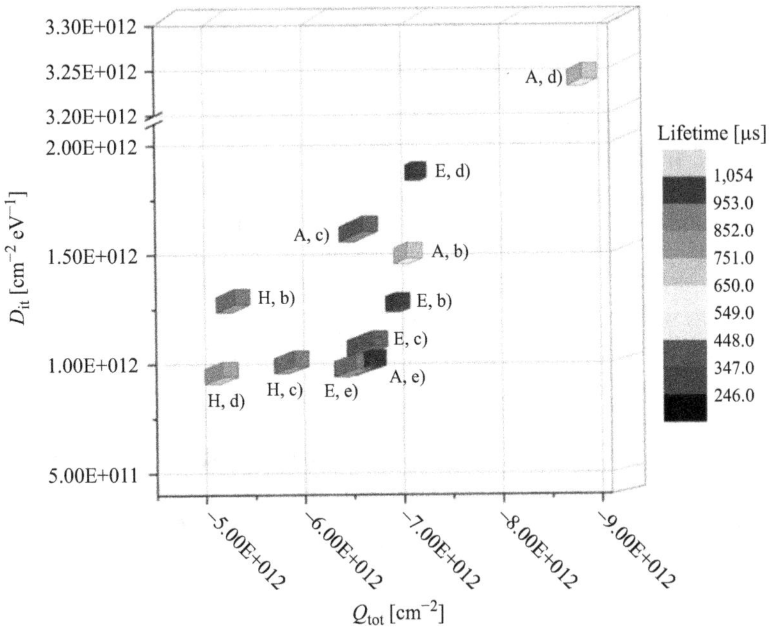

Figure 5.17 Contactless capacitance-voltage (COCOS, corona oxide characterization of semiconductor) measurements of differently prepared surfaces applied to different deposition parameters. Optimum processing leads to the lowest D_{it} and the highest negative Q_{tot} values. For example, following the results for samples with 1×10^{12} cm^{-2} eV^{-1}, one can find increasing lifetimes with increasing negative charge density [25]

Similar to the variation of the O_2 flux, also other process parameters can play a crucial role on the passivation results. However, it is possible to find robust high-quality processing setups.

COCOS measurements were performed to investigate the physical mechanisms behind the passivation results. Figure 5.17 provides an overview. The exact details of the process conditions are not given in this paper. However, one can see the range of results. In Figure 5.17, the lifetime results are additionally represented in the colors and heights of the data points. The surface preparation process variation is labeled with a small letter, the deposition process variation with a capital letter. Deposition process H allows excellent surface passivation almost independent of the preconditioning of the wafer surface (within the applied frame here). The passivation quality of layers deposited by process A is significantly depending on the surface preparation.

Thus, one can find combinations of surface preconditioning and passivation that allow high quality.

5.4 Performance comparison with other techniques (P&M)

5.4.1 Comparison between MW-, ICP-PECVD, and ALD

As a result of the ICP-PECVD investigations, the AlO_x layer thickness could be lowered to ranges below 5 nm with very good passivation results as shown in Figure 5.18 which were comparable to thick ICP-PECVD AlO_x layers of about 18 nm thickness and to a variety of ALD passivation results from the literature [27–29]. MW-PECVD AlO_x layers were reported to show significant lifetime decrease starting at ~10 nm [30]. This shows that PECVD processes can be designed in such a way that very thin aluminum oxide layers with excellent surface passivation properties are achievable.

5.4.2 Solar cell results

The most relevant test for a technology dedicated for solar cell is the application and test on the device. In this section, thin aluminum oxide layer (5 nm) deposited by ALD, aluminum oxide layer deposited by PECVD (MW-PECVD, see Section 5.2) and thermally grown silicon dioxide layers are compared when applied

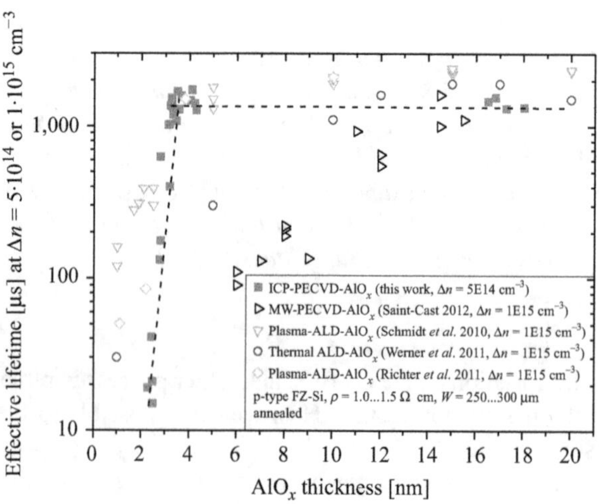

Figure 5.18 *Carrier lifetimes of c-Si wafers with AlO_x layers of thicknesses in the range up to 20 nm on both sides after annealing (FZ-Si substrates 1.0… 1.5 Ω cm, p-type, 250… 300 µm). A slight variation of the O_2 flux provided the data behind the red ICP-PECVD symbols shown. It seems evident that the deposition process is quite robust in this parameter range. With changing O_2 flux, still excellent passivation properties were possible. Lifetimes of about 1.7 ms reflect a SRV of about 5 cm/s. For comparison, also ALD-AlO_x and PECVD-AlO_x passivation layers on comparable substrate types from literature [27–30] are shown [25]*

Table 5.3 Illuminated current voltage measurement results, the measurement was performed in standard illumination conditions (1,000 W/m^2 AM1.5)

Rear surface passivation	V_{OC} (mV)	j_{SC} (mA cm^2)	*FF* (%)	η (%)
Thin ALD $Al_2O_3 + SiO_X$ (best cell*)	681	39.3	79.6	21.3
Mean/Std. dev. 7 cells	677/2	39.6/0.4	78.8/1.6	21.1/0.5
PECVD Al_2O_3 (best cell*)	684	39.4	79.8	21.5
Mean/Std. dev. 7 cells	680/1	39.9/0.2	77.8/1.1	21.1/0.4
Thermal SiO_2 (best cell*)	685	39.2	79.8	21.4
Mean/Std. dev. 7 cells	683/1	39.7/0.1	78.7/1.3	21.3/0.4

Calibrated measurement in CalLab.

on the rear surface of *p*-type rear passivated solar cells. The cells concerned are lab size 2 × 2 cm^2, the front surface has inverted pyramid, an homogeneous emitter passivated by thermal silicon oxide and the front contact is obtain by photolithography and Ag plating. The rear passivation is varied as mentioned, the rear electrode is from evaporated Al and the contacts are formed by laser-fired contact. The results of the illuminated current voltage measurement in standard irradiation conditions (1,000 W/m^2 AM1.5) are presented in Table 5.3. Very similar cells performances are obtained for each of the passivation layers used in comparison. Each one of these rear passivation layers allowed for an efficiency of more than 21% and a V_{OC} of more than 680 mV; they are therefore a valid alternative to thermally grown SiO_2 for the rear surface passivation of *p*-type solar cells. Bias-light-dependent spectral response shows that aluminum oxide provides a very good passivation even at low injection conditions this is a direct consequence of the stability of aluminum oxide the passivation quality at low injection (see Figure 5.5). This could increase the solar cells perform in low light conditions and represent an advantage of using aluminum oxide as rear passivation.

The high throughput, the flexibility, and the scalability of PECVD techniques are strong advantage for AlO_x PECVD layers as far as industrial application is concerned.

References

[1] R. Hezel and K. Jaeger, Journal of the Electrochemical Society, 136/2 (1989), 518.

[2] G. Agostinelli, P. Vitanov, Z. Alexieva, *et al.*, Proceedings of the 19th European Photovoltaic Solar Energy Conference and Exhibition, Paris, 2004.

[3] G. Agostinelli, A. Delabie, P. Vitanov, *et al.*, Solar Energy Materials and Solar Cells, 90 (2006), p. 3438.

[4] J. Schmidt, A. Merkle, R. Brendel, B. Hoex, M.C.M. van de Sanden, and W.M.M. Kessels, Progress in Photovoltaics: Research and Applications, 16(6), 461–466 (2008).

[5] J. Benick, B. Hoex, M.C.M. van de Sanden, W.M.M. Kessels, O. Schultz, and S. Glunz, Applied Physics Letters, 92 (2008), 253504.

[6] S. Miyajima, J. Irikawa, A. Yamada, and M. Konagai, Proceedings of the 23rd European Photovoltaic Solar Energy Conference, Valencia, Spain, 2008, p. 1029.

[7] P. Saint-Cast, D. Kania, M. Hofmann, J. Benick, J. Rentsch, and R. Preu, Applied Physics Letters, 95/15 (2009), 151502.

[8] T.-T. Li and A. Cuevas, Physica Status Solidi RRL, 3/5 (2009), 160.

[9] F. Werner, B. Veith, V. Tiba, *et al.*, Applied Physics Letters 97/16 (2010), 162103.

[10] L.E. Black, K.M. Provancha, and K.R. McIntosh, Proceedings of the 26th European Photovoltaic Solar Energy Conference, Hamburg, Germany, 2011, p. 1120.

[11] B. Veith, T. Dullweber, M. Siebert, *et al.*, Energy Procedia, 27/0 (2012), 379.

[12] G. Dingemans, M.C.M. van de Sanden, and W.M.M. Kessels, Plasma Processes and Polymers, 9/8 (2012), 761.

[13] Roth & Rau, *Plasma Source LPS1, Operating Instructions*, 2003, Roth & Rau AG, p. 34.

[14] A. Richter, S.W. Glunz, F. Werner, J. Schmidt, and A. Cuevas, Physical Review B, 86 (2012), 165202.

[15] B. Hoex, J. Schmidt, P. Pohl, M.C.M. van de Sanden, and W.M.M. Kessels, Journal of Applied Physics, 104 (2008), 044903.

[16] R. Castagne and A. Vapaille, Surface Science, 28 (1971), 157.

[17] D.E. Kane and R.M. Swanson, Proceedings of the 18th IEEE Photovoltaic Specialists Conference, Las Vegas, Nevada, USA, 1985.

[18] W. Shockley and W.T.J. Read, Physical Review, 87 (1952), 835–842.

[19] R.N. Hall, Physical Review, 87 (1952), 387.

[20] A.B. Sproul and M.A. Green, Journal of Applied Physics, 70 (1991), 846–854.

[21] K.R. McIntosh and P.P. Altermatt, Proceedings of the 35th IEEE Photovoltaic Specialists Conference, Honolulu, Hawaii, USA, 2010.

[22] M.A. Lieberman and A.J. Lichtenberg, Principles of plasma discharges and materials processing: John Wiley & Sons, New York, USA, 1994.

[23] B.F.P. Roos, T. Dippell, O. Hohn, *et al.* 27th European Photovoltaic Solar Energy Conference, Frankfurt, Germany, 2012, p. 1684–1687.

[24] A. Pflug, M. Siemers, T. Melzig, L. Schäfer, and G. Bräuer, Surface and Coatings Technology, 260 (2014), 411–416.

[25] M. Hofmann, M. Jäcklein, P. Saint-Cast, *et al.*, 32nd European Photovoltaic Solar Energy Conference, Munich, Germany, 2016, pp. 345–349.

[26] J.W. Coburn, and M. Chen, Journal of Applied Physics, 51 (6) (1980), 3134–3136.

[27] J. Schmidt, B. Veith, F. Werner, D. Zielke, and R. Brendel, Proceedings of the 35th IEEE Photovoltaic Specialist Conference, pp. 885–890 (2010).

[28] F. Werner, B. Veith, D. Zielke, *et al.* Journal of Applied Physics, 109 (2011), 113701.

[29] A. Richter, J. Benick, M. Hermle, M. Glunz, and W. Stefan, Physica Status Solidi RRL 5, No. 5-6 (2011), 202–204.

[30] P. Saint-Cast, Passivation of Si surfaces By PECVD aluminum oxide, PhD Thesis, University of Konstanz, 2012.

Chapter 6

Atmospheric pressure chemical vapor deposition of aluminum oxide for silicon surface passivation—background and materials science

Kristopher O. Davis[1]

6.1 Background on atmospheric pressure chemical vapor deposition

The most recent research [1–8] on atmospheric pressure chemical vapor deposition (APCVD) of Al_2O_3 was carried out using an in-line APCVD system manufactured by SCHMID Thermal Systems, Inc. (STI), formerly SierraTherm, Inc. This system consists of a Ni alloy (Inconel 601) transport belt, or ceramic rollers in newer designs; heating zones to control the substrate temperature, since this is a thermal CVD process; precursor delivery systems for both gas-phase and liquid-phase precursors; CVD injector(s) separated from the rest of the system by N_2 gas curtains; and exhaust management. An illustration of the CVD injector is shown in Figure 6.1, where N_2-diluted precursors are delivered to a CVD injector zone and exposed to the substrate, which is at an elevated temperature. Because the deposition is performed at atmospheric pressure and the CVD injectors are separated by N_2 gas curtains, multiple CVD injectors can be included in one system, allowing for multilayer film stacks to be deposited within a single process run.

The origins of this APCVD system can be traced back to Watkins-Johnson Company, who developed and produced these systems primarily for integrated circuit applications [9]. Much of the work focused on the deposition of undoped and doped SiO_x, using N_2-diluted SiH_4 and O_2 or tetraethoxysilane and O_3 as precursors [9–12]. These systems have also been used to deposit titanium oxide (TiO_2) [13,14], silicon nitride [15], titanium nitride [16], tin and fluorine-doped indium oxide [17], tungsten [18], and tungsten silicide [19].

APCVD oxides have been used for a wide range of different functions in crystalline silicon (c-Si) solar cells. Historically, APCVD was widely used in the PV industry to deposit TiO_2 antireflection coatings. TiO_2 was the main material used for this application from the 1970s through to the early 2000s [20], when it was replaced by plasma-enhanced chemical vapor deposition (PECVD)

[1]Department of Materials Science and Engineering, University of Central Florida, USA

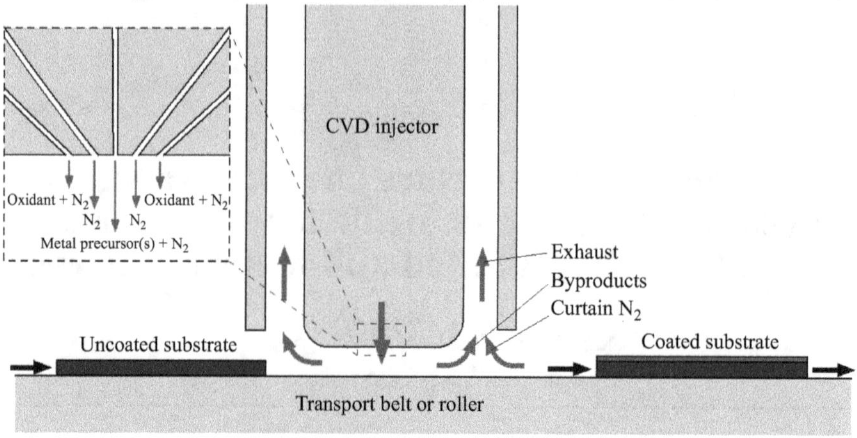

Figure 6.1 Diagram highlighting the dynamic in-line APCVD process used in this work

Table 6.1 Examples of how APCVD oxides have been used in c-Si solar cells

Oxide material	Functions within c-Si solar cells
Aluminum oxide (Al_2O_3, AlO_x)	• Surface passivation layer [1–8,22] • Top layer antireflection coating (ARC) in double-layer ARC structures [23]
Phosphosilicate glass (PSG), i.e., phosphorus-doped SiO_x (SiO_x:P)	• Phosphorus dopant source [24–34]
Borosilicate glass (BSG), i.e., boron-doped SiO_x (SiO_x:B)	• Boron dopant source [26,32,33,35–38]
Silicon oxide (SiO_2, SiO_x)	• Intermediate rear reflector in rear passivated cells [7] • Capping layer during PSG and BSG doping processes
Titanium oxide (TiO_2, TiO_x)	• Surface passivation layer [14,39–41] • Optical coating in single- and double-layer ARC structures [13,23,41–43] • Rear reflector and capping layer in rear passivated cells to protect the passivation layer from the Al during firing [7]
Boron-doped aluminum oxide (AlO_x:B)	• Boron dopant source and surface passivation layer [44,45]

hydrogenated silicon nitride (SiN_x:H) due to the bulk and surface passivation provided by SiN_x:H films [21]. APCVD has also been used to deposit phosphorus- and boron-containing layers as dopant diffusion sources. An overview of these and other functionalities explored in the literature is included in Table 6.1. The majority of recent work has focused on Al_2O_3 for surface passivation, the focus of this chapter.

6.2 Composition and structure of the APCVD Al_2O_3–Si interface

Several theories exist for the origin of the low D_{it} and large negative Q_{tot} for Al_2O_3 passivated Si. With respect to the low D_{it}, the primary reasons found in the literature are the formation of a thin SiO_x interlayer between the Si and Al_2O_3, diffusion of H to the Si surface, and film relaxation [46]. For the large negative Q_{tot}, defects like Al vacancies and O interstitials and/or local atomic coordination of the Al are thought to be the primary factors [46,47]. Thankfully, significant research has been performed on CVD and ALD Al_2O_3 films on Si using a variety of different characterization methods, which can be leveraged to better understand the role of composition and structure on the electronic properties of Al_2O_3 passivation layer.

With respect to the formation of a SiO_x interlayer, significant evidence in the literature suggests this interlayer is actually a compound containing Si, Al, and O (e.g., aluminum silicate). Klein *et al.* showed evidence of aluminum silicate formation during LPCVD of Al_2O_3 on Si [48]. Here, a substrate deposition temperature of 400 °C was used with TEDA-TSB as the Al source and O_2 as the oxidant. Nuclear reaction resonance profiling was used to obtain concentration profiles of Al with subnanometer resolution. Following the bulk Al_2O_3 film, decreasing Al concentration was found in an intermediate layer between the Al_2O_3 and the Si substrate, believed to be an aluminum silicate interlayer. Transmission electron microscopy (TEM) images also showed the presence of an interlayer, and X-ray photoelectron spectroscopy (XPS) data showed evidence of Al–O–Si bonding further supporting the evidence of an aluminum silicate. XPS data also showed an O/Al ratio of 1.64, slightly oxygen rich, similar the results from Davis *et al.* for APCVD Al_2O_3, or AlO_x [8]. Krug *et al.* later showed more details of the transport of Si, Al, and O and the chemical stability of the Al_2O_3 films deposited on silicon using ALD with TMA and H_2O [49]. Using nuclear reaction resonance profiling to measure samples in the as-deposited state and after annealing in ^{18}O-enriched O_2, density functions profiles were calculated for all of the species under investigation. A notable conclusion was that Si was found to migrate into the Al_2O_3 region forming a compound containing Si, Al, and O (e.g., aluminum silicate) or a SiO_2-Al_2O_3 mixture. Kimoto *et al.* also explored the Al_2O_3-Si interface using electron energy loss spectroscopy (EELS) in a TEM and interpreted the results using first-principles calculations [50]. They also found evidence of the aluminum silicate formation at this interface layer and believe an increase in tetrahedral-to-octahedral Al coordination (Figure 6.2) near is due to this aluminum silicate interlayer. Later work by Hoex *et al.* also found an increase in the ratio of tetrahedral-to-octahedral Al peak intensity near the Al_2O_3–Si interface for both thermal and plasma-assisted ALD Al_2O_3 upon annealing at 400 °C compared to the as-deposited state [51]. Using Fourier-transform infrared spectroscopy (FTIR), Black showed a strong correlation between the tetrahedral-to-octahedral peak intensity, measured via FTIR and the Q_{tot} for APCVD Al_2O_3 films [52]. This was later supported by Kühnhold *et al.* on PECVD Al_2O_3 films [53].

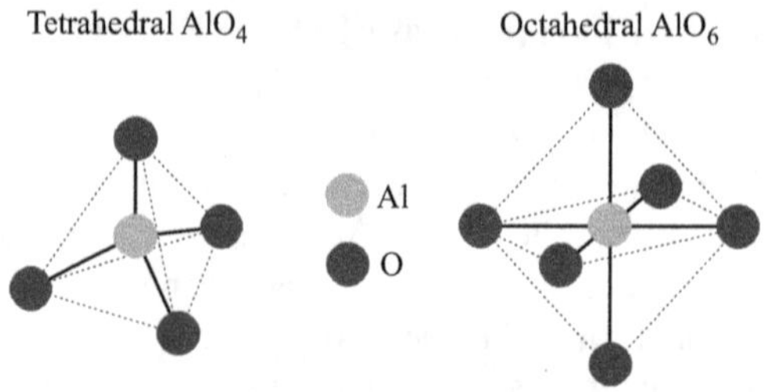

Figure 6.2 Illustration of tetrahedral AlO_4 and octahedral AlO_6

Recently, process-structure-property studies have been carried out on APCVD Al_2O_3 by Zhang *et al.* [54] and Looney *et al.* [55]. Al_2O_3 was deposited at ≈450 °C using the APCVD TMA-O_2 process from [8]. Following the deposition, a subset of the wafers were annealed at 500 °C for 30 min and some at 800 °C for 60 min. Changes in APCVD Al_2O_3-based films for different annealing conditions were studied using aberration-corrected high-resolution transmission electron microscopy, annular dark field imaging, and EELS. Measurements of S_{eff}, mid-gap D_{it}, and Q_{tot} were also performed. Following the 500 °C anneal, S_{eff} drops from ≈25 cm/s (as-deposited state) to ≈9 cm/s due to an order of magnitude reduction in the mid-gap D_{it} from 2.1×10^{12} to 2.8×10^{11} $ev^{-11} \cdot cm^{-2}$. Almost no change in Q_{tot} was observed. The TEM results show a slight increase in the interfacial layer from 1.5 to 2.3 nm, and the annular dark field images show a dramatic reduction in strain at the Al_2O_3–Si interface (Figure 6.3). The notion that stress can influence D_{it} is consistent with prior work by Lucovsky and Phillips [56].

For the films annealed at 800 °C, the mid-gap D_{it} remains in the 10^{11} $ev^{-1} \cdot cm^{-2}$ range, but the Q_{tot} drops in half resulting in a S_{eff} of ≈15 cm/s. This higher temperature anneal results in the partial crystallization and densification of the Al_2O_3 films evidenced by γ-Al_2O_3 crystallites dispersed within the bulk amorphous Al_2O_3 film. The depth-dependent EELS spectra for the different films show a change in the short-range coordination of Al from predominantly octahedral near the Si surface in the as-deposited state to both octahedral and tetrahedral for the 500 °C anneal. This suggests the SiO_x interlayer near the Si surface is actually an aluminum silicate, consistent with previous literature. Detailed analysis of the electronic properties of APCVD Al_2O_3–Si interfaces by Black *et al.* provides further evidence that the interface layer formed is electrically distinct from SiO_2–Si interfaces [5]. For the 800 °C anneal, the EELS spectra within the interfacial layer suggest that a SiO_2 layer is formed near the surface and the aluminum silicate glass layer now present further from the Si surface at the Al_2O_3–SiO_2 interface. The low mid-gap D_{it} and the reduction in the Q_{tot} may be due to the interface becoming more similar to a SiO_2–Si-type of interface.

Many of the conclusions found in [54,55] for APCVD Al_2O_3 are consistent with results from established theory and previous experiments on ALD Al_2O_3. However, more detailed process-structure-property studies of ALD, PECVD, and

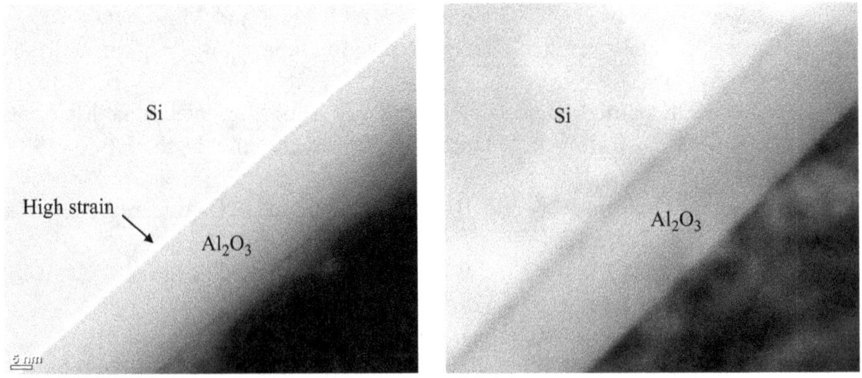

(a) (b)

Figure 6.3 Annular dark field images of the APCVD Al_2O_3 films and the Al_2O_3–Si interface in (a) the as-deposited state and (b) upon annealing at 500 °C. The Al_2O_3 films are ≈20 nm thick and the bright region at the Al_2O_3-Si interface for the as-deposited sample is indicative of high strain, as described by Looney et al. [55]. The mid-gap D_{it} for the as-deposited sample is approximately an order of magnitude higher than the annealed sample

APCVD Al_2O_3 would be beneficial, particularly since most of the subnanometer-scale interface analysis carried out on Al_2O_3-Si did not connect recombination-related electronic properties to the compositional and structural information.

References

[1] L.E. Black, T. Allen, A. Cuevas, K.R. McIntosh, B. Veith, and J. Schmidt, Solar Energy Materials and Solar Cells, 120 (2013), 339–345.

[2] L.E. Black, T. Allen, K.R. McIntosh, and A. Cuevas, Journal of Applied Physics, 115 (2014), 093707.

[3] L.E. Black, T.C. Kho, K.R. McIntosh, and A. Cuevas, Energy Procedia, 55 (2014), 750–756.

[4] L.E. Black, and K.R. McIntosh, Applied Physics Letters, 100 (2012), 202107-202101-202107-202105.

[5] L.E. Black, and K.R. McIntosh, IEEE Journal of Photovoltaics, 3 (2013), 936–943.

[6] L.E. Black, K.M. Provancha, and K.R. McIntosh, Surface passivation of crystalline silicon by APCVD aluminium oxide, 26th European Photovoltaic Solar Energy Conference and Exhibition, Hamburg, Germany, 2011, pp. 1120–1124.

[7] K.O. Davis, K. Jiang, C. Demberger, *et al.*, IEEE Journal of Photovoltaics, 3 (2013), 641–648.

[8] K.O. Davis, K. Jiang, M. Wilson, *et al.*, Physica Status Solidi (RRL) – Rapid Research Letters, 7 (2013), 942–945.

[9] L.D. Bartholomew, N.M. Gralenski, M.A. Richie, and M.L. Hersh, Atmospheric pressure chemical vapor deposition apparatus, Watkins-Johnson Company, USA, 1989.
[10] W. Kern, Dielectric Materials, in: R.A. Levy (Ed.) Microelectronic Materials and Processes, Springer Netherlands, Dordrecht, the Netherlands, 1989, pp. 247–273.
[11] J.P. Nieto, L. Jeannerot, and B. Caussat, Chemical Engineering Science, 60 (2005), 5331–5340.
[12] G.U. Pignatel, J.C. Sisson, and W.P. Weiner, Journal of The Electrochemical Society, 138 (1991), 1723–1728.
[13] M.C. Talló, and K.R. McIntosh, Permeability of TiO_2 anti-reflection coatings to damp heat, 24th European Photovoltaic Solar Energy Conference and Exhibition, Hamburg, Germany, 2009, pp. 2037–2040.
[14] A.F. Thomson, K.R. McIntosh, and B.S. Richards, Surface recombination in oxide-passivated Si after the deposition of APCVD TiO_2, 22nd European Photovoltaic Solar Energy Conference and Exhibition, Milan, Italy, 2007, pp. 1305–1308.
[15] X. Lin, D. Endisch, X. Chen, and A. Kaloyeros, MRS Proceedings, 495 (2011).
[16] J.N. Musher, and R.G. Gordon, Journal of Materials Research, 11 (2011), 989–1001.
[17] B. Mayer, Thin Solid Films, 221 (1992), 166–182.
[18] N. Gralenski, and L. Bartholomew, Selective and blanket tungsten by APCVD, Tungsten and other refractory metals for VLSI applications II, Materials Research Society, 1987, pp. 131–136.
[19] D. Dobkin, L. Bartholomew, G. McDaniel, and J. DeDontney, Journal of The Electrochemical Society, 137 (1990), 1623–1626.
[20] B.S. Richards, Novel uses of titanium dioxide for silicon solar cells, Centre for Photovoltaic Engineering, vol. Doctor of Philosophy, University of New South Wales, New South Wales, Australia, 2002.
[21] A.G. Aberle, Solar Energy Materials and Solar Cells, 65 (2001), 239–248.
[22] R. Hezel, and K. Jaeger, Journal of the Electrochemical Society, 136 (1989), 518–523.
[23] K.O. Davis, K. Jiang, D. Habermann, and W.V. Schoenfeld, IEEE Journal of Photovoltaics, 5 (2015), 1265–1270.
[24] R.R. King, R.A. Sinton, and R.M. Swanson, IEEE Transactions on Electron Devices, 37 (1990), 365–371.
[25] A. Metz, D. Adler, S. Bagus, *et al.*, Solar Energy Materials and Solar Cells, 120 (2014), 417–425.
[26] R. King, P.E. Gruenbaum, R.A. Sinton, and R.M. Swanson, Passivated emitters in silicon solar cells, Photovoltaic Specialists Conference, 1990, Conference Record of the Twenty First IEEE, 1990, vol. 221, pp. 227–232.
[27] A. Slaoui, B. Hartiti, J.C. Muller, R. Stuck, M. Loghmarti, and P. Siffert, Rapid thermal diffusion of phosphorus into silicon from doped oxide films, 22nd IEEE Photovoltaic Specialists Conference, Las Vegas, NV, 1991, vol. 441, pp. 445–449.

[28] S. Bourdais, G. Beaucarne, A. Slaoui, J. Poortmans, B. Semmache, and C. Dubois, Solar Energy Materials and Solar Cells, 65 (2001), 487–493.

[29] S. Sivoththaman, W. Laureys, P. De Schepper, J. Nijs, R. Mertens, IEEE Electron Device Letters, 21 (2000), 274–276.

[30] S. Sivoththaman, W. Laureys, J. Nijs, and R. Mertens, Applied Physics Letters, 67 (1995), 2335–2337.

[31] K.O. Davis, K. Jiang, C. Demberger, *et al.*, Physica Status Solidi (RRL) – Rapid Research Letters, 7 (2013), 319–321.

[32] R.A. Sinton, Y. Kwark, J.Y. Gan, and R.M. Swanson, IEEE Electron Devices Letters, 7 (1986), 567–569.

[33] R.A. Sinton, Y. Kwark, S. Swirhun, and R.M. Swanson, IEEE Electron Device Letters, 6 (1985), 405–407.

[34] F. Book, H. Knauss, C. Demberger, F. Mutter, and G. Hahn, Phosphorous doping from APCVD deposited PSG, 32nd European Photovoltaic Solar Energy Conference and Exhibition, 2016, pp. 824–827.

[35] R.R. King, and R.M. Swanson, IEEE Transactions on Electron Devices, 38 (1991), 1399–1409.

[36] P. Rothhardt, S. Meier, S. Maier, K. Jiang, A. Wolf, and D. Biro, IEEE Journal of Photovoltaics, 4 (2014), 827–833.

[37] P. Rothhardt, S. Meier, R. Hoenig, A. Wolf, D. and Biro, Solar Energy Materials and Solar Cells, 153 (2016), 25–30.

[38] K. Ryu, E. Cho, A. Rohatgi, and Y.-W. Ok, Progress in Photovoltaics: Research and Applications, 24 (2016), 1109–1115.

[39] K.R. McIntosh, S.C. Baker-Finch, N.E. Grant, A.F. Thomson, S. Singh, and I.D. Baikie, Journal of The Electrochemical Society, 156 (2009), G190–G195.

[40] A.F. Thomson, and K.R. McIntosh, Progress in Photovoltaics: Research and Applications, 20 (2011), 343–349.

[41] R.S. Bonilla, K.O. Davis, E.J. Schneller, W.V. Schoenfeld, and P.R. Wilshaw, IEEE Journal of Photovoltaics, 7 (2017), 1603–1610.

[42] D. Hocine, M.S. Belkaid, M. Pasquinelli, L. Escoubas, P. Torchio, and A. Moreau, Physica Status Solidi (c), 12 (2015), 323–326.

[43] B.S. Richards, Solar Energy Materials and Solar Cells, 79 (2003), 369–390.

[44] K.O. Davis, W.V. Schoenfeld, K. Jiang, and D. Habermann, Silicon material substrate doping method, structure and applications, 2013.

[45] K.O. Davis, Atmospheric pressure chemical vapor deposition of functional oxide materials for crystalline silicon solar cells, CREOL, the College of Optics & Photonics, University of Central Florida, Orlando, FL, 2015.

[46] G. Dingemans, and W.M.M. Kessels, Journal of Vacuum Science & Technology A: Vacuum, Surfaces, and Films, 30 (2012), 040802-040801-040802-040827.

[47] B. Hoex, J.J.H. Gielis, M.C.M. van de Sanden, and W.M.M. Kessels, Journal of Applied Physics, 104 (2008), 113703-113701-113703-113707.

[48] T.M. Klein, D. Niu, W.S. Epling, *et al.*, Applied Physics Letters, 75 (1999), 4001.

[49] C. Krug, E.B.O. da Rosa, R.M.C. de Almeida, *et al.*, Physical Review Letters, 85 (2000), 4120–4123.
[50] K. Kimoto, Y. Matsui, T. Nabatame, *et al.* Applied Physics Letters, 83 (2003), 4306.
[51] B. Hoex, M. Bosman, N. Nandakumar, and W.M.M. Kessels, Physica Status Solidi (RRL) – Rapid Research Letters, 7 (2013), 937–941.
[52] L.E. Black, New perspectives on surface passivation: Understanding the Si-Al_2O_3 interface, vol. Ph.D., The Australian National University, 2015.
[53] S. Kühnhold, P. Saint-Cast, B. Kafle, M. Hofmann, F. Colonna, and M. Zacharias, Journal of Applied Physics, 116 (2014), 054507.
[54] W. Zhang, K.O. Davis, W.V. Schoenfeld, P. Looney, and F. Wang, STEM-EELS Studies of the local structure and coordination of Al_2O_3/Si interfaces in Si solar cells, Microscopy and Microanalysis, vol. 20, Cambridge University Press, 2014, pp. 396–397.
[55] J.P. Looney, W. Zhang, F. Wang, K.O. Davis, K. Jiang, and W.V. Schoenfeld, Interfacial Structure and Passivation Properties of Al_2O_3 on Silicon, 43rd IEEE Photovoltaic Specialists Conference, Portland, OR, 2016.
[56] G. Lucovsky, and J.C. Phillips, Journal of Physics: Condensed Matter, 16 (2004) S5139–S5151.

Chapter 7

Al_2O_3 by atmospheric pressure chemical vapour deposition

Lachlan E. Black[1]

Thin film, amorphous Al_2O_3 possesses some remarkable properties that make it well suited to passivating silicon surfaces in photovoltaic devices. It is a wide-bandgap material, with little or no absorption in the visible range. It may be deposited by various methods at moderate temperatures, well below those required for thermal oxidation. It is capable of forming a high-quality interface with crystalline silicon, with a low density of interface states. Most distinctively, it possesses a large, stable negative charge that makes it ideally suited to passivate highly and lowly doped *p*-type silicon surfaces.

It is this ability to effectively passivate *p*-type surfaces that has generated so much recent interest in Al_2O_3 for photovoltaic applications. Conventional SiN_x passivation layers are not well suited to this role, because their positive charge makes them susceptible to parasitic shunting on undiffused *p*-type surfaces [1] and induces depletion at diffused *p*-type surfaces that significantly increases surface recombination [2]. Al_2O_3 thus has applications both as a replacement for the Al-alloyed 'back surface field' at the rear of conventional *p*-type silicon cells (which increasingly limits efficiency), and as an enabling passivation layer for the boron-diffused $p+$ region in new high-efficiency cells based on *n*-type material, with its desirable material properties [3].

Much recent work on Al_2O_3 for surface passivation has utilised ALD, either thermal or plasma-assisted, as a deposition method, with trimethylaluminium (TMA) as the chemical precursor. However, for industrial photovoltaic applications, conventional ALD is generally regarded as too slow to be economically viable [4,5]. Consequently, significant attention in recent years has focussed on alternative high-throughput deposition methods for Al_2O_3, including large-scale batch or spatial ALD [6–10], plasma-enhanced chemical vapour deposition (PECVD) [11–13], and reactive sputtering [14,15]. Al_2O_3 films deposited by all of these methods have been shown to be capable of providing excellent passivation in recent years. Various other deposition techniques, including sol-gel [16–18] and

[1]Department of Applied Physics, Eindhoven University of Technology, the Netherlands

screen-printed pastes [19,20], have also been investigated, though so far with less impressive results.

Given such interest in low-cost, high-throughput deposition techniques for Al_2O_3, surprisingly little attention has been paid to what would seem an obvious candidate, namely atmospheric pressure chemical vapour deposition (APCVD). This technique possesses a number of inherent advantages that make it particularly suited for such a role. Deposition occurs at atmospheric pressure under ambient conditions, using simple thermal activation at moderate temperatures. Therefore, no load-lock system, vacuum pump, or plasma is required. Deposition is continuous, rather than self-limiting, allowing in-line operation with only a single deposition zone. A wider range of aluminium precursors can be used than for ALD, many at lower cost than TMA. The use of non-pyrophoric (non-explosive) precursors in particular significantly reduces costs associated with shipping, storage, and handling of such chemicals. All of these factors contribute to a low cost of ownership for APCVD Al_2O_3.

As a CVD process, deposition by APCVD is based on the *chemical* reaction of one or more precursor species delivered in the *vapour* phase, where this process occurs at *atmospheric* pressure. The deposition reaction occurs at the substrate surface, or nearby, and generally requires an external source of energy. For APCVD that energy is purely thermal, coming from the elevated temperature of the substrate (PECVD, on the other hand, supplements this thermal energy by providing energy to reactive species in a plasma). The deposition reaction may be based on one or more of several mechanisms, including pyrolysis (thermal dissociation), hydrolysis (reaction with water), oxidation, nitridation, or reduction. The reaction may involve a single, or two or more reactant species (common secondary species include H_2O, O_2, CO_2, and NH_3). The primary reactant, or precursor, is often a metal-organic molecule. In this case the process may also be referred to as metal-organic chemical vapour deposition.

A wide variety of Al-containing chemical species can be used as precursors for Al_2O_3 deposition by APCVD. Precursors for which (non-plasma) Al_2O_3 CVD processes have been reported include halide ($AlCl_3$) [21], alkyl (e.g. trimethylaluminium, TMA or $Al(CH_3)_3$) [22], alkoxide (e.g. aluminium tri-isopropoxide, ATI) [23], and mixed alkyl-alkoxide compounds (e.g. dimethylaluminium isopropoxide, DMAI) [24], as well as more complex structures such as aluminium acetyl acetonate ($Al(acac)_3$) [25] and aluminium 2-ethylhexanoate [26]. Deposition may be performed at temperatures ranging from 200 to 1,000 °C, depending on the precursor and the desired film properties.

APCVD was used in the very first investigation of Al_2O_3 as a material for electronic device applications by Aboaf [23] in 1967. Deposition was performed by pyrolysis of aluminium tri-isopropoxide (ATI). It was reported that unlike SiO_2, which was well known to increase the *n*-type character of a surface (i.e. induce a negative surface charge due to its positive fixed charge), Al_2O_3 applied to Si or Ge surfaces could, for certain deposition conditions, have the opposite effect, producing a strongly *p*-type surface regardless of the substrate type. This was the first evidence of Al_2O_3's large negative fixed charge. While this negative charge was considered useful for some electronic device applications, interest in Al_2O_3 as a possible replacement for SiO_2 was stimulated mainly by its higher dielectric constant, effectiveness as a diffusion barrier for ionic impurities, better radiation resistance, and the possibility of low temperature deposition by various methods [27–31]. Subsequent investigations of the Si–Al_2O_3 interface revealed a high-quality interface, with a low density of interface states [27,30,32].

APCVD was again used in the first investigation of Al_2O_3 as a passivation layer for photovoltaic applications by Jaeger and Hezel in the late 1980s [33,34]. Deposition was again performed using ATI as the precursor, following Aboaf. The results (which included the first silicon solar cells incorporating Al_2O_3 passivation) were quite encouraging. Hezel and Jaeger [34] reported a low mid-gap interface defect density D_{it} of 8×10^{10} eV^{-1} cm^{-2} and a large fixed charge concentration Q_f/q of -3.2×10^{12} cm^{-2}. They also measured a surface generation velocity S_0 of 210 cm/s by the pulsed MIS capacitor technique. The latter result has usually been misunderstood and hence misrepresented in the more recent literature, where the S_0 value reported by [34] has been conflated with the more familiar effective surface recombination velocity S_{eff} [12,14,35]. The latter includes the effect of charge-assisted passivation, while the former does not, being measured under conditions of free-carrier depletion. By misleadingly reporting the S_0 value measured by [34] as though it were an S_{eff}, this result, and by association the quality of APCVD Al_2O_3 passivation in general, were made to appear significantly worse in comparison with those of other techniques.

The results of Jaeger and Hezel with Al_2O_3, while promising, were not followed up at the time, as attention shifted to PECVD SiN_x (also investigated as a passivation layer by the same authors), which was better suited to passivating the phosphorus-diffused front-side of conventional *p*-type cells. Interest revived in the mid-2000s as the need for a passivation material for *p*-type surfaces became apparent, with reports by Agostinelli *et al.* [35,36] and Hoex *et al.* [37] highlighting the excellent passivation achievable using Al_2O_3 from ALD. Subsequent years saw an explosion of studies on Al_2O_3 for photovoltaic applications. However, while the pioneering work of Jaeger and Hezel was acknowledged, most of these new studies employed ALD, as the technique with which the most recent excellent results had been achieved.

The first work on silicon surface passivation by APCVD Al_2O_3 since that of Jaeger and Hezel was reported in 2011 by Black *et al.* [38]. Instead of ATI, this work employed triethyldialuminium-tri-(*sec*-butoxide) (TEDA-TSB) [39] as the aluminium precursor together with H_2O as co-reactant. TEDA-TSB has the advantage of being liquid at room temperature, unlike ATI, and does not suffer from the same stability issues [40]. S_{eff} values of 1.2 cm/s and 12.5 cm/s were reported on 1 Ωcm n-type and 0.8 Ωcm p-type <100> Si surfaces, respectively, with a high negative fixed charge of -3.0×10^{12} cm^{-2} and reasonably low interface state density of 3.3×10^{11} eV^{-1} cm^{-2} using a non-optimised process. These good results together with the further improved values reported in follow-up works [41] showed unambiguously that APCVD Al_2O_3 was capable of providing excellent passivation of silicon surfaces.

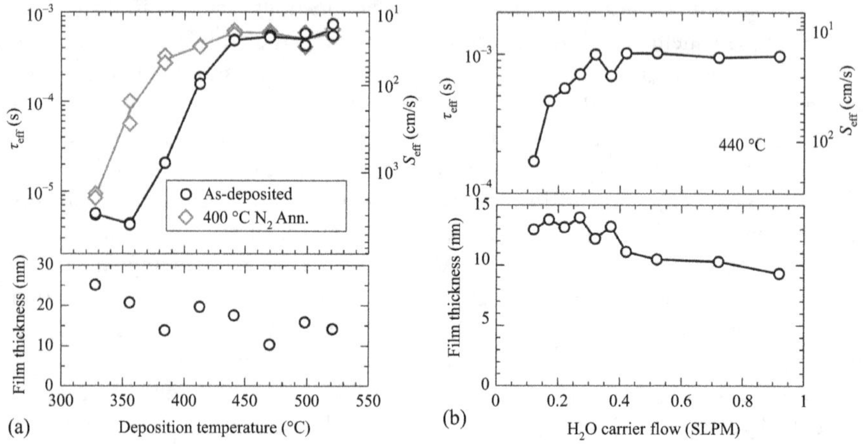

Figure 7.1 *(a) Effective excess carrier lifetime τ_{eff} and film thickness vs. deposition temperature for APCVD Al_2O_3 deposited from TEDA-TSB on 325 μm thick 0.8 Ωcm p-type 100 FZ Si, as-deposited and after annealing at 400 °C in N_2 for 30 min. (b) τ_{eff} and film thickness vs. H_2O co-reactant N_2 carrier flow rate (in standard litres per minute) for the same process at a deposition temperature of 440 °C on 500 μm thick 1 Ωcm p-type 100 FZ Si. H_2O carrier flow rate for a) was 0.25 SLPM*

Further development and optimisation of the TEDA-TSB/H_2O process demonstrated the importance of the deposition process parameters to surface passivation performance. Passivation was generally found to improve with increasing deposition temperature in the range of 325–520 °C (Figure 1a), due primarily to reducing interface state density (Figure 2), saturating at temperatures above ~440 °C [41]. Films deposited at these higher temperatures provided good passivation already in their as-deposited state, benefiting only marginally from additional post-deposition annealing at 400–425 °C. This is in contrast to the well-known need for such post-deposition annealing in order to 'activate' the passivation of ALD Al_2O_3, which is typically deposited at significantly lower temperatures. The concentration of the H_2O co-reactant was also found to be a critical deposition parameter for passivation, with passivation increasing and then saturating with increasing H_2O vapour flow rate (Figure 1b), due mainly to a reduction of D_{it} [42]. Deposition using TEDA-TSB with O_2 as co-reactant instead of H_2O was also investigated [42], but this resulted in films with a high density of interface states and resulting high surface recombination velocities, regardless of the O_2 flow rate or deposition temperature.

The best passivation results for APCVD Al_2O_3 so far reported have been achieved through the use of optimised deposition conditions together with a very short (1 s peak) rapid thermal anneal at 500–550 °C [43]. An S_{eff} of 7 cm/s was determined for 10–12 nm thick films prepared in this way on p-type 1 Ωcm <100> Si, corresponding to an exceptionally low J_0 of 7 fA cm^{-2}. It was found that this was due to a significantly increased fixed charge density (-3.3×10^{12} cm^{-2}) compared to that obtained following

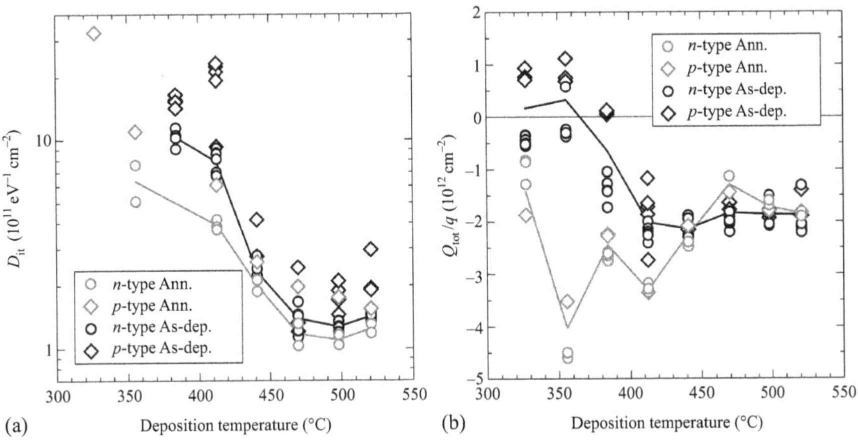

Figure 7.2 (a) Interface state density D_{it} and (b) fixed charge Q_f for the same deposition and annealing conditions as in Figure 7.1(a)

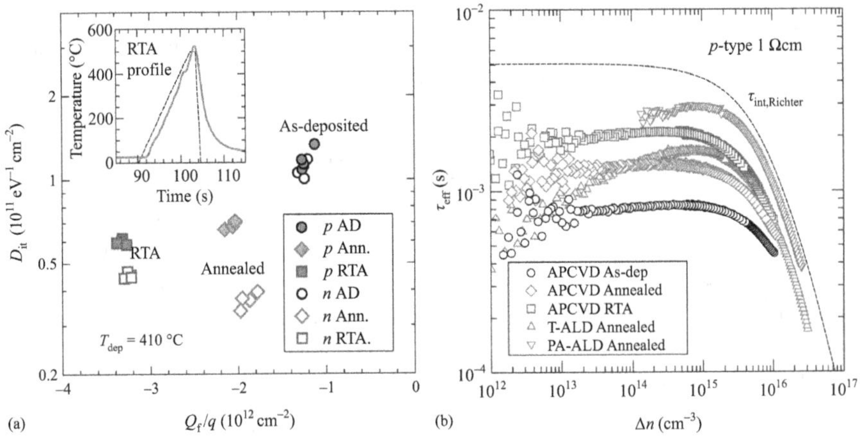

Figure 7.3 (a) D_{it} vs. Q_f and (b) τ_{eff} vs. excess carrier concentration Δn for APCVD Al_2O_3 (from TEDA-TSB) as-deposited at 410 °C and after either a standard furnace anneal at 425 °C for 30 min, or a rapid thermal anneal (RTA) at 500 °C for 1 s. τ_{eff} for the same substrates passivated by Al_2O_3 from standard thermal and plasma ALD processes is also shown. The inset of (a) shows the set (dashed) and measured (solid) RTA temperature profiles

a conventional extended (30 min) anneal at 425 °C, combined with a similarly low density of interface states (~5×10^{10} eV^{-1} cm^{-2}) (Figure 3a). This level of passivation was found to be as good as or better than that provided by a standard thermal ALD Al_2O_3 film on the same substrates (Figure 3b), demonstrating the outstanding passivation potential of Al_2O_3 films prepared by even such a relatively simple CVD process.

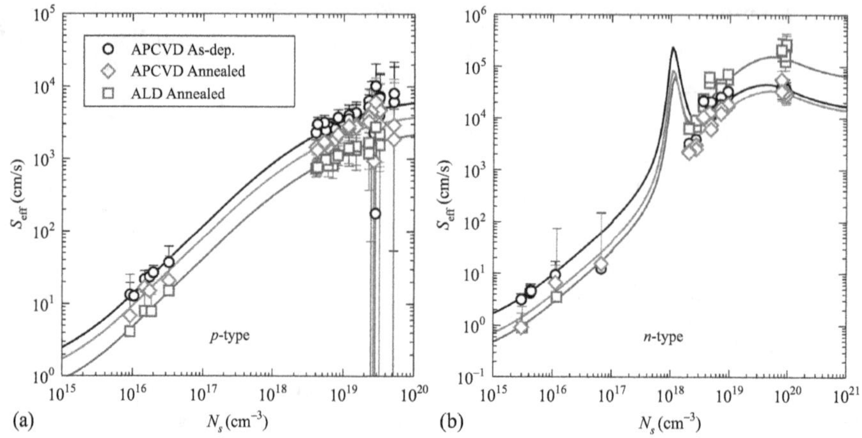

Figure 7.4 S_{eff} *vs. surface dopant concentration* N_s *for (a) boron- and (b) phosphorus-doped <100> Si surfaces passivated by* Al_2O_3 *from APCVD (as-deposited and annealed) and thermal ALD (annealed). Lines show theoretical predictions based on an experimentally derived model of the Si–*Al_2O_3 *interface properties (see Figure 5a) together with measured values of* D_{it} *and* Q_f

The thermal stability of Al_2O_3 passivation layers has been a significant concern in recent work on this material, due to the necessity for compatibility with the high-temperature ‘firing’ step employed in industry-standard screen-printed metallisation. Investigations into this question for films deposited by various methods have resulted in conflicting reports regarding the stability of the Al_2O_3 passivation and the necessity for and effectiveness of SiN_x capping layers [44–51]. APCVD Al_2O_3 has displayed outstanding quality in this regard, with lifetimes found to be stable or actually improve after a standard firing step, with or without a capping SiN_x layer [52]. A direct comparison of Al_2O_3 passivation layers deposited by different methods [52] showed that the post-firing lifetimes of the samples with APCVD Al_2O_3 was higher than for Al_2O_3 films deposited by inductively coupled plasma (ICP) PECVD, and not far below that for state-of-the-art plasma-assisted ALD films, with an S_{eff} of 3 cm/s on *p*-type 1.35 Ωcm surfaces, and less than 2 cm/s on 1.2 Ωcm *n*-type surfaces.

The influence of surface dopant concentration on the passivation performance of both APCVD and ALD Al_2O_3 films has also been investigated through a study of these films on both boron- and phosphorus-diffused surfaces with a wide range of dopant profiles (Figure 4) [42,53]. Excellent passivation of heavily doped *p*-type surfaces was observed, comparable to that reported for Al_2O_3 by other deposition techniques. Numerical modelling of the measured profiles was used to separate the surface recombination rate from the Auger recombination in the heavily doped regions. It was shown that, unlike what has been reported for thermal SiO_2, the interface properties of both APCVD and ALD Al_2O_3 are independent of the surface dopant concentration. Furthermore, this work was able to demonstrate a direct quantitative correlation between the interface properties determined by capacitance–voltage measurements and the surface recombination rate derived from photoconductance measurements.

Systematic investigations of APCVD Al_2O_3 passivation of Si <100>, <111>, and random-textured surfaces [54] have revealed only a weak influence of the surface orientation and geometry on surface recombination. J_0 was found to increase only by a factor of 1.0–1.5 for <111> vs. <100> surfaces, and 2.0–3.3 for textured vs. <100>, with the increase for textured (<111> facets) vs. planar <111> attributable simply to the increased surface area. This is consistent with similar comparisons performed for ALD and PECVD Al_2O_3 [55–58], and is in contrast to the much larger dependence on orientation and morphology observed for thermal SiO_2 and PECVD SiN_x [55]. Investigations of the interface properties of <100> and <111> Si surfaces passivated by APCVD Al_2O_3 using C–V measurements have shown that the difference in passivation between surfaces of different orientation is due solely to differences in the fixed charge density, not to different interface state densities or properties as might be expected [42]. This relatively weak dependence on orientation and morphology makes Al_2O_3 particularly suitable for the passivation of textured silicon surfaces.

APCVD Al_2O_3 has also been employed in fundamental investigations [42,59] of the Si–Al_2O_3 interface state properties. Apparent interface state-capture cross sections for both majority and minority carriers were determined by (frequency-dependent) measurements of the metal–insulator–semiconductor parallel conductance, together with an extended analysis including minority-carrier contributions (Figure 5a). The properties of the dominant recombination-active defect were then disentangled from the apparent values by a careful analysis based on an appreciation of the distinction between real and apparent capture cross sections. The recombination rate predicted by the resulting model of the Si–Al_2O_3 interface was subsequently compared with that determined by independent photoconductance measurements for a wide range of samples (Figure 5b), with excellent quantitative agreement. This model of the interface state properties has been shown to be generally applicable to Si–Al_2O_3 interfaces, regardless of the Al_2O_3 deposition method [42].

While most work to date on APCVD of Al_2O_3 for surface passivation has employed TEDA-TSB as precursor (with H_2O), alternative precursors have also received some attention. Davis *et al.* [60] have investigated the passivation performance of Al_2O_3 deposited by APCVD using TMA with O_2 as co-reactant. This process has the advantage of employing the same precursor used in the great majority of investigations of Al_2O_3 passivation by ALD and PECVD, which is thus both familiar to the photovoltaic community and widely available. High static deposition rates of 70–150 nm/min (assuming a 156 mm substrate) were reported for films deposited at 500 °C. Passivation was found to improve with increasing O_2/TMA flow ratio, with S_{eff} as low as 8 cm/s reported for 15–20 nm thick films on 10–20 Ωcm *p*-type <100> Si after firing. This process has been commercialised by Schmid Group in their APCVD systems.

Aluminium tri-(*sec*-butoxide) (ATSB) has also been investigated [61] as a lower-cost alternative to TEDA-TSB or TMA. Like the former, it is liquid at room

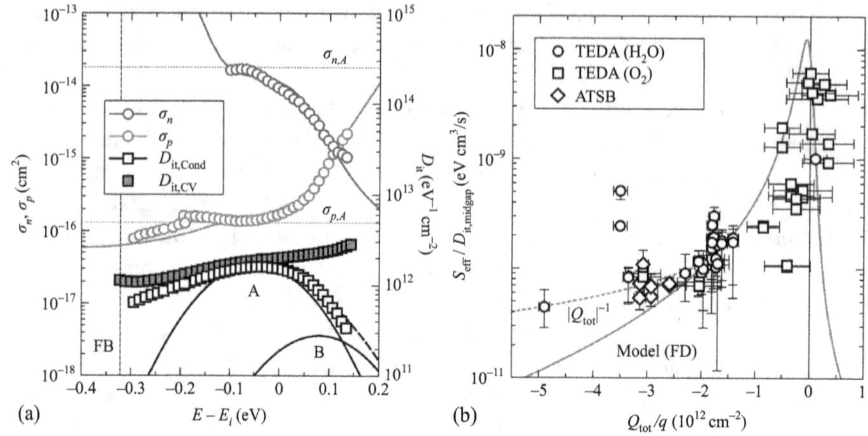

Figure 7.5 (a) Apparent capture cross-sections σn and σp for electrons and holes at the Si–Al_2O_3 interface as a function of energy relative to the intrinsic Fermi energy E_i determined by the extended conductance method for a p-type Si substrate, together with D_{it} determined by the same method and by capacitance – voltage measurements. Solid lines show simulated apparent capture cross-sections used to fit the data, based on a model including two Gaussian-distributed interface states A and B with constant capture cross-sections, where the dashed lines show modelled σn and σp for state A, which is the dominant recombination-active state under most conditions at the Si–Al_2O interface. (b) Measured (symbols) and modelled (solid lines) surface recombination velocity S_{eff} normalised to midgap D_{it} vs total interface charge Q_{tot} at $\Delta n = 1015\ cm^{-3}$ for 0.8 Ωcm p-type Si passivated by APCVD Al_2O_3 films prepared under various conditions. Modelling is based on the interface state model shown in Figure 5a

temperature, and nonpyrophoric (in contrast to TMA). Its high oxygen content (O:Al ratio 3:1) makes possible single-source (without additional co-reactants) deposition of Al_2O_3 via pyrolysis. This allows a considerable simplification of the APCVD vapour delivery system that would reduce system costs and also facilitate the higher delivery line temperatures needed to compensate for the relatively low vapour pressure of ATSB, which is its principal drawback. Preliminary investigations using this precursor have demonstrated promising passivation quality using a non-optimised single-source process, with S_{eff} values as low as 18 cm/s measured for ~20 nm thick unannealed films on 1.1 Ωcm *p*-type <100> Si (compared to 9 cm/s for annealed thermal ALD films on the same substrates) (Figure 6) [61].

Several results for solar cells incorporating APCVD Al_2O_3 have recently been published, which tend to show the comparability of APCVD Al_2O_3 with Al_2O_3 from other techniques like ALD and PECVD on a device level. Metz *et al.* [62] reported an efficiency of 20.7% for large-area screen-printed PERC cells based on

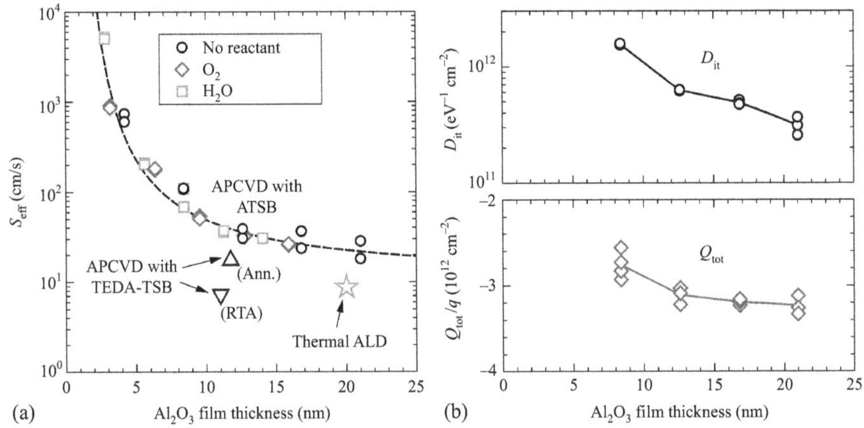

Figure 7.6 *(a)* S_{eff} *and (b)* D_{it} *and* Q_{tot} *vs. film thickness for* Al_2O_3 *deposited by APCVD with ATSB as precursor, with either no additional reactants or with either* O_2 *or* H_2O. S_{eff} *for the same substrates with* Al_2O_3 *by APCVD with TEDA-TSB (with either a conventional anneal or an optimised RTA) or by thermal ALD (annealed) are also shown for comparison*

p-type Cz Si with rear passivation provided by APCVD Al_2O_3 from TMA and O_2 [60], with cell parameters found to be identical to those of similarly processed cells with PECVD Al_2O_3. APCVD Al_2O_3 films deposited from TEDA-TSB and H_2O have also recently been demonstrated as surface-passivating layers in two different solar cell structures. Best efficiencies of 21.6% were achieved for small-area front-junction *n*-type cells fabricated at ANU [63], and 21.3% for 6 inch rear-junction *n*-type cells fabricated in collaboration by ANU and IMEC [42] (the design of these cells has been reported in detail by [64]), both incorporating Al_2O_3 passivation of the boron-doped $p+$ region. In the latter case, the cells could be directly compared to similarly processed cells featuring passivation by thermal SiO_2 or ALD Al_2O_3, exhibiting a similar V_{oc} to the latter (significantly increased compared to SiO_2). These results provide a proof of concept for the implementation of APCVD Al_2O_3 into industrial solar cells.

References

[1] S. Dauwe, L. Mittelstädt, A. Metz, and R. Hezel, "Experimental evidence of parasitic shunting in silicon nitride rear surface passivated solar cells," *Progress in Photovoltaics*, vol. 10, pp. 271–278, 2002.

[2] M. J. Kerr, *Surface, emitter and bulk recombination in silicon and development of silicon nitride passivated solar cells*. PhD thesis, The Australian National University, 2002.

[3] D. Macdonald and L. J. Geerligs, "Recombination activity of interstitial iron and other transition metal point defects in p- and n-type crystalline silicon," *Applied Physics Letters*, vol. 85, no. 18, pp. 4061–4063, 2004.

[4] G. Dingemans and W. M. M. Kessels, "Status and prospects of Al_2O_3-based surface passivation schemes for silicon solar cells," *Journal of Vacuum Science and Technology A*, vol. 30, no. 4, 2012.

[5] J. Schmidt, F. Werner, B. Veith, *et al.*, "Industrially relevant Al_2O_3 deposition techniques for the surface passivation of Si solar cells," in *Proceedings of the 25th European Photovoltaic Solar Energy Conference* (Valencia, Spain), pp. 1130–1133, 6–10 September 2010.

[6] I. Cesar, E. Granneman, P. Vermont, *et al.*, "Excellent rear side passivation on multi-crystalline silicon solar cells with 20 nm uncapped Al_2O_3 layer: Industrialization of ALD for solar cell applications," in *Proceedings of the 35th IEEE Photovoltaic Specialists Conference*, pp. 000044–000049, June 2010.

[7] E. H. A. Granneman, P. Vermont, V. Kuznetsov, M. Coolen, and K. Vanormelingen, "High-throughput, in-line ALD Al_2O_3 system," in *Proceedings of the 25th European Photovoltaic Solar Energy Conference* (Valencia, Spain), 6–10 September 2010.

[8] P. Poodt, A. Lankhorst, F. Roozeboom, K. Spee, D. Maas, and A. Vermeer, "High-speed spatial atomic-layer deposition of aluminum oxide layers for solar cell passivation," *Advanced Materials*, vol. 22, pp. 3564–3567, 2010.

[9] B. Vermang, H. Goverde, A. Lorenz, *et al.*, "On the blistering of Al_2O_3 passivation layers for Si solar cells," in *Proceedings of the 26th European Photovoltaic Solar Energy Conference* (Hamburg, Germany), p. 1129–1131, 5–9 September 2011.

[10] F. Werner, W. Stals, R. Görtzen, B. Veith, R. Brendel, and J. Schmidt, "High-rate atomic layer deposition of Al_2O_3 for the surface passivation of Si solar cells," *Energy Procedia*, vol. 8, pp. 301–306, 2011.

[11] S. Miyajima, J. Irikawa, A. Yamada, and M. Konagai, "Hydrogenated aluminum oxide films deposited by plasma enhanced chemical vapor deposition for passivation of p-type crystalline silicon," in *Proceedings of the 23rd European Photovoltaic Solar Energy Conference* (Valencia, Spain), 1–5 September 2008.

[12] P. Saint-Cast, D. Kania, M. Hofmann, J. Benick, J. Rentsch, and R. Preu, "Very low surface recombination velocity on p-type c-Si by high-rate plasma-deposited aluminum oxide," *Applied Physics Letters*, vol. 95, 151502, 2009.

[13] S. Miyajima, J. Irikawa, A. Yamada, and M. Konagai, "High quality aluminum oxide passivation layer for crystalline silicon solar cells deposited by parallel-plate plasma-enhanced chemical vapor deposition," *Applied Physics Express*, vol. 3, 012301, 2010.

[14] T.-T. Li and A. Cuevas, "Effective surface passivation of crystalline silicon by rf sputtered aluminum oxide," *Physica Status Solidi RRL*, vol. 3, pp. 160–162, 2009.

[15] X. Zhang and A. Cuevas, "Plasma hydrogenated, reactively sputtered aluminium oxide for silicon surface passivation," *Physica Status Solidi RRL*, vol. 7, no. 9, pp. 619–622, 2013.

[16] P. Vitanov, X. Loozen, A. Harizanova, T. Ivanova, and G. Beaucarne, "A study of sol-gel deposited Al_2O_3 films as passivating coatings for solar cells application," in *Proceedings of the 23rd European Photovoltaic Solar Energy Conference* (Valencia, Spain), p. 1596–1599, 1–5 September 2008.

[17] P. Vitanov, A. Harizanova, T. Ivanova, and T. Dimitrova, "Chemical deposition of Al_2O_3 thin films on Si substrates," *Thin Solid Films*, vol. 517, pp. 6327–6330, 2009.

[18] H.-Q. Xiao, C.-L. Zhou, X.-N. Cao, *et al.*, "Excellent passivation of p-type Si surface by sol-gel Al_2O_3 films," *Chinese Physics Letters*, vol. 26, 088102, 2009.

[19] J.-Y. Hung, J.-C. Wang, S.-W. Chen, *et al.*, "Back-side AlO_x passivation material and technology for the application of high efficiency (20%) and low cost PERC solar cells," in *Proceedings of the 40th IEEE Photovoltaic Specialists Conference*, pp. 3308–3312, June 2014.

[20] Y.-S. Lin, J.-Y. Hung, T.-C. Chen, *et al.*, "Effect of post deposition annealing of printed AlO_x film on PERC solar cells," in *Proceedings of the 40th IEEE Photovoltaic Specialists Conference*, pp. 0615–0618, 2014.

[21] I. Nasution, A. Velasco, and H.-j. Kim, "Atmospheric pressure chemical vapor deposition mechanism of Al_2O_3 film from $AlCl_3$ and O_2," *Journal of Crystal Growth*, vol. 311, pp. 429–434, 2009.

[22] R. S. Ehle, B. J. Baliga, and W. Katz, "Low temperature aluminum oxide deposition using trimethylaluminum," *Journal of Electronic Materials*, vol. 12, no. 3, pp. 587–601, 1983.

[23] J. A. Aboaf, "Deposition and properties of aluminum oxide obtained by pyrolytic decomposition of an aluminum alkoxide," *Journal of the Electrochemical Society*, vol. 114, pp. 948–952, 1967.

[24] D. Barreca, G. A. Battiston, R. Gerbasi, and E. Tondello, "Al_2O_3 thin films from aluminium dimethylisopropoxide by metal–organic chemical vapour deposition," *Journal of Materials Chemistry*, vol. 10, no. 9, pp. 2127–2130, 2000.

[25] T. Maruyama and S. Arai, "Aluminum oxide thin films prepared by chemical vapor deposition from aluminum acetylacetonate," *Applied Physics Letters*, vol. 60, no. 3, pp. 322–323, 1992.

[26] T. Maruyama and T. Nakai, "Aluminum oxide thin films prepared by chemical vapor deposition from aluminum 2-ethylhexanoate," *Applied Physics Letters*, vol. 58, no. 19, pp. 2079–2080, 1991.

[27] A. Waxman and K. H. Zaininger, "Al_2O_3-silicon insulated gate field effect transistors," *Applied Physics Letters*, vol. 12, no. 3, pp. 109–110, 1968.

[28] C. A. T. Salama, "RF sputtered aluminum oxide films on silicon," *Journal of the Electrochemical Society*, vol. 117, no. 7, pp. 913–917, 1970.

[29] T. Tsujide, S. Nakanuma, and Y. Ikushima, "Properties of aluminum oxide obtained by hydrolysis of $AlCl_3$," *Journal of the Electrochemical Society*, vol. 117, no. 5, pp. 703–708, 1970.

[30] M. T. Duffy and A. G. Revesz, "Interface properties of Si-(SiO_2)-Al_2O_3 structures," *Journal of the Electrochemical Society*, vol. 117, no. 3, pp. 372–377, 1970.
[31] M. Duffy, J. Carnes, and D. Richman, "Dielectric and interface properties of pyrolytic aluminum oxide films on silicon substrates," *Metallurgical Transactions*, vol. 2, no. 3, pp. 667–672, 1971.
[32] T. Tanaka and S. Iwauchi, "Interface characteristics of the reactively sputtered Al_2O_3-Si structure," *Japanese Journal of Applied Physics*, vol. 8, no. 6, pp. 730–738, 1969.
[33] K. Jaeger and R. Hezel, "A novel thin silicon solar cell with Al_2O_3 as surface passivation," in *Proceedings of the 18th IEEE Photovoltaic Specialists Conference* (Las Vegas, USA), pp. 1752–1753, 21–25 October 1985.
[34] R. Hezel and K. Jaeger, "Low-temperature surface passivation of silicon for solar cells," *Journal of the Electrochemical Society*, vol. 136, pp. 518–523, February 1989.
[35] G. Agostinelli, P. Vitanov, Z. Alexieva, *et al.*, "Surface passivation of silicon by means of negative charge dielectrics," in *Proceedings of the 19th European Photovoltaic Solar Energy Conference* (Paris, France), pp. 132–134, 7–11 June 2004.
[36] G. Agostinelli, A. Delabie, P. Vitanov, *et al.*, "Very low surface recombination velocities on p-type silicon wafers passivated with a dielectric with fixed negative charge," *Solar Energy Materials and Solar Cells*, vol. 90, pp. 3438–3443, 23 November 2006.
[37] B. Hoex, S. B. S. Heil, E. Langereis, M. C. M. v. d. Sanden, and W. M. M. Kessels, "Ultralow surface recombination of c-Si substrates passivated by plasma-assisted atomic layer deposited Al_2O_3," *Applied Physics Letters*, vol. 89, 24 July 2006.
[38] L. E. Black, K. M. Provancha, and K. R. McIntosh, "Surface passivation of crystalline silicon by APCVD aluminium oxide," in *Proceedings of the 26th European Photovoltaic Solar Energy Conference* (Hamburg, Germany), pp. 1120–1124, 2011.
[39] R. G. Gordon, K. Kramer, and X. Liu, "Chemical vapor deposition and properties of amorphous aluminum oxide films," *Materials Research Society Symposium Proceedings*, vol. 446, pp. 383–388, 1997.
[40] M.-M. Sovar, D. Samélor, A. N. Gleizes, and C. Vahlas, "Aluminium tri-isopropoxide: Shelf life, transport properties, and decomposition kinetics for the low temperature processing of aluminium oxide-based coatings," *Surface and Coatings Technology*, vol. 201, no. 22–23, pp. 9159–9162, 2007.
[41] L. E. Black and K. R. McIntosh, "Surface passivation of c-Si by atmospheric pressure chemical vapor deposition of Al_2O_3," *Applied Physics Letters*, vol. 100, 202107, 2012.
[42] L. E. Black, *New perspectives on surface passivation: understanding the Si–Al_2O_3 interface*. PhD thesis, The Australian National University, 2015.

[43] L. E. Black, T. Allen, K. R. McIntosh, and A. Cuévas, "Improved silicon surface passivation of APCVD Al_2O_3 by rapid thermal annealing," *Energy Procedia*, vol. 92, pp. 317–325, 2016.

[44] J. Benick, A. Richter, M. Hermle, and S. W. Glunz, "Thermal stability of the Al_2O_3 passivation on p-type silicon surfaces for solar cell applications," *Physica Status Solidi RRL*, vol. 3, pp. 233–235, 2009.

[45] B. Vermang, X. Loozen, C. Allebé, *et al.*, "Characterization and implementation of thermal ALD Al_2O_3 as surface passivation for industrial Si solar cells," in *Proceedings of the 24th European Photovoltaic Solar Energy Conference* (Hamburg, Germany), p. 1051–1054, 21–25 September 2009.

[46] J. Schmidt, B. Veith, and R. Brendel, "Effective surface passivation of crystalline silicon using ultrathin Al_2O_3 films and Al_2O_3/SiN_x stacks," *Physica Status Solidi RRL*, vol. 3, pp. 287–289, 2009.

[47] B. Veith, F. Werner, D. Zielke, R. Brendel, and J. Schmidt, "Comparison of the thermal stability of single Al_2O_3 layers and Al_2O_3/SiN_x stacks for the surface passivation of silicon," *Energy Procedia*, vol. 8, pp. 307–312, 17–20 2011.

[48] D. Schuldis, A. Richter, J. Benick, and M. Hermle, "Influence of different post deposition treatments on the passivation quality and interface properties of thermal ALD Al_2O_3 capped by PECVD SiN_x," in *Proceedings of the 27th European Photovoltaic Solar Energy Conference* (Frankfurt, Germany), pp. 1933–1937, 2012.

[49] G. Dingemans, P. Engelhart, R. Seguin, *et al.*, "Stability of Al_2O_3 and Al_2O_3/a-SiN_x:H stacks for surface passivation of crystalline silicon," *Journal of Applied Physics*, vol. 106, 114907, 2009.

[50] B. Veith, T. Dullweber, M. Siebert, *et al.*, "Comparison of ICP-AlO_x and ALD-Al_2O_3 layers for the rear surface passivation of c-Si solar cells," *Energy Procedia*, vol. 27, pp. 379–384, 2012.

[51] P. Saint-Cast, D. Kania, R. Heller, *et al.*, "High-temperature stability of c-Si surface passivation by thick PECVD Al_2O_3 with and without hydrogenated capping layers," *Applied Surface Science*, vol. 258, no. 21, pp. 8371–8376, 2012.

[52] L. E. Black, T. Allen, A. Cuevas, K. R. McIntosh, B. Veith, and J. Schmidt, "Thermal stability of silicon surface passivation by APCVD Al_2O_3," *Solar Energy Materials and Solar Cells*, vol. 120, Part A, pp. 339–345, 2014.

[53] L. E. Black, T. Allen, K. R. McIntosh, and A. Cuevas, "Effect of boron concentration on recombination at the *p*-Si–Al_2O_3 interface," *Journal of Applied Physics*, vol. 115, 093707, 2014.

[54] L. E. Black, T. C. Kho, K. R. McIntosh, and A. Cuevas, "The influence of orientation and morphology on the passivation of crystalline silicon surfaces by Al_2O_3," *Energy Procedia*, vol. 55, pp. 750–756, 2014.

[55] S. C. Baker-Finch and K. McIntosh, "The contribution of planes, vertices, and edges to recombination at pyramidally textured surfaces," *IEEE Journal of Photovoltaics*, vol. 1, pp. 59–65, 2011.

[56] W. Liang, K. J. Weber, D. Suh, *et al.*, "Surface passivation of boron-diffused p-type silicon surfaces with (100) and (111) orientations by ALD Al_2O_3 layers," *IEEE Journal of Photovoltaics*, vol. 3, pp. 678–683, 2013.

[57] S. Duttagupta, F. Lin, K. D. Shetty, A. G. Aberle, and B. Hoex, "Excellent boron emitter passivation for high-efficiency Si wafer solar cells using AlO_x/SiN_x dielectric stacks deposited in an industrial inline plasma reactor," *Progress in Photovoltaics: Research and Applications*, vol. 21, pp. 760–764, 2012.

[58] A. Richter, J. Benick, and M. Hermle, "Boron emitter passivation with Al_2O_3 and Al_2O_3/SiN_x stacks using ALD Al_2O_3," *IEEE Journal of Photovoltaics*, vol. 3, pp. 236–245, 2013.

[59] L. E. Black and K. R. McIntosh, "Modeling recombination at the Si–Al_2O_3 interface," *IEEE Journal of Photovoltaics*, vol. 3, pp. 936–943, 2013.

[60] K. O. Davis, K. Jiang, M. Wilson, *et al.*, "Influence of precursor gas ratio and firing on silicon surface passivation by APCVD aluminium oxide," *Physica Status Solidi RRL*, vol. 7, no. 11, pp. 942–945, 2013.

[61] L. E. Black, T. Allen, and K. R. McIntosh, "Safe and inexpensive Al_2O_3 deposited by APCVD with single-source precursor," in *Proceedings of the 28th European Photovoltaic Solar Energy Conference* (Paris, France), pp. 1068–1072, October 2013.

[62] A. Metz, D. Adler, S. Bagus, *et al.*, "Industrial high performance crystalline silicon solar cells and modules based on rear surface passivation technology," *Solar Energy Materials and Solar Cells*, vol. 120, Part A, pp. 417–425, 2014.

[63] Y. Wan, C. Samundsett, T. Kho, *et al.*, "Towards industrial advanced front-junction n-type silicon solar cells," in *Proceedings of the 40th IEEE Photovoltaic Specialist Conference*, pp. 0862–0865, June 2014.

[64] E. Cornagliotti, A. Sharma, L. Tous, *et al.*, "Large area n-type c-Si solar cells featuring rear emitter and efficiency beyond 21%," in *Proceedings of the 6th World Conference Photovoltaic Energy Conversion* (Kyoto, Japan), 23–27 November 2014.

Chapter 8

Surface passivation of industrial PERC solar cells

Thorsten Dullweber[1] and Jan Schmidt[1,2]

8.1 Introduction

Since more than two decades, around 90% of the solar cells worldwide have been manufactured applying silicon wafers and a screen-printed full-area aluminum (Al) layer contacting the complete rear wafer surface. Thereby, the recombination of photo-generated charge carriers at the aluminum rear contact is only marginally suppressed by the local Al doping of the rear wafer surface, the so-called aluminum back surface field (Al-BSF). Also, the full-area Al layer partly absorbs the infrared light which reduces the light absorption of the solar cell. Due to these two loss mechanisms, the efficiency of today's industrial silicon solar cells with full-area Al-BSF is limited to around 20%. A cell concept which overcomes these limitations was introduced already in 1989 on laboratory scale by Blakers *et al.* [1] and named "Passivated Emitter and Rear Cell" (PERC). This first lab-type PERC solar cell applied a silicon dioxide layer at the rear silicon wafer surface and the evaporated Al rear layer only locally contacted the rear silicon surface, hence minimizing the recombination of charge carriers at the rear surface and increasing the internal rear reflectivity and absorption of infrared light. The local point contact openings in the silicon dioxide passivation layer were then realized by means of photolithography.

It took 25 years until research institutes and companies developed suitable low-cost and high-performance production processes, in particular, for the rear passivation layer and the rear contact definition which enabled to achieve an economic benefit, not just an efficiency benefit, of the PERC technology over the conventional Al-BSF technology. Since a few years, the new production processes are available with sufficient quality and throughput and now several solar cell manufacturers are introducing PERC solar cells into industrial mass production. The International Technology Roadmap for Photovoltaics (ITRPV) predicts a strong take-up of the PERC concept into industrial production and forecasts market shares of the PERC technology of 45% within the next years [2] as shown in

[1]Department of Photovoltaics, Institute for Solar Energy Research Hamelin (ISFH), Germany
[2]Department of Solar Energy, Institute of Solid-State Physics, Leibniz University Hanover, Germany

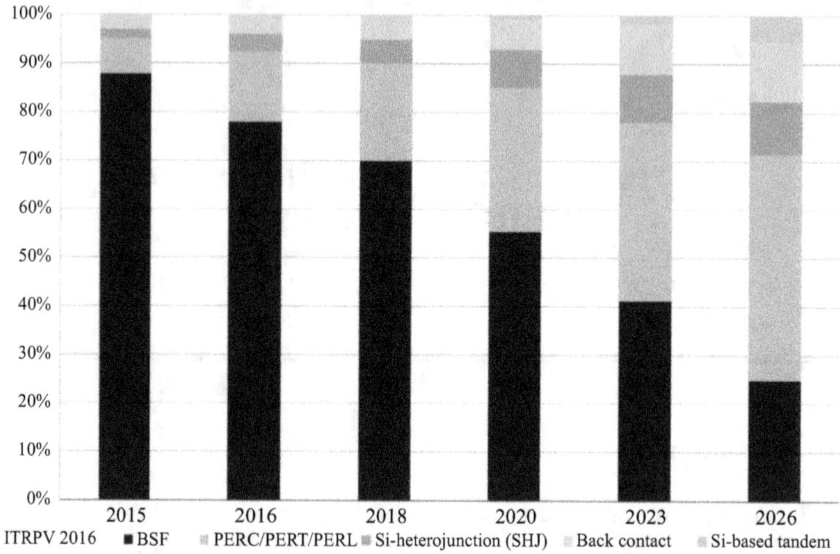

Figure 8.1 Forecasted worldwide market share of different c-Si solar cell technologies according to the International Technology Roadmap for Photovoltaics (ITRPV) as published in 2016 [2]. PERC cells are expected to gain approximately 45% market share within the next years

Figure 8.1. Several publications [3–5] provide a good overview of the historic development and the present status of the PERC technology.

In this chapter, we summarize the main developments and improvements within the last 25 years which enabled the industrialization of the lab-type PERC solar cell concept. In particular, we describe the historic development and current status of key industrial PERC manufacturing process steps such as the rear passivation layer, typically applying an AlO_x/SiN_y layer stack, as well as the local Al rear contact formation by laser ablation, Al screen printing and subsequent firing. At the end of this chapter, an outlook is provided on future development opportunities such as further efficiency increases and bifacial PERC applications.

8.2 Brief historical review

The PERC structure was introduced by Blakers *et al.* [1] in the research group of M. A. Green at the University of New South Wales (UNSW), Australia, in 1989. At that time, the goal was not to demonstrate a production-feasible solar cell process but to increase the conversion efficiency to a new world record value of 22.8% applying mostly laboratory-type processes and wafer materials. As substrate material 0.2 Ωcm p-type float-zone silicon wafers were used. The rear passivation consisted of a thermally grown (~1,000 °C) silicon dioxide layer which was locally

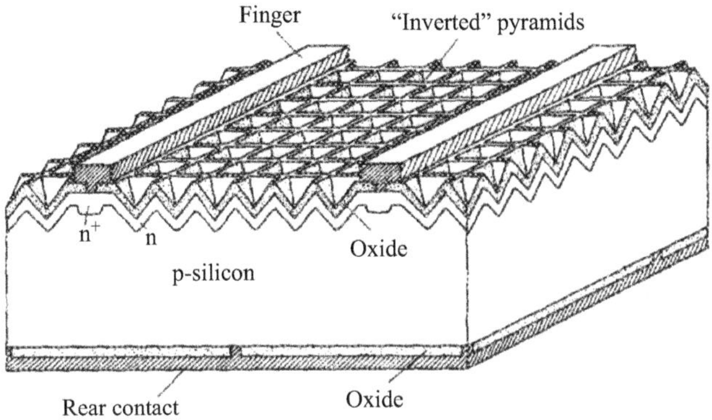

Figure 8.2 Schematic drawing of the laboratory-type PERC solar cell published by Blakers et al. in 1989 [1] applying a thermal silicon dioxide rear passivation, photolithographically defined point contact opening for the rear contacts and an evaporated aluminum rear metal layer. Figure 8.2 is reproduced from Ref. [1], with the permission of AIP Publishing

opened by photolithography and a masked etch forming point-shaped local rear contacts. The aluminum rear layer was evaporated followed by a 15 min anneal in forming gas. The processing of the front side involved several photolithography and oxide masking steps forming inverted pyramids, a selective phosphorus emitter and evaporated and subsequently plated TiPdAg front contacts. The resulting 4 cm^2 laboratory-type PERC solar cell of 1989 is shown schematically in Figure 8.2.

In the following years, the PERC concept was further developed into the passivated emitter and rear locally diffused (PERL) solar cell resulting in an efficiency of 25% by present standards [6] which remained the world-record efficiency for silicon solar cells for almost two decades. The PERC and PERL solar cell designs of the UNSW provided the foundation and basic concept of today's industrial PERC solar cells. However, a broad variety of process and technology developments were required to develop an industrial PERC manufacturing process sequence applying novel low-cost and high-throughput processes with sufficient quality for high conversion efficiencies. Particular technological breakthroughs were necessary concerning surface passivation, local contact definition, and metallization.

As an important step toward industrialization, a lot of efforts were directed to evaluate low-cost high-throughput processes for the deposition of the rear passivation layer replacing the lab-type high-temperature thermal oxidation process. For quite a while silicon nitride (SiN_x) deposited by plasma-enhanced chemical vapor deposition (PECVD) seemed to be a promising candidate [7–11] as it was successfully applied already at that time as a front side passivation layer of the phosphorus-doped emitter in industrial solar cells. However, it was shown that on p-type rear silicon surfaces of PERC cells SiN_x passivation layers induce an inversion of the surface due the positive fixed charges within the SiN_x layer leading to an enhanced carrier recombination which was identified and denoted as "parasitic shunting" by Dauwe *et al.* [12] in 2002. As alternative material, aluminum oxide (AlO_x) was found to be a very suitable candidate as it exhibits a high negative fixed charge density avoiding any parasitic shunting on p-type silicon surfaces [13,14,16]. Whereas most initial research work on AlO_x passivation was performed on lab-type low-throughput atomic layer deposition (ALD) of AlO_x layers [15,16], from 2009 onwards high-throughput tools and processes for industrial mass production have been developed such as PECVD [17–21] and spatial ALD [22,23]. An overview is provided in [24]. The addition of a SiN_y layer on top of the AlO_x layer made the AlO_x/SiN_y layer stack [25] more firing stable and resistant to Al pastes during firing. Also, the SiN_y capping layer improves the optical reflectivity of the rear surface thereby increasing the absorption of infrared light in the solar cell due to improved light trapping [26].

Another important step toward industrialization of the PERC concept was introduced in the year 2000 by Preu *et al.* [9] demonstrating that laser ablation of the rear passivation layer is a promising industrial process to define the rear contact geometries instead of the previously used laboratory-type photolithographic rear contact structuring. This approach was further developed by Agostinelli *et al.* [11,27] in 2005 into the so-called i-PERC concept which proposed for the first time to combine laser ablation of the rear passivation layer with screen-printed Al layers, thereby forming local Al-BSF regions by alloying of the silicon and the aluminum through the laser contact openings (LCOs) during the furnace firing step. This approach resulted then in large-area i-PERC solar cells with efficiencies up to 17.2% [11] which outperformed the corresponding full-area Al-BSF cells by $0.7\%_{abs}$. Gatz *et al.* [28] were the first to apply optimized line-shaped LCOs in combination with improved aluminum pastes which resulted in PERC cell conversion efficiencies of 19.4% in 2011.

In order to upgrade a conventional full-area Al-BSF industrial production process into a PERC process, besides introducing a rear passivation layer and a LCO, a third process requirement has to be fulfilled: the phosphorus diffusion should only dope the wafer front surface and not the rear surface in order to avoid a floating rear junction or shunting of the electrodes. Two industrial solutions have been developed to solve this issue. The company Centrotherm developed an industrial PERC process flow where the rear passivation layer (in this case a PECVD-SiO_x/SiN_y stack) is deposited prior to the phosphorus diffusion thereby acting as a diffusion barrier and protecting the rear side from being phosphorus doped [29]. This so-called Centaurus process was the first turn-key solution for industrial PERC production lines becoming available in 2011 [30]. Another industrial PERC process very similar to the Centaurus process was developed by Dullweber *et al.* [31,32] at ISFH in 2011 and applies a SiN_y layer which protects the rear wafer surface against the following alkaline texture and $POCl_3$ diffusion. A second industrial PERC process sequence, which achieves a single-sided phosphorus doping applies wet chemical removal of the phosphorus-doped rear surface after $POCl_3$ diffusion. Early work of RENA and IMEC on this concept [33,34] wet chemically polished the rear wafer surface post double-sided texturing and subsequently applied a wet chemical junction isolation post $POCl_3$ diffusion. This process sequence enabled almost planar wafer rear surfaces by using long polishing times and an aggressive polishing chemistry removing up to 20 μm of silicon from the rear surface [35]. However, it was shown by Kranz *et al.* [35] that in particular when applying an AlO_x/SiN_y passivation stack the rear surface roughness can be quite high without detrimental impact on the rear passivation quality and hence conversion efficiency. Accordingly, further work by RENA and ISFH [36,37] simplified the PERC polishing process sequence to texturing—$POCl_3$ diffusion—wet chemical rear polishing by developing a less aggressive wet chemical polishing chemistry.

Several different metallization technologies have been evaluated for industrial PERC solar cells. As alternative to the previously introduced LCO contact formation, in 2001 Schneiderlöchner *et al.* [38] introduced the so-called laser-fired contacts (LFC) where the local Al contact is established by melting the printed and fired Al paste with short laser pulses thereby locally removing the passivation layer and alloying the Al-Si contact. PERC cells with LFC contacts were the first to obtain conversion efficiencies above 18% [39] in combination with screen-printed contacts. Several companies further developed the LFC approach obtaining excellent PERC conversion efficiencies up to 20.2% [40,41]. As alternative to the widely used Ag screen-printed front contacts, nickel copper (NiCu) contacts deposited by plating have been developed [42,43] demonstrating an approximately $0.4\%_{abs.}$ higher efficiency potential compared to Ag screen-printed contacts. In terms of cell interconnection technology for module assembly, a so-called metal wrap through (MWT) technology has been developed which guides the front contacts through laser-drilled holes to the rear side of the solar cell [44,45]. Hence, there are no front busbars reflecting the sunlight which provides an efficiency gain potential of approximately $0.5\%_{abs.}$ compared to conventional 3 or 5 busbar layouts. Efficiencies up to 20.6% on float zone wafers have been obtained when applying the MWT concept to PERC cells [45]. Even though all of the alternative metallization technologies mentioned above demonstrated promising results, so far none of these technologies is applied in high-volume industrial PERC cell production. In case of NiCu plating, the increased process complexity compared to screen printing could be a reason as argued in [42], whereas in case of LFC contacts the local Al-BSF thickness is rather thin [40] resulting in higher contact recombination compared to LCO contacts. In case of MWT, a possible reason preventing large-scale production is that the required changes to the module interconnection technology have not been economically successful so far.

Whereas in the past two decades many options for industrial PERC processes have been evaluated, it seems that today a mainstream technology has evolved for the production of industrial PERC solar cells [3]: as rear passivation mostly AlO_x/SiN_y layer stacks are used deposited by PECVD with high deposition rates of several nm/s. The rear contacts are typically defined by laser ablation (LCO) of the rear passivation forming line-shaped or dashed-line-shaped local rear contacts. The full-area rear aluminum layer is deposited by screen-printing of aluminum

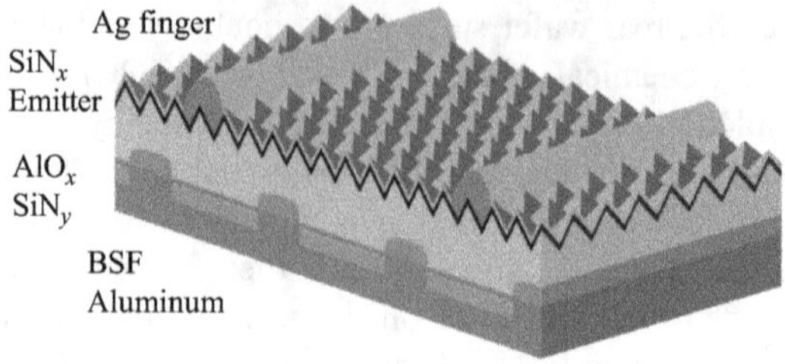

Figure 8.3 Schematic drawing of an industry-typical PERC solar cell with AlO_x/SiN_y rear passivation and local screen-printed aluminum rear contacts formed by LCO of the rear passivation layer. Picture taken from Ref. [46]

paste which is specifically optimized for PERC applications. The local Al-Si contact is formed during furnace firing for a few seconds at temperatures around 750–850 °C, where the silicon locally alloys with the aluminum forming a local Si-Al eutectic layer with a surrounding Al-BSF. Front-side processing includes formation of random pyramids by acidic or alkaline anisotropic wet chemical etching, phosphorus diffusion of the emitter, rear emitter removal by wet chemical polishing, SiN_x antireflective coating and front surface passivation by PECVD as well as screen printing of the silver front contacts with a three, four, or five busbar layout, which is compatible with conventional stringing module interconnection technology. The resulting industrial PERC cell structure is shown in Figure 8.3.

Figure 8.4 shows the evolution of record conversion efficiencies over the past 6 years obtained with industrial PERC solar cells (large area >148 cm^2 p-type monocrystalline Czochralski-grown (Cz) silicon wafers, printed metal front and rear contacts with conventional busbar layout) [28,40,47–54]. Starting with an efficiency of 19.2% [47] reported by Centrotherm in 2010, the benchmark in 2012 has been 21.0% [49] as demonstrated by Schott Solar. In 2014, the efficiency has been further increased to 21.2% by ISFH [50] and in July 2015 to 21.7% by SolarWorld [52]. The actual champion efficiency of 22.6% for an industrial PERC cell has been reported by Trina Solar in December 2016 [54]. As can be derived from Figure 8.4, the record efficiencies of industrial PERC cells increase by approximately $0.5\%_{abs}$ per year as average value over the past 6 years. Simply extrapolating this trend predicts that in 2018 record PERC cell efficiencies will exceed 23% applying industrially feasible production processes.

An attractive technology option is bifacial silicon solar cells which absorb the direct sunlight from front surface and in addition the diffuse light from the rear surface of the solar cell. When applied in PV power plants the produced electricity can be increased by up to 20% using bifacial instead of monofacial solar modules [55,56] depending on the detailed installation conditions, in particular, the albedo of the ground and the mounting height. Accordingly, the photovoltaic technology roadmap ITRPV predicts a market share of bifacial solar modules of 30% until 2026 [2]. Whereas the lab-type PERC cell of Blakers *et al.* as well as typical

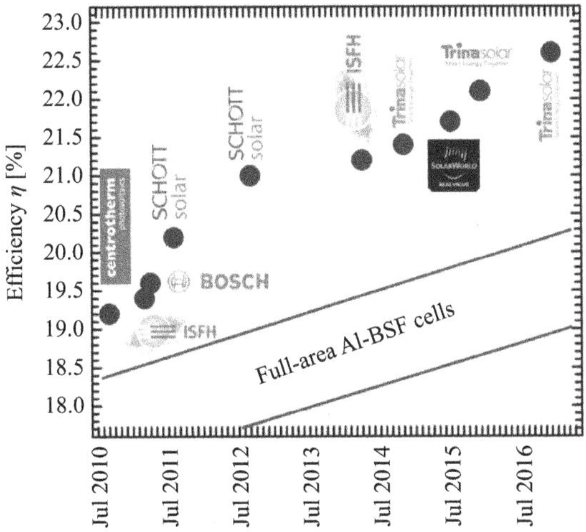

Figure 8.4 Independently confirmed record conversion efficiencies of industrial-type PERC solar cells with printed metal contacts using monocrystalline large area p-type Cz silicon wafers. Typical efficiencies of full-area Al-BSF solar cells are indicated as well. Graph updated from Ref. [3]

industrial PERC solar cells as outlined above and shown in Figures 8.2 and 8.3 are monofacial due to the full-area Al rear layer, in 2015, a bifacial industrial PERC solar cell named "PERC+" was introduced by Dullweber *et al.* [57,58] by screen printing an Al finger grid instead of the full-area Al rear layer. The Al fingers are printed on top of the LCOs with a finger widths around 150 μm accounting for approximately 10% to 20% rear side shadowing [58,59]. Hence, the PERC+ conversion efficiency when illuminated form the rear side is a few percent absolute lower compared to the front side efficiency resulting in around 75% bifaciality [59]. Hence, bifacial PERC+ cells have the potential to increase the energy yield of photovoltaic power plants between 5% and 25% [59,60] compared with monofacial PERC cells depending on the installation details. In particular, bifacial PERC+ modules mounted on 1-axis-trackers increase the energy yield by up to 24% compared to fixed-mounted monofacial PERC modules [61]. In 2015, SolarWorld and later several additional solar cell manufacturers started mass production of bifacial PERC+ cells [59,62].

8.3 Emitter passivation

Phosphorus-diffused n^+ emitters are very effectively passivated by means of silicon dioxide (SiO_2) grown at high temperature (typically >900 °C) [63–65]. SiO_2 was hence used as standard surface passivation scheme in the early high-efficiency laboratory-type silicon solar cells developed in the 1980s and 1990s [66–68].

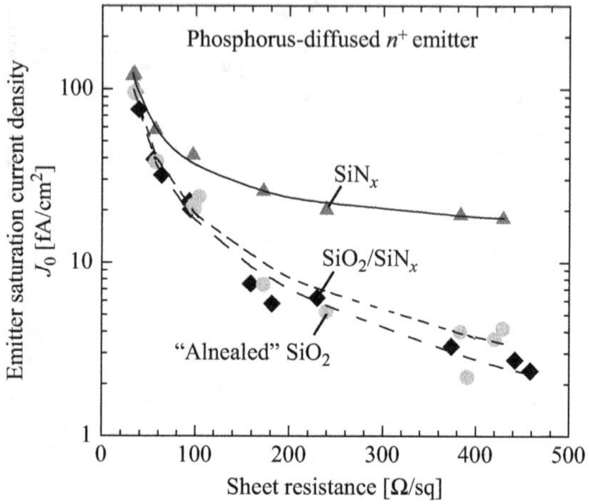

Figure 8.5 Measured emitter saturation current densities J_0 as a function of the sheet resistance for phosphorus-diffused n^+ emitters passivated by (i) "alnealed" thermally grown SiO_2; (ii) PECVD-SiN_x; and (iii) SiO_2/SiN_x stacks (data taken from Ref. [76])

However, the high-temperature oxidation turned out to be less suitable for industrial solar cells because of several practical issues, such as too high production cost and the sensitivity of the silicon bulk material properties to high temperatures. Hence, an alternative low-temperature passivation scheme, plasma-enhanced chemical-vapor-deposited (PECVD) silicon nitride (SiN_x), which was routinely applied in microelectronics then, was in the 1980s transferred to photovoltaics [69,70]. In the 1990s, numerous studies analyzed and optimized the surface passivation properties of SiN_x layers on undiffused and diffused silicon surfaces [71–73] and it turned out that an unexpectedly high level of surface passivation can be achieved, even after high-temperature firing as used for the contact formation of screen-printed metal contacts in industrial solar cells [74,75]. In addition, the refractive index of SiN_x films can be easily tuned by adjusting the film composition and the SiN_x films are hence also suited as excellent antireflection coatings when applied to the front of silicon solar cells. All these factors together made SiN_x films the standard front coating in today's P-diffused industrial silicon solar cells.

Figure 8.5 shows an exemplary set of measured emitter saturation current densities J_0 on planar PECVD-SiN_x-passivated n^+ emitters as a function of the sheet resistance in comparison with two high-temperature SiO_2-based passivation schemes [76]. In order to realize the best-case passivation scenario as a reference, the single-layer SiO_2-passivated samples included in Figure 8.5 received a so-called 'alneal', which consists of the deposition of a thin aluminum layer on top of the SiO_2, an annealing step at ~400 °C and a subsequent removal of the aluminum. This procedure generates atomic hydrogen by a reaction between residual water molecules within the SiO_2 layer and the aluminum [77]. The atomic hydrogen is then able to effectively passivate dangling bonds at the c-Si/SiO_2 interface, thereby drastically reducing the interface state density and consequently improving the passivation quality. The alnealed thermally grown SiO_2 is regarded as the best-case reference scenario. In addition to the single-layer SiO_2 and SiN_x passivation layers, a stack consisting of a thin (~10 nm) thermally grown SiO_2 plus a thicker (~60 nm) PECVD-SiN_x capping layer is included in Figure 8.5 as well. This improved passivation behavior of the SiO_2/SiN_x stack has been attributed to the very high hydrogen content in the PECVD-SiN_x layer (>10 at. %) [76]. The atomic hydrogen diffuses through the thin SiO_2 layer and passivates dangling bonds as effective as in the case of the alnealing treatment. Typical sheet resistances of industrial n^+ emitters are in the range between 80 and 140 Ω/sq. In this most relevant range, the emitter with thermally grown SiO_2 and SiO_2/SiN_x stacks shown in Figure 8.5 achieve J_0 values of 12–25 fA/cm^2, while the emitter with a single-layer SiN_x exhibits J_0 values of 30–40 fA/cm^2. Hence, SiN_x-passivated n^+ emitters show a high potential, which is, however, outperformed by SiO_2/SiN_x stacks, which perform as excellently as the alnealed SiO_2 reference. Single layers of PECVD-SiN_x are the standard in the n^+ emitter passivation in today's industrial solar cells, however, as other recombination channels, such as bulk and rear surface recombination, are more and more reduced, the front emitter is becoming one of the performance-limiting factors in the next generation of industrial solar cells. Hence, the introduction of SiO_2/SiN_x stacks is an attractive option to further enhance the potential of industrial n^+ front emitters, in particular in solar cells with selective emitters.

In order to reduce the front reflection and to improve light trapping, the solar cell front is typically textured, e.g. by random pyramids in the case of monocrystalline silicon solar cells, and the emitter is diffused into the textured surface. The effect of texturing on the surface passivation properties of SiO_2, SiN_x, and SiO_2/SiN_x stacks has been studied [78,79] and it was found that for typical industrial emitters, J_0 increases approximately by the fraction of the surface area increase, i.e. for random pyramids J_0 increases by a factor of 1.73. This leads for the best random-pyramid-textured state-of-the-art industrial selective emitters with a sheet resistance of ~120 Ω/sq in the shallow-diffused n^+ region to J_0 values of ~25 fA/cm^2, which have recently indeed been realized on state-of-the-art >22% efficient industrial PERC devices [80].

8.4 Rear passivation technologies

In high-efficiency laboratory-type PERC silicon solar cells, the rear surface recombination is very effectively suppressed by means of silicon dioxide (SiO_2) grown at high temperatures (≳900 °C) in a quartz-tube furnace [1]. One reason why high-temperature oxidation was not transferred into industrial solar cell production is the high sensitivity of the silicon bulk charge carrier lifetime to high-temperature processes. In particular, in the case of multi-crystalline silicon wafers, thermal processes above 900 °C typically lead to a significant degradation in the bulk lifetime. Hence, low-temperature surface passivation alternatives were required for future industrial high-efficiency silicon solar cells. Hezel and Jaeger [13] introduced silicon nitride (SiN_x) grown by plasma-enhanced chemical vapor deposition (PECVD) at temperatures around 400 °C into silicon-photovoltaics. Another advantage of PECVD is the high deposition rate resulting in short process times of just several seconds per wafer enabling high throughput and low production costs. As was shown by Aberle *et al.* [81], on p-type silicon surfaces optimized PECVD-SiN_x can result in low surface recombination velocities (SRVs) <10 cm/s as it is typically only obtained by highest-quality high-temperature SiO_2. However, when PECVD-SiN_x was applied to the rear of PERC solar cells the short-circuit current density was found to be strongly reduced compared to the SiO_2-passivated cell. Dauwe *et al.* [12] attributed this effect to the high density of fixed positive charges within the SiN_x layer when deposited onto a c-Si surface, inducing an inversion layer in the crystalline silicon underneath the SiN_x. The coupling of this inversion layer to the base contact leads to a significant loss in the short-circuit current density, a detrimental effect named 'parasitic shunting'. One way of overcoming the parasitic shunting is to introduce a thermally grown typically ~10 nm thick SiO_2 layer between the p-type silicon surface and the SiN_x layer, which was successfully demonstrated by Schmidt *et al.* on laboratory-type PERC devices [82]. As the fixed positive charge density in SiO_2 on c-Si is considerably lower compared to SiN_x, parasitic shunting is absent on p-type silicon surfaces passivated with SiO_2/SiN_x stacks. Furthermore, it was demonstrated [82] that the surface passivation by SiO_2/SiN_x stacks improves considerably during high-temperature annealing, which was attributed to the diffusion of hydrogen from the hydrogen-rich PECVD-SiN_x layer to the SiO_2/c-Si interface, where the hydrogen effectively passivates silicon dangling bonds. Interestingly, a very similar beneficial behavior was also observed when the thermally grown SiO_2 within the SiO_2/SiN_x stack was replaced by PECVD-SiO_x or SiO_xN_y, which was first examined by Agostinelli *et al.* [83] and later by Hofmann *et al.* [84] and Petres *et al.* [85], who reported surface recombination velocities of 10 cm/s on p-type silicon wafers using SiO_xN_y/SiN_z stacks deposited in a commercial PECVD system after firing. Note that this type of stack is used e.g. in the Centrotherm "Centaurus" turn-key PERC production process [29] and hence is still applied in several present-day industrial PERC production lines.

Another alternative passivation scheme is intrinsic hydrogenated amorphous silicon (a-Si) deposited by PECVD in the temperature range between 200 and 250 °C [86,87]. For lab-type PERC cells with evaporated contacts and a-Si rear passivation, efficiencies up to 21.7% have been demonstrated [88]. Despite the fact that no parasitic shunting occurs in the case of an a-Si passivated cell rear, new problems arise from the high sensitivity of the a-Si passivation to thermal processes, such as furnace firing, which makes this passivation scheme largely incompatible with the current industrial screen-printing technology.

In recent years, the negative-charge dielectric aluminum oxide (Al_2O_3) has proven capable of providing an outstanding level of passivation on crystalline silicon surfaces [13–17,83,89–95]. Al_2O_3 dielectric layers were first demonstrated by Hezel and Jaeger [13] in 1989 to effectively passivate crystalline silicon

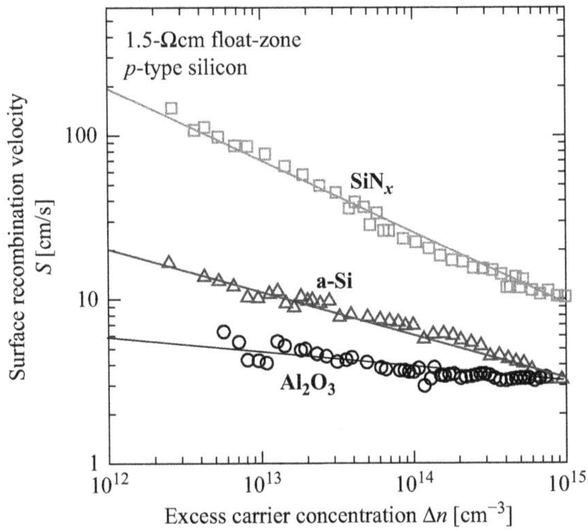

Figure 8.6 Comparison of the injection-dependent effective surface recombination velocity S(Δn) measured on 1.5-Ωcm p-type float-zone silicon wafers passivated by (i) SiN_x deposited by remote-PECVD; (ii) intrinsic a-Si deposited in a parallel-plate PECVD reactor; and (iii) Al_2O_3 deposited by means of plasma-assisted ALD (data taken from Schmidt et al., Ref. [94])

surfaces. However, these first promising results were not taken up by the photovoltaic community then, until almost two decades later Agostinelli *et al.* [83] demonstrated surface recombination velocities below 10 cm/s on p-type silicon wafers, which was the beginning of the "revival" of Al_2O_3 as a dielectric passivation layer with striking properties, which later turned out to be perfectly suited for the application to the rear surface passivation of PERC solar cells. Al_2O_3 can be deposited by various techniques, such as ALD leading to the best passivation properties [14–16], plasma-enhanced chemical vapor deposition (PECVD) [17,18], inductively coupled plasma (ICP) deposition [19,20], atmospheric pressure chemical vapor deposition [13,96], or reactive sputtering [90]. It was demonstrated to be ideally suited to the rear passivation of PERC solar cells, as the above-mentioned parasitic shunting is completely absent – thanks to the fixed *negative* charges [16].

Figure 8.6 shows a comparison of the injection-dependent effective SRVs measured on 1.5-Ωcm *p*-type float-zone silicon wafers passivated by (i) SiN_x deposited by microwave remote-PECVD [97], (ii) intrinsic a-Si deposited in a parallel-plate PECVD reactor [87], and (iii) Al_2O_3 deposited by means of plasma-assisted ALD [15]. The measurements are shown in the injection range between 10^{12} and 10^{15} cm^{-3}, which is the most relevant range for the rear of one-sun silicon solar cells. Within this injection range it becomes obvious from Figure 8.6 that the Al_2O_3 passivation outperforms the well-optimized SiN_x as well as the a-Si passivation. Importantly, the Al_2O_3 passivation results in a very weak injection level dependence in the injection range between 10^{12} and 10^{15} cm^{-3}, and the SRV

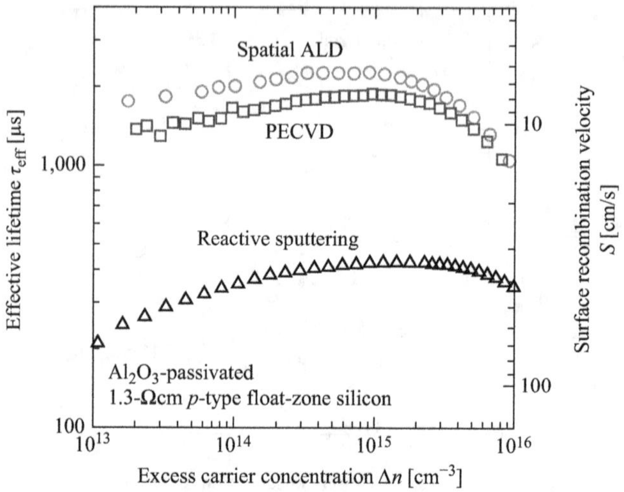

Figure 8.7 Effective lifetime τ_{eff} (left scale) and corresponding surface recombination velocity S (right scale) as a function of the excess carrier concentration, measured on 1.3-Ω cm p-type float-zone silicon passivated by Al_2O_3 deposited by spatial ALD, PECVD, and reactive sputtering (data taken from Schmidt, Ref. [93])

stays well below 10 cm/s even at low injection densities $<10^{13}$ cm^{-3}. The strong injection level dependence of the effective SRV measured on SiN_x-passivated *p*-type Si surfaces can be attributed to recombination within the space charge region induced by the high fixed positive charge density within the SiN_x layer.

In a conventional ALD process, the separation of the half-reactions is implemented by alternate dosing of the process gases. Exposure times of only a few milliseconds are sufficient to ensure complete saturation of the growth surface. In between both precursor doses, however, the reactor chamber is purged by an inert gas and subsequently pumped to remove the residual process gas and reaction products. To prevent parasitic CVD processes and ensure a true ALD process, pumping times of the order of a few seconds are required, severely limiting the deposition rate to approximately 2 nm/min. This makes conventional ALD unsuitable for high-throughput industrial solar cell production. However, Poodt *et al.* proposed a high-rate ALD concept based on spatially separated ALD ("spatial ALD") [98], enabling deposition rates of 1 nm/second. In contrast to the conventional sequential separation, both half-reactions are *spatially* separated, thus eliminating the need for intermediate pumping steps. By now, high-throughput spatial ALD reactors are commercially available by the companies SoLayTec and Levitech. Figure 8.7 compares the effective lifetimes measured on 1.3-Ωcm *p*-type float-zone silicon wafers passivated by Al_2O_3 films deposited using three different industrially feasible Al_2O_3 deposition techniques: (i) spatial ALD, (ii) PECVD, and (iii) RF magnetron sputtering [93]. Spatial ALD and PECVD provide *S* values <10 cm/s clearly outperforming the sputtered Al_2O_3.

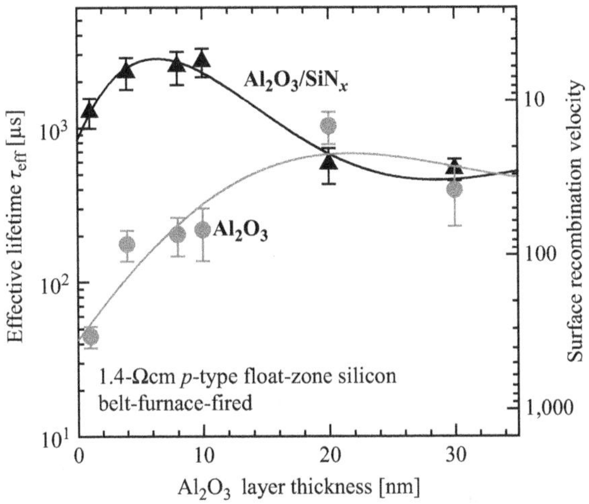

Figure 8.8 Effective lifetime τ_{eff} measured at a fixed excess carrier concentration of $\Delta n = 10^{15}$ cm^{-3} as a function of Al_2O_3 layer thickness after firing. Shown are the results for single Al_2O_3 layers and Al_2O_3 layers with SiN_x capping layer. The Al_2O_3 layers were deposited using plasma-assisted atomic-layer-deposition (ALD), whereas the SiN_x films were deposited using plasma-enhanced chemical vapor deposition (PECVD) (data taken from Veith, Ref. [99])

Industrial-type solar cells use screen-printing of metal pastes and firing in a belt-line furnace at high temperatures (750–850 °C) for contact formation, which makes an excellent thermal stability of the surface passivation indispensable. As demonstrated by Schmidt *et al.*, ultrathin Al_2O_3 layers tend to degrade during firing, however, a hydrogen-containing SiN_x capping layer deposited by PECVD improves the firing stability considerably [91]. Comparable results were later also obtained for Al_2O_3/SiN_x layers deposited by ICP and PECVD [19] when optimizing the thermal budget of the SiN_x capping layer deposition [20], thereby reducing blistering of the passivation layer stack. Please note that the SiN_x capping layer also serves as a protection layer, since the Al_2O_3 is not stable against the aluminum paste used at the rear of PERC cells. Figure 8.8 shows an exemplary result of measurements of the effective lifetime after firing for different layer systems as a function of Al_2O_3 layer thickness (taken from Ref. [99]). For ALD-Al_2O_3 single layers thinner than 20 nm, a pronounced decrease of the effective lifetime after firing is observed, whereas Al_2O_3/SiN_x stacks with Al_2O_3 layers below 20 nm show lifetimes between 1 and 3 ms after firing. Hence, these stacks are perfectly suited for the application to the rear surface passivation of screen-printed PERC solar cells and were hence an important milestone on the path toward the industrialization of the PERC cell. Meanwhile, many solar cell manufacturers have introduced Al_2O_3/SiN_x stacks as rear passivation of industrial PERC solar cells into mass production.

8.5 Local aluminum rear contacts

In addition to the rear passivation, the local aluminum (Al) rear contacts strongly influence the current voltage parameters of industrial PERC solar cells. In particular, recombination of charge carriers at the local Al rear contacts has to be suppressed by the local aluminum p+ doping called the aluminum back-surface-field (Al-BSF) surrounding the contact. Also, the specific contact resistance should be as low as possible. Finally the contact area and contact spacing has to be chosen as the optimum balance between resistive losses (bulk spreading resistance and contact resistance) on the one hand and rear surface recombination losses (recombination at the passivated and contacted rear surface) on the other hand. Whereas lab-type PERC cells often use photolithography to define the rear contact geometries and evaporated aluminum as rear metallization, industrial PERC cells apply the much cheaper laser contact opening (LCO) and screen-printing of aluminum paste. In the LCO approach, the rear passivation layer is locally removed by laser ablation at areas where later the screen-printed aluminum alloys with the silicon wafer thereby forming local rear contacts [9,27]. Typically, pulsed lasers are used and one laser pulse ablates a circular spot with a diameter of around 30 μm. The PERC cell rear contact geometry, for example, continuous lines or points of a certain diameter, is hence composed of adjacent laser-ablated spots where a certain overlap of neighboring laser spots ensures that the full contact area is ablated. Afterwards, the full PERC rear side is coated with aluminum. For industrial PERC applications, screen-printing of aluminum paste is the preferred process technology since it is a widely used, cheap high-throughput process technology proven already for full-area Al-BSF solar cells without rear passivation. After drying of the aluminum paste at temperatures of around 200 °C in a belt furnace, the PERC process is completed by firing the wafers in an infrared belt furnace for a few seconds at a peak temperature between 750 and 850 °C. Thereby, the firing temperature profile is mainly optimized for the requirements of the front silver paste which has a much narrower firing temperature process window compared with the aluminum paste on the rear side. When the wafer temperature during furnace firing exceeds 660 °C, the aluminum melts and according to the Al–Si phase diagram [100] silicon from the wafer dissolves in the aluminum at areas where the rear passivation has been removed by LCO. If the firing temperature and time are sufficiently high, the silicon concentration in the aluminum melt exceeds the eutectic concentration of 12.6 wt.% of silicon in aluminum. During cool down of the firing process excess silicon beyond the eutectic concentration epitaxially recrystallizes at the silicon wafer surface thereby incorporating aluminum up to its solid solubility limit which is in the range of 5×10^{18} cm^3. This local Al-BSF helps to suppress the recombination of charge carriers at the rear contacts due to the p/p+ built-in electric field. Whereas the basic Al–Si contact alloying mechanism as described above was already well known from full-area Al-BSF solar cells, the specific implications of the locally confined Al-Si contact area of PERC solar cells on the alloying mechanism required further research, which will be outlined below.

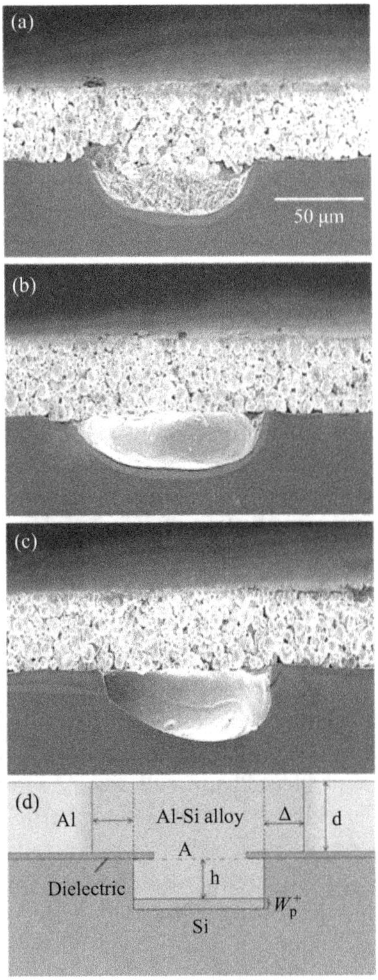

Figure 8.9 Typical scanning electron microscope (SEM) cross-sectional images of local Al contacts processed with LCO of the passivation layer, Al screen printing and firing. (a) a solid Al-Si eutectic layer and 7 μm deep Al-BSF is formed; (b) a large void occurs instead of the Al-Si eutectic and a 6 μm deep Al-BSF is present; (c) a large void is visible but no Al-BSF; and (d) schematic drawing of the alloyed Al-Si contact indicating the areas of Al-Si exchange during furnace firing

Figure 8.9 shows typical scanning electron microscope (SEM) images of local Al contacts with (a) a solid Al-Si eutectic and 7 μm deep Al-BSF, (b) a large void within the Al-Si eutectic and a 6 μm deep AlBSF, and (c) a large void and no Al-BSF. Figure 8.9(d) shows a schematic drawing of the Al rear contact indicating the locally alloyed Al-Si region which extends significantly beyond the laser-ablated contact area. The Al-BSF with thickness W_p^+ is indicated as well. Whereas the i-PERC approach by Agostinelli *et al.* in 2005 [11,27] provided the basic concept for local LCO rear contacts, much more technology development was necessary to improve the quality of the local Al rear contact and an in-depth characterization and modeling was required to understand the Al–Si alloying mechanism of locally confined rear contacts in particular with respect to the formation of the local Al–Si eutectic and Al-BSF. In 2011 and in the following years, several papers were published by different research groups which greatly helped to understand the physics and technology involved in the Al–Si contact formation and its implications on the contact recombination and conversion efficiency of PERC solar cells. This section summarizes the main findings of these papers. Whereas the lab-type PERC cell of Blakers *et al.* [1] applied point-shaped local rear contacts, in 2011 it was shown by Lauermann *et al.* [101] that point contacts in industrial-type PERC cells (i.e., processed with local removal of the passivation layer, Al screen-printing and firing) cause severe void formation similar to Figure 8.9(b) and 8.9(c) resulting in a scattering of the Al-BSF depth from 0 to 6 μm. In contrast, according to [101], line-shaped local rear contacts decrease the amount of voids and increase the Al-BSF depth to 4 μm to 7 μm thereby reducing contact recombination and increasing PERC conversion efficiencies. Further investigations applying carrier lifetime measurements demonstrated that compared with point contacts line contacts achieve a lower rear contact recombination velocity down to 800 cm/s [102]. Gatz *et al.* [103] revealed that the Al-BSF depth increases with increasing LCO line width thereby reducing the contact recombination velocity down to 300 cm/s. The application of an LCO line width of 80 μm, full-area Al screen printing and firing to industrial-type PERC solar cells resulted in a record efficiency of 19.4% in 2011 [28]. In the same year, Urrejola *et al.* [104] systematically investigated the diffusion of silicon from the wafer through local line-shaped contact openings into the aluminum paste during furnace firing. One main result was that the maximum spread of silicon in the aluminum paste ranges between 300 and 1,000 μm depending on the peak firing temperature and on the opening width. Hence, the silicon diffusion and following Al–Si alloy formation extends several hundred μm beyond the local rear contact edges into the aluminum paste. Taking this finding into account, in 2012, Müller *et al.* [105] proposed an analytical model which calculates the Al-BSF thickness in dependence of the contact geometry and several other process parameters. The model assumes a box-shaped silicon distribution in the aluminum paste with eutectic concentration after firing. The width of the box is used as a fit parameter to adjust the model to experimental Al-BSF values. The assumption of a box profile rather than the more realistic Gaussian silicon concentration profile greatly simplifies the model since no differential equations for the silicon diffusion have to be solved. The model further assumes that the excess silicon in the aluminum paste above the eutectic concentration recrystallizes as Al-BSF during cool down of the firing step. The model reproduces the experimental finding that a certain minimum line-width or a minimum point diameter is required to obtain several μm deep Al-BSFs as required for low contact recombination and high-efficiency PERC cells. According to the model, the physical reason is that in case of thin lines or small points the large silicon spread beyond the LCO edges in the aluminum paste significantly reduces the silicon concentration in the aluminum paste and hence the amount of silicon which is available for the Al-BSF formation. In the extreme situation of very small contacts, the silicon concentration in the aluminum paste is reduced below the eutectic concentration and hence no silicon is available for the Al-BSF growth. Another finding by experiment and modeling is that compared with points smaller line widths are sufficient to obtain a deep Al-BSF. Here, the physical reason is that in case of lines the spread of silicon in aluminum occurs only in one dimension whereas for points the spread occurs in two dimensions thereby further reducing the silicon concentration in the aluminum. Recently, Lauermann *et al.* [106] extended the model of Müller *et al.* by including differential equations to describe the silicon diffusion in the aluminum paste. This extended model accurately describes the distribution of the silicon concentration and allows e.g. to determine the diffusivity of silicon in the respective aluminum paste.

The analytical models of Müller *et al.* and Lauermann *et al.* as presented above provide an accurate description of the Al-BSF depth depending on the contact geometry and other process parameters. Nevertheless, they do not describe the impact of voids on the Al-BSF formation and depth. Two recent papers [107,108] demonstrate that the appearance of voids increases the scattering of the measured Al-BSF depths ranging from no Al-BSF (e.g., Figure 8.9(c)) up to similarly deep Al-BSFs as compared to void-free contacts (e.g., compare Figure 8.9(a) and 8.9(b)). As physical mechanism for the formation of voids Urrejola *et al.* [104] proposed the Kirkendall effect [109] which states that voids build up at the interface of two materials A and B if the diffusivity of material A in material B is higher than the diffusivity of material B in material A. The different diffusivities result in a net mass transport of material A into material B forming voids at the interface of material A. However, this mechanism of void formation would start from the beginning of the firing process and hence during cool down there would be no Al-Si melt being in contact with the silicon interface and hence no Al-BSF could epitaxially regrow. In contrast, experimentally quite often a deep Al-BSF is observed despite the occurrence of a void (see, e.g., Figure 8.9(b)) which suggests that the void forms rather at the end of the firing process during cool down shortly before the eutectic Al–Si composition solidifies. One alternative physical explanation for the formation of voids is that the surface energy of the screen-printed Al particles is minimized during firing by pulling the liquid Al–Si melt out of the area adjacent to the silicon wafer surface which then wets the large surface area of the Al particles. Experimental evidence for this mechanism is presented by Kranz *et al.* and by Großer *et al.* [108,110]. The mechanism of surface energy explains why the voids form at the end of the firing process since then sufficient volume of the liquid AlSi is available to wet the Al particle surface. Minimizing surface energies also describes why in case of voids the Al–Si eutectic always forms planar surfaces, see e.g. Figure. 8.8(b) and 8.8(c). Recently, an analytical model has been proposed by Kranz *et al.* [111], which quantitatively describes void formation based on minimizing the surface energies of the liquid Al, the silicon wafer surface and the screen-printed Al particle surface.

All investigations of the Al-BSF formation as described above favor either wide line-shaped rear contacts or point-shaped rear contacts with very large diameter in order to fabricate a deep Al-BSF and hence to reduce the surface recombination velocity at the Al rear contacts. Nevertheless, for optimum PERC conversion efficiencies the metallized area fraction should be small in order to keep the absolute contact recombination small. Also, the wafer bulk spreading resistance increases with wider contact spacing, hence limiting the efficiency. These considerations favor rear contact geometries which achieve a sufficiently deep Al-BSF with as small as possible contact area. An interesting approach to achieve this was introduced in 2013 by the company Sunrise [112]. They proposed a dashed-line rear contact geometry which potentially achieves a deep Al-BSF due to the line-shaped local contacts and simultaneously a small metallized area fraction in combination with low bulk resistance contribution due to the arrangement of the dashed-lines similar to point-shaped contact patterns [112]. Another reason why industrial PERC cells favor as small as possible LCO contact areas is to limit the LCO process time to a few seconds per wafer enabling throughputs of several thousand wafers per hour per LCO production tool. This is the main reason why a typical industrial LCO line width today is in the range of 20–30 μm [113]. Thanks to the tremendous progress in optimizing PERC specific Al pastes as outlined in the following section, today several μm deep Al-BSFs can be obtained with industrially feasible 30 μm narrow LCO openings.

Besides the LCO contact geometry, the composition of the aluminum paste has a very strong influence on the alloying process and hence the Al-BSF formation. In 2006, Meemongkolkiat *et al.* [114] demonstrated that a key requirement is to use a low content of glass frits in the Al paste in order to prevent etching of the rear passivation layer during furnace firing. Also, it was shown that the addition of certain materials to the Al paste resulted in a strongly increased Al-BSF depth [114]. It was, however, not disclosed neither what type of materials were added to the Al paste nor what the physical mechanism was that enabled the deeper Al-BSF. In 2011, Rauer *et al.* [115] demonstrated that the addition of up to 24 wt% silicon to the Al paste increases the Al-BSF depth due to the higher Si concentration in the Al-Si melt during furnace firing. However, the same author also reported that a drawback of the Si addition is that it significantly increases the specific resistivity of the Al paste and that it lowers the internal reflectance of the local rear contacts [116]. In any case, the development of Al pastes which were specifically optimized for local LCO rear contacts greatly contributed to reducing the rear contact recombination [117] and improving PERC conversion efficiencies. By analyzing the maximum spread of silicon in different Al pastes after furnace firing, Lauermann *et al.* [106] found that the diffusivity of Si in Al strongly varies between different Al pastes. This indicates that another approach for PERC Al paste optimization could be to apply additives which reduce the spread of Si in Al during furnace firing thereby enabling the formation of deeper Al-BSFs.

Industrial PERC cells typically apply silver pads on the rear side where later during module fabrication the copper ribbons are soldered to the PERC cells rear side applying a conventional module fabrication process. The silver pads are printed first, afterwards the full-area aluminum is printed on the rear side except for the areas covered by the Ag pads. In order to preserve the PERC rear passivation, specific Ag pad pastes are used with low glass frit content, which do not fire through the rear passivation layer. Hence, the Ag pads contact the silicon wafer only at areas where the rear passivation has been removed by laser ablation. For typical Ag pad and LCO geometries the area where the Ag pad is in direct contact with the silicon wafer accounts for less than 0.5% of the wafer rear side [118]. Even when assuming a very high surface recombination velocity of 10^7 cm/s at the Ag-Si interface the Ag pads impose a reduction of the open circuit voltage of less than 2 mV due to the small area fraction according to analytical modeling as well as experimental PERC cell results [118]. Hence, the missing Al-BSF in the Ag pad region only has a minor impact on the efficiency of industrial PERC solar cells.

8.6 Industrial PERC process flows

As described in Section 8.2, the first PERC solar cell fabricated at the UNSW in 1989 [1] applied mostly lab-type processes such as photolithography for structuring, thermal oxidation for masking and surface passivation as well as evaporation of metals for contact deposition. One important step towards an industrial PERC production sequence was introduced by IMEC in 2005 [11,27] proposing a combination of PECVD SiN_x as rear surface passivation, contact structuring by LCO or etching paste, and rear metallization via Al screen printing. However, from today's perspective also the PERC process sequence proposed by IMEC applied several lab-type plasma-processes e.g. for texturing and rear emitter removal [11] which are not state of the art in industrial manufacturing. This section describes today's most widely used industrial PERC production process sequences and how they evolved. In order to change a conventional industrial full-area Al-BSF production process to a PERC production process, besides adding a rear passivation layer and a LCO process another technological criterion has to be met: only the wafer front surface has to be phosphorus-doped, not the rear surface in order to avoid a floating rear junction and/or shunting of the electrodes. The two industrial solutions which have been developed to solve this process requirement will be introduced in Sections 8.6.1 and 8.6.2.

8.6.1 PERC process with rear protection layer

At ISFH, in 2011, an industrial PERC process was developed by Dullweber *et al.* [31] applying a dielectric protection layer at the rear wafer surface after KOH damage etch. Hence, the following alkaline texture and $POCl_3$ diffusion affect only the wafer front surface whereas the wafer rear surface remains planar and non-diffused. The protection layer is removed afterwards and the wafer rear side is coated with a passivation layer stack, e.g. AlO_x/SiN_y. After LCO of line-shaped rear contacts, the rear-side processing is completed by full-area Al screen printing and firing. The full-process sequence is described in detail in Ref. [32]. This PERC baseline process has since then been continuously improved up to a record efficiency of 21.2% in 2014 [50]. The ISFH PERC baseline process served as an R&D platform for the development of production processes in collaboration with several PV companies such as solar cell producers, equipment manufacturers, and material suppliers.

In parallel to the aforementioned efforts at ISFH, the company Centrotherm developed a similar industrial PERC process flow where the rear passivation layer itself acts as a barrier against phosphorus diffusion [29]. The complete Centrotherm PERC process sequence is shown in column 2 of Table 8.1, "PERC with protection layer." In this process sequence, after texturing the rear side is wet chemically polished in order to improve the surface passivation quality of the following PECVD SiO_xN_y/SiN_z layer stack [29]. The $POCl_3$ diffusion only dopes the front surface since the SiO_xN_y/SiN_z layer stack prevents that phosphorus atoms diffuse into the rear wafer surface. After phosphorus silicate glass (PSG) removal and front SiN_x deposition, the PERC process is completed by rear LCO, Ag and Al screen-printing and firing. Compared to the ISFH PERC process, the Centrotherm PERC process does not require a sacrificial protection layer making it more attractive from a cost perspective. This so-called Centaurus PERC process was one of the first turn-key solutions for industrial PERC production lines becoming available in 2011 [30]. The Centaurus process sequence requires that the rear passivation layer maintains its passivation quality after being exposed to a high temperature $POCl_3$ diffusion. So far, this capability has been demonstrated only for PECVD SiO_xN_y/SiN_z layer stacks [29] but not for AlO_x/SiN_y stacks. Compared to an industry-typical full-area Al-BSF solar cell process, the Centaurus PERC process requires just two additional process steps as shown in Table 8.1.

Table 8.1 Typical production process flows of industrial silicon solar cells applying a full-area Al-BSF as well as PERC cells with rear protection [29,30] or PERC cells with polishing of the rear surface [36,37]. The additional PERC process steps as well as the adapted polishing process are highlighted in gray

Full-area Al-BSF	**PERC with protection layer**	**PERC with polishing**
Texturing	Texturing	Texturing
	Rear polishing	
	Rear passivation	
$POCl_3$ diffusion	$POCl_3$ diffusion	$POCl_3$ diffusion
Junction isolation		Rear polishing
PSG etch	PSG etch	PSG etch
		Rear passivation
Front SiN_x	Front SiN_x	Front SiN_x
	Laser ablation	Laser ablation
Ag screen printing	Ag screen printing	Ag screen printing
Al screen printing	Al screen printing	Al screen printing
Firing	Firing	Firing
8 process steps	**10** process steps	**10** process steps

8.6.2 PERC process with rear polishing

A second industrial PERC process sequence which achieves a single-sided phosphorus doping applies wet chemical removal of the phosphorus-doped rear surface after $POCl_3$ diffusion. Early work of RENA and IMEC on this concept [33,34] applied the process sequence double-sided texturing— wet chemical polish of the rear wafer surface—$POCl_3$ diffusion—wet chemical junction isolation of the rear wafer surface. This process sequence enabled almost planar wafer rear surfaces by using long polishing times and an aggressive polishing chemistry removing up to 20 μm of silicon from the rear surface [35]. A second benefit of this approach is that during polishing the wafer front surface was not yet phosphorus doped. Hence, the aggressive gas phase of the polishing bath did not affect the $POCl_3$ emitter properties.

However, it was shown by Kranz *et al.* [35] that, in particular, when applying an AlO_x/SiN_y passivation stack the rear surface roughness can be quite high without detrimental impact on the rear passivation quality and hence conversion efficiency as shown in Figure 8.10. Accordingly, further work by RENA and ISFH [36,37] simplified the PERC polishing process sequence to texturing—$POCl_3$ diffusion—wet chemical rear polishing by developing a less aggressive wet chemical polishing chemistry. This adapted polishing process removes only 2 to 5 μm of silicon thereby only marginally smoothening the rear pyramids while removing the rear phosphorus doping. A second benefit of the less aggressive polishing recipe is that the gas phase is less reactive hence leading to only marginally increased emitter

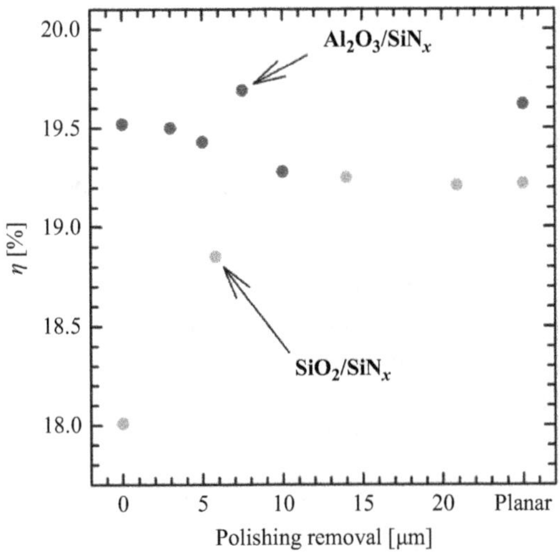

Figure 8.10 Dependence of the conversion efficiency η of industrial PERC cells on the silicon removal during the polishing process. Higher silicon removals reduce the surface roughness. Hence, PERC cells with Al_2O_3/SiN_x rear passivation are rather insensitive to the rear surface roughness in contrast to SiO_2/SiN_x-passivated PERC cells. Graph taken from [35]

sheet resistances at the cell front side after polishing [36]. Hence, the wet chemical polishing recipe which is now typically used for industrial PERC cell production is regarding its chemical composition in-between a classical junction isolation recipe and a polishing recipe. The complete PERC process flow with rear polishing is shown in column 3 of Table 8.1, "PERC with polishing." With 10 process steps in total, it is as lean as the Centaurus PERC process flow. However, one advantage of the "PERC with polishing" process sequence is that it is compatible with the favorable AlO_x/SiN_y rear passivation. This maybe one reason why it became the most widely used process sequence for industrial PERC solar cell production.

8.7 Outlook

As described in the previous sections, a collaborative effort by many institutes and PV companies successfully transferred the PERC silicon solar cell from a lab-type cell concept into mass-production by developing new PERC-specific high-throughput and high-quality process steps for rear emitter removal and polishing, deposition of the rear passivation layer, and LCO for establishing the local Al rear contacts. Nevertheless, research and development (R&D) of industrial PERC solar cells is not finished yet but in contrary there exist still many research opportunities for future improvements of industrial PERC solar cells.

The probably most important R&D topic remains the further increase of the conversion efficiency of industrial PERC solar cells as it is one of the biggest leverages to further decrease the levelized cost of electricity of photovoltaic power supply systems. A recent numerical-based power loss analysis [119,120] of industry-typical PERC solar cells revealed that the four dominant loss mechanisms are (i) front grid shadowing, (ii) emitter recombination, (ii) SiN front surface reflection, and (iv) Al rear contact recombination. Several technologies are already in development that have the potential to minimize these power losses. Promising technology approaches to reduce the front grid shadowing loss are 5 busbar front grid designs [46] and multiwire module interconnection technologies [121,122] which facilitate screen-printed fineline front side silver finger grids. The emitter recombination can be reduced by applying selective emitters, see, e.g. the review paper by Hahn [123], or by advanced homogeneously doped emitters [124,125] with reduced phosphorus concentration at the silicon surface. The Al rear contact recombination can be suppressed by deeper Al-BSFs obtained, e.g. by printing Al finger dots or grids instead of a full-area Al layer [58,107,111] or by further developed Al pastes with improved alloying properties. With these and other potential improvements being already in development, we expect that within the next few years the efficiency of industrial PERC solar cells will exceed the efficiency of 22.8% of the lab-type PERC solar cell of Blakers *et al.* presented in 1989 [1]. Beyond that, the rear-side efficiency and hence energy yield of bifacial PERC+ solar cells can be further improved, e.g. by printing narrower Al fingers and by applying multiwire module interconnection technologies which reduces rear side shadowing losses and resistive losses of the Al finger grid [59].

References

[1] A. W. Blakers, A. Wang, A. M. Milne, J. Zhao, and M. A. Green, "22.8% efficient solar cell," *Appl. Phys. Lett.*, vol. 55, pp. 1363–1365, 1989.

[2] International Technology Roadmap for Photovoltaic (ITRPV.net), 2015 Results, October 2016. Available at http://www.itrpv.net/Reports/Downloads/

[3] T. Dullweber and J. Schmidt, "Industrial silicon solar cells applying the Passivated Emitter and Rear Cell (PERC) concept—a review", *IEEE Journal of Photovoltaics*, vol. 6, pp. 1366–1381, 2016.

[4] A. Metz, D. Adler, S. Bagus, *et al.*, "Industrial high performance crystalline silicon solar cells and modules based on rear surface passivation technology", *Solar Energy Mat. Solar Cells*, vol. 120, pp. 417–425, 2014.

[5] M.A. Green, "The Passivated Emitter and Rear Cell (PERC): From conception to mass production", *Solar Energy Mat. Solar Cells*, vol. 143, pp. 190–197, 2015.

[6] J. Zhao, A. Wang, M. A. Green, and F. Ferrazza, "Novel 19.8% efficient "honeycomb" textured multicrystalline and 24.4% monocrystalline silicon solar cells", *Appl. Phys. Letters*, vol. 73, pp. 1991–1993, 1998.

[7] A. Hübner, A.G. Aberle, and R. Hezel, "Novel cost-effective bifacial silicon solar cells with 19.4% front and 18.1% rear efficiency", *Appl. Phys. Lett.*, vol. 70, pp. 1008–1010, 1997.

[8] J. Schmidt, M. Kerr, and A. Cuevas, "Surface passivation of Si solar cells using plasma-enhanced chemical vapor deposited SiN films and thermal SiO_2/plasma SiN stacks", *Semiconductor Science and Technology*, vol. 16, pp. 164–170, 2001.

[9] R. Preu, S. W. Glunz, S. Schäfer, R. Lüdemann, W. Wettling, and W. Pfleging, "Laser ablation – a new low-cost approach for passivated rear contact formation in crystalline silicon solar cell technology," in *Proc. 16th Eur. Photovoltaic Solar Energy Conf.*, Glasgow, United Kingdom, 2000, pp. 1181–1184.

[10] S. Steckemetz, A. Metz, and R. Hezel, "Thin Cz-silicon solar cells with rear silicon nitride passivation and screen printed contacts", in *Proc. 17th Eur. Photovoltaic Solar Energy Conf.*, Munich, Germany, 2001, pp. 1902–1905.

[11] G. Agostinelli, P. Choulat, H. F. W. Dekkers, S. de Wolf, and G. Beaucarne, "Screen printed large area crystalline silicon solar cells on thin substrates," in *Proc. 20th Eur. Photovoltaic Solar Energy Conf.*, Barcelona, Spain, 2005, pp. 647–650.

[12] S. Dauwe, L. Mittelstädt, A. Metz, and R. Hezel, "Experimental evidence of parasitic shunting in silicon nitride rear surface passivated solar cells," *Prog. Photovolt: Res. Appl.*, vol. 10, pp. 271–278, 2002.

[13] R. Hezel and K. Jaeger, "Low-temperature surface passivation of silicon for solar cells", *J. Electrochem. Soc.*, vol. 136, pp. 518–523, 1989.

[14] G. Agostinelli, A. Delabie, P. Vitanov, *et al.*, "Very low surface recombination velocities on p-type silicon wafers passivated with a dielectric with fixed negative charge", *Sol. En. Mat. Sol. Cells*, vol. 90, pp. 3438–3443, 2006.

[15] B. Hoex, J. Schmidt, P. Pohl, M. C. M. van de Sanden, and W. M. M. Kessels, "Silicon surface passivation by atomic-layer-deposited Al_2O_3", *J. Appl. Phys.*, vol. 104, 044903/1-12, 2008.

[16] J. Schmidt, A. Merkle, R. Brendel, M. C. M van de Sanden, and W. M. M. Kessels, "Surface passivation of high-efficiency silicon solar cells by atomic-layer-deposited Al_2O_3", *Prog. Photovoltaics: Res. Appl.*, vol. 16, pp. 461–466, 2008.

[17] P. Saint-Cast, D. Kania, M. Hofmann, J. Benick, J. Rentsch, and R. Preu, "Very low surface recombination velocity on p-type c-Si by high-rate plasma-deposited aluminum oxide", *Appl. Phys. Letters*, vol. 95, pp. 151502-1–151502-3, 2009.

[18] H.-P. Sperlich, D. Decker, P. Saint-Cast, E. Erben, and L. Peters, "High productive solar cell passivation on Roth & Rau MAiAs MW-PECVD inline machine—a comparison of Al_2O_3, SiO_2 and SiN_x–H process conditions and performance", in *Proc. 25th Eur. Photovoltaic Solar Energy Conf.*, Valencia, Spain, 2010, pp. 1352–1357.

[19] B. Veith, T. Dullweber, M. Siebert, *et al.*, "Comparison of ICP-AlO_x and ALD-Al_2O_3 layers for the rear surface passivation of c-Si Solar cells, *Energy Procedia*, vol. 27, pp. 379–384, 2012.

[20] T. Dullweber, C. Kranz, B. Beier, *et al.*, "Inductively coupled plasma chemical vapour deposited AlO_x/SiN_y layer stacks for applications in high-efficiency industrial-type silicon solar cells", *Solar Energy Mat. Solar Cells*, vol. 112, pp. 196–201, 2013.

[21] F. Schwarz, R. Beckmann, N. Kohn, *et al.*, "Advanced anti-reflection and passivation layer systems produced by high-power plasma in the new Manz PECVD system", in *Proc. 27th European Photovoltaic Solar Energy Conf.*, Frankfurt, Germany, 2012, pp. 1938–1941.

[22] F. Werner, W. Stals, R. Goertzen, B. Veith, R. Brendel, and J. Schmidt, "High-rate atomic layer deposition of Al_2O_3 for the surface passivation of Si solar cells", *Energy Procedia*, vol. 8, pp. 301–306, 2011.

[23] P. G. Vermont, V. Kuznetsov, and E. H. A. Granneman, "High-throughput solar cell passivation on in-line Levitrack ALD Al_2O_3 system—demonstration of process performance", in *Proc. 26th Eur. Photovoltaic Solar Energy Conf.*, Hamburg, Germany, 2011, pp. 1644–1647.

[24] J. Schmidt, F. Werner, B. Veith, *et al.*, "Industrially relevant Al_2O_3 deposition techniques for the surface passivation of Si solar cells", in *Proc. 25th European Photovoltaic Solar Energy Conf. and 5th World Conf. Photovolt. Energy Conv.*, Valencia, Spain, 2010, pp. 1130–1133.

[25] J. Schmidt, B. Veith, and R. Brendel, "Effective surface passivation of crystalline silicon using ultrathin Al_2O_3 films and Al_2O_3/SiN_x stacks", *Phys. Status Solidi RRL*, vol. 3, pp. 287–289, 2009.

[26] D. Kray, M. Hermle, and S. W. Glunz, "Theory and experiments on the back side reflectance of silicon wafer solar cells", *Prog. Photovolt: Res. Appl.*, vol. 16, pp. 1–15, 2008.

[27] G. Agostinelli, J. Szlufcick, P. Choulat, and G. Beaucarne, "Local contact structures for industrial PERC-type solar cells", in *Proc. 20th Eur. Photovoltaic Solar Energy Conf.*, Barcelona, Spain, 2005, pp. 942–945.

[28] S. Gatz, H. Hannebauer, R. Hesse, *et al.*, "19.4%-efficient large-area fully screen-printed silicon solar cells," *Phys. Stat. Sol. RRL*, vol. 5, pp. 147–149, 2011.

[29] K.A. Münzer, M. Hein, J. Schöne, *et al.*, "Technical performance and industrial implementation in favour of centaurus technology," *Energy Procedia*, vol. 27, pp. 631–637, 2012.

[30] K.A. Münzer, J. Schöne, M. Hein, *et al.*, "Development and implementation of 19% rear passivation and local contact centaurus technology," in *Proc. 26th Eur. Photovoltaic Solar Energy Conf.*, Hamburg, Germany, 2011, pp. 2292–2297.

[31] T. Dullweber, S. Gatz, H. Hannebauer, *et al.*, "19.4%-efficient large area rear-passivated screen-printed silicon solar cells", in *Proc. 26th Eur. Photovoltaic Solar Energy Conf.*, Hamburg, Germany, 2011, pp. 811–816.

[32] T. Dullweber, S. Gatz, H. Hannebauer, *et al.*, "Towards 20% efficient large-area screen-printed rear-passivated silicon solar cells", *Prog. Photovolt: Res. Appl.*, vol. 20, pp. 630–638, 2012.

[33] S. Queisser, K. De Keersmaecker, T. Borgers, *et al.*, "Inline single side polishing and junction isolation for rear side passivated solar cells", in *Proc. 24th Eur. Photovoltaic Solar Energy Conf.*, Hamburg, Germany, 2009, pp. 1792–1794.

[34] E. Cornagliotti, M. Ngamo, L. Tous, *et al.*, "Integration of inline single-side wet emitter etch in PERC cell manufacturing", *Energy Procedia*, vol. 27, pp. 624–630, 2012.

[35] C. Kranz, S. Wyczanowski, S. Dorn, *et al.*, "Impact of the rear surface roughness on industrial-type PERC solar cells", in *Proc. 27th European Photovoltaic Solar Energy Conf.*, Frankfurt, Germany, 2012, pp. 557–560.

[36] C. Kranz, S. Wyczanowski, U. Baumann, *et al.*, "Wet chemical polishing for industrial type PERC solar cells", *Energy Procedia*, vol. 38, pp. 243–249, 2013.

[37] D. Pysch, C. Schmitt, B. Latzel, *et al.*, "Implementation of an ALD-Al_2O_3 PERC-Technology into a multi- and monocrystalline industrial pilot production", in *Proc. 29th Eur. Photovoltaic Solar Energy Conf.*, Amsterdam, Netherlands, 2014, pp. 612–616.

[38] E. Schneiderlöchner, R. Preu, R. Lüdemann, and S. W. Glunz, "Laser-fired rear contacts for crystalline silicon solar cells", *Prog. Photovolt: Res. Appl.*, vol. 10, pp. 29–34, 2002.

[39] L. Gautero, M. Hofmann, J. Rentsch, *et al.*, "All-screen-printed 120-μm-thin large-area silicon solar cells applying dielectric rear passivation and laser-fired contacts reaching 18% efficiency," in *Proc. 34th IEEE Photovoltaic Specialists Conf.*, Philadelphia, CA, USA, 2009, pp. 1888–1893.

[40] T. Boscke, R. Hellriegel, T. Wutherich, *et al.*, "Fully screen-printed PERC cells with laser-fired contacts – an industrial cell concept with 19.5%

efficiency", in *Proc. 37th IEEE Photovolt. Spec. Conf.*, Seattle, USA, 2011, pp. 3663–3666.

[41] P. Engelhart, D. Manger, B. Klöter, *et al.*, "QUANTUM – Q-Cells next generation high-power silicon cell & module concept", in *Proc. 26th Eur. Photovoltaic Solar Energy Conf.*, Hamburg, Germany, 2011, pp. 821–826.

[42] L. Tous, R. Russell, J. Beckers, *et al.*, "A simple copper based plating process resulting in efficiencies above 20.5% using pilot processing equipment," in *Proc. 28th European Photovoltaic Solar Energy Conf.*, Paris, France, 2013, pp. 1008–1012.

[43] J. Horzel, N. Bay, M. Passig, *et al.*, "Low cost metallization based on Ni/Cu plating enabling high efficiency industrial solar cells," in *Proc. 29th Eur. Photovoltaic Solar Energy Conf.*, Amsterdam, Netherlands, 2014, pp. 507–512.

[44] I. Romijn, A. A. Mewe, M. W. P. E. Lamers, E. Kossen, E. Bende, and A. W. Weeber, "An overview of MWT cells and evolution to the ASPIRe concept: a new integrated mc-Si cell and module design for high-efficiencies," in *Proc. 23rd European Photovoltaic Solar Energy Conf.*, Valencia, Spain, 2008, pp. 1000–1005.

[45] E. Lohmüller, B. Thaidigsmann, M. Pospischil, *et al.*, "20% efficient passivated large-area metal wrap through solar cells on boron-doped Cz silicon," *IEEE Electron Device Letters*, vol. 32(12), pp. 1719–1721, 2011.

[46] T. Dullweber, H. Hannebauer, U. Baumann, *et al.*, "Fine-line printed 5 busbar PERC solar cells with conversion efficiencies beyond 21%," in *Proc. 29th Eur. Photovoltaic Solar Energy Conf.*, Amsterdam, Netherlands, 2014, pp. 621–626.

[47] K. A. Münzer, J. Schöne, A. Teppe, *et al.*, "Advanced rear side technology for industrial high efficiency solar cells", in *Proc. 25th Eur. Photovoltaic Solar Energy Conf.*, Valencia, Spain, 2010, pp. 2314–2318.

[48] Schott Solar AG, press release August 24th 2011. Available at http://www.presseportal.de/pm/23114/2101260

[49] A. Lachowicz, K. Ramspeck, P. Roth, *et al.*, "Nox-free solution for emitter etch-back", in *Proc. 27th Eur. Photovoltaic Solar Energy Conf.*, Frankfurt, Germany, 2012, pp. 1846–1850.

[50] H. Hannebauer, T. Dullweber, U. Baumann, T. Falcon, and R. Brendel, "21.2%-efficient fineline-printed PERC solar cell with 5 busbar front grid", *Physica Status Solidi RRL*, vol. 8, pp. 675–679, 2014.

[51] Trina Solar, press release, November 17, 2014. Available at http://www.prnewswire.com/news-releases/trina-solar-announces-new-efficiency-records-for-silicon-solar-cells-282907361.html

[52] SolarWorld AG, press release July 17, 2015. Available at http://www.solarworld.de/en/group/investor-relations/news-announcements/corporate-news/single-ansicht/article/solarworld-ag-sets-new-world-record-for-solar-cell-efficiency/

[53] Trina Solar, press release, December 16, 2015. Available at http://www.prnewswire.com/news-releases/trina-solar-announces-new-efficiency-record-of-2213-for-mono-crystalline-silicon-solar-cell-300193727.html

[54] Trina Solar, press release, December 19, 2016. Available at http://www.prnewswire.com/news-releases/trina-solar-announces-new-efficiency-record-of-2261-for-mono-crystalline-silicon-perc-cell-300380906.html

[55] C. Comparotto, M. Noebels, L. Popescu, *et al.*, "Bifacial n-type solar modules: indoor and outdoor evaluation", in *Proc. 29th European Photovoltaic Solar Energy Conf.*, 2014, pp. 3248–3250.

[56] G. J. M. Janssen, B. B. Van Aken, A. J. Carr, and A. A. Mewe, "Outdoor performance of bifacial modules by measurements and modelling", *Energy Procedia*, vol. 77, pp. 364–373, 2015.

[57] T. Dullweber, C. Kranz, R. Peibst, *et al.*, "The PERC+ cell: a 21%-efficient industrial bifacial PERC solar cell" in *Proc. 31st European Photovoltaic Solar Energy Conf.*, 2015, pp. 341–350.

[58] T. Dullweber, C. Kranz, R. Peibst, *et al.*, "PERC+: industrial PERC solar cells with rear Al grid enabling bifaciality and reduced Al paste consumption", *Prog. Photovolt: Res. Appl.*, vol. 24, pp. 1487–1498, 2016.

[59] T. Dullweber, H. Schulte-Huxel, S. Blankemeyer, *et al.*, "Present status and future perspectives of bifacial PERC+ solar cells and modules", *Jap. Journ. Appl. Phys.*, vol. 57, 08RA01, 2018.

[60] K. Krauß, F. Fertig, J. Greulich, S. Rein, and R. Preu, "biPERC silicon solar cells enabling bifacial applications for industrial solar cells with passivated rear sides," *Phys. Status Solidi A*, vol. 213(1), pp. 68–71, 2016.

[61] L. C. Boon (Trina Solar), presented at the PV Module Tech Conference, Malaysia (2017).

[62] SolarWorld press release, 2015. Available at http://www.pvtech.org/news/intersolar_europe_solarworld_to_launch_glass_glass_bifacial_modules

[63] R. R. King, R. A. Sinton, and R. M. Swanson, "Studies of diffused phosphorus emitters: saturation current, surface recombination velocity, and quantum efficiency", *IEEE Trans. Electron Dev.*, vol. 37, pp. 365–371 (1990).

[64] S. W. Glunz, S. Sterk, R. Steeman, W. Warta, J. Knobloch, and W. Wettling, "Emitter dark saturation currents of high-efficiency solar cells with inverted pyramids", in *Proc. 13th European Photovoltaic Solar Energy Conf.*, Nice, France (1995), p. 409.

[65] A. Cuevas, P. A. Basore, G. Giroult-Matlakowski, and C. Dubois, "Surface recombination velocity of highly doped *n*-type silicon", *J. Appl. Phys.*, vol. 80, p. 3370 (1996).

[66] R. M. Swanson, S. K. Beckwith, R. A. Crane, *et al.*, "Point-contact silicon solar cells", *IEEE Trans. Electron Dev.* ED-31, 661–664 (1984).

[67] A. W. Blakers, A. Wang, A. M. Milne, J. Zhao, and M. A. Green, "22.8% efficient silicon solar cell", *Appl. Phys. Lett.*, vol. 55, p. 1363 (1989).

[68] S. Sterk, S. W. Glunz, J. Knobloch, and W. Wettling, "High efficiency (>22%) Si-solar cells with optimized emitter", in *Proc. 1st World Conf. Photovolt. Energy Conv.*, Hawaii, USA (1994), pp. 1303–1306.

[69] R. Hezel, "Silicon nitride for the improvement of silicon inversion layer solar cells", *Solid-State Electron.*, vol. 24, pp. 863–868 (1981).

[70] R. Hezel and K. Jaeger, "Low-temperature surface passivation of silicon for solar cells", *J. Electrochem. Soc.*, vol. 136, pp. 518–523 (1989).

[71] A. G. Aberle, T. Lauinger, J. Schmidt, and R. Hezel, "Injection-level dependent surface recombination velocities at the silicon-plasma silicon nitride interface", *Appl. Phys. Lett.*, vol. 66, p. 2828 (1995).

[72] C. Leguijt, P. Lölgen, J. A. Eikelboom, *et al.*, "Low-temperature surface passivation for silicon solar cells", *Sol. En. Mater. Sol. Cells*, vol. 40, pp. 297–345 (1996).

[73] T. Lauinger, J. Schmidt, A. G. Aberle, and R. Hezel, "Record low surface recombination velocities on 1 Ωcm *p*-silicon using remote plasma silicon nitride passivation", *Appl. Phys. Lett.*, vol. 68, p. 1232 (1996).

[74] B. Lenkeit, *Elektronische und strukturelle Eigenschaften von Plasma-Siliziumnitrid zur Oberflächenpassivierung von siebgedruckten, bifacialen Silizium-Solarzellen*, PhD thesis, University of Hanover (2002).

[75] J. Schmidt, J. D. Moschner, J. Henze, S. Dauwe, and R. Hezel, "Recent progress in the surface passivation of silicon solar cells using silicon nitride", in *Proc. 19th European Photovoltaic Solar Energy Conf.*, Paris, France (2004), pp. 391–396.

[76] J. Schmidt, M. Kerr, and A. Cuevas, "Surface passivation of Si solar cells using plasma-enhanced chemical vapor deposited SiN films and thermal SiO_2/plasma SiN stacks", *Semicond. Sci. Technol.*, vol. 16, pp. 164–170 (2001).

[77] M. L. Reed and J. D. Plummer, "Chemistry of Si-SiO_2 interface trap annealing", *J. Appl. Phys.*, vol. 63, pp. 5776–5793 (1988).

[78] M. J. Kerr, J. Schmidt, A. Cuevas, and J. H. Bultman, "Surface recombination velocity of phosphorus-diffused silicon solar cell emitters passivated with plasma enhanced chemical vapor deposited silicon nitride and thermal silicon oxide", *J. Appl. Phys.*, vol. 89, pp. 3821–3826 (2001).

[79] K. R. McIntosh and L. P. Johnson, "Recombination at textured silicon surfaces passivated with silicon dioxide", *J. Appl. Phys.*, vol. 105, p. 124520 (2009).

[80] F. Ye, W. Deng, W. Guo, *et al.*, "22.13% efficient industrial *p*-type mono PERC solar cell", in *Proc. 43rd IEEE Photovoltaic Specialists Conf.*, Portland, USA (2016), pp. 3360–3365.

[81] A. G. Aberle, T. Lauinger, J. Schmidt, and R. Hezel, "Injection-level dependent surface recombination velocities at the silicon-plasma silicon nitride interface", *Appl. Phys. Lett.* vol. 66, pp. 2828–2830, 1995.

[82] J. Schmidt, M. Kerr, and A. Cuevas, "Surface passivation of Si solar cells using plasma-enhanced chemical vapor deposited SiN films and thermal SiO_2/plasma SiN stacks", *Semiconductor Science and Technology*, vol. 16, pp. 164–170, 2001.

[83] G. Agostinelli, P. Choulat, H. F. W. Dekkers, E. Vermarien, and G. Beaucarne, "Rear surface passivation for industrial solar cells on thin substrates", in *Proc. 4th World Conf. on Photovoltaic Energy Conversion*, Hawaii, USA, 2006, pp. 1004–1007.

[84] M. Hofmann, S. Kambor, C. Schmidt, *et al.*, "PECVD-ONO: A new deposited firing stable rear surface passivation layer system for crystalline silicon solar cells", *Advances in OptoElectronics*, vol. 2008, Article ID 485467, 2008.

[85] R. Petres, V. D. Mihailetchi, J. Maier, S. Keller, T. Pernau, and P. Fath, "Silicon surface passivation by industrial low frequency PECVD films – properties and performance of SiC_x and SiO_xN_y", in *Proc. 24th European Photovoltaic Solar Energy Conf.*, Hamburg, Germany, 2009, pp. 1613–1616.

[86] T. Sawada, N. Terada, S. Tsuge *et al.*, "High-efficiency a-Si/c-Si heterojunction solar cell", in *Proc. 24th IEEE Photovoltaic Specialists Conf.*, Waikoloa, HI, USA, 1994, pp. 1219–1226.

[87] S. Dauwe, J. Schmidt, and R. Hezel, "Very low surface recombination velocities on p- and n-type silicon wafers passivated with hydrogenated amorphous silicon films", in *Proc. 29th IEEE Photovoltaic Specialists Conf.*, New Orleans, USA, 2002, pp. 1246–1249.

[88] M. Hofmann, S. Glunz, R. Preu, and G. Willeke, "21%-efficient silicon solar cells using amorphous silicon rear side passivation," in *Proc. 21st European Photovoltaic Solar Energy Conf.*, Dresden, Germany, 2006, pp. 609–613.

[89] S. Miyajima, J. Irikawa, A. Yamada, and M. Konagai, "Hydrogenated aluminum oxide films deposited by plasma enhanced chemical vapor deposition for passivation of p-type crystalline silicon", in *Proc. 23rd European Photovoltaic Solar Energy Conf.*, Valencia, Spain, 2008, pp. 1029–1032.

[90] T. T. Li and A. Cuevas, "Effective surface passivation of crystalline silicon by rf sputtered aluminum oxide", *Phys. Status Solidi RRL*, vol. 3, pp. 160–162, 2009.

[91] J. Schmidt, B. Veith, and R. Brendel, "Effective surface passivation of crystalline silicon using ultrathin Al_2O_3 films and Al_2O_3/SiN_x stacks", *Phys. Status Solidi RRL*, vol. 3, pp. 287–289, 2009.

[92] G. Dingemans, R. Seguin, P. Engelhart, M. C. M. van de Sanden, and W. M. M. Kessels, "Silicon surface passivation by ultrathin Al_2O_3 films synthesized by thermal and plasma atomic layer deposition", *Phys. Status Solidi RRL*, vol. 4, 10–12, 2010.

[93] J. Schmidt, F. Werner, B. Veith, *et al.*, "Industrially relevant Al_2O_3 deposition techniques for the surface passivation of Si solar cells", in *Proc. 25th European Photovoltaic Solar Energy Conf. and 5th World Conf. Photovolt. Energy Conv.*, Valencia, Spain, 2010, pp. 1130–1133.

[94] J. Schmidt, A. Merkle, R. Bock, *et al.*, "Progress in the surface passivation of silicon solar cells", *Proc. 23rd European Photovoltaic Solar Energy Conf.*, Valencia, Spain, 2008, pp. 974–981.

[95] G. Dingemans, M. C. M. van de Sanden, and W. M. M. Kessels, "Influence of the deposition temperature on the c-Si surface passivation by Al_2O_3 films synthesized by ALD and PECVD", *Electrochem. Solid-State Lett.*, vol. 13, pp. H76–H79, 2010.

[96] L. E. Black, T. Allen, A Cuevas, K. R. McIntosh, B. Veith, and J. Schmidt, "Thermal stability of surface passivation by APCVD Al_2O_3", *Sol. En. Mat. Sol. Cells*, vol. 120, pp. 339–345, 2014.

[97] S. Dauwe, "Low-temperature surface passivation of crystalline silicon and its application to the rear side of solar cells", Ph.D. thesis, ISFH, University of Hanover, Germany, 2004.

[98] P. Poodt, A. Lankhorst, F. Roozeboom, K. Spee, D. Maas, and A. Vermeer, "High-speed spatial atomic layer deposition of aluminum oxide layers for solar cell passivation", *Adv. Mat.*, vol. 22, pp. 3564–3567, 2010.

[99] B. Veith, F. Werner, D. Zielke, R. Brendel, and J. Schmidt, "Comparison of the thermal stability of single Al_2O_3 layers and Al_2O_3/SiN_x stacks for the surface passivation of silicon", *Energy Procedia*, vol. 8, pp. 307–312, 2011.

[100] J. L. Murray and A. J. McAlister, "The Al-Si (Aluminum-Silicon) System", *Bull. Alloy Phase Diagrams*, vol. 5, pp. 74–84, 1984.

[101] T. Lauermann, A. Zuschlag, S. Scholz, G. Hahn, and B. Terheiden, "The influence of contact geometry and sub-contact passivation on the performance of screen-printed Al_2O_3-passivated solar cells", in *Proc. 26th Eur. Photovoltaic Solar Energy Conf.*, Hamburg, Germany, 2011, pp. 1137–1143.

[102] J. Müller, K. Bothe, S. Gatz, H. Plagwitz, G. Schubert, and R. Brendel, "Contact formation and recombination at screen-printed local aluminum-alloyed silicon solar cell base contacts", *IEEE Trans. Electron Devices*, vol. 58, pp. 3239–3245, 2011.

[103] S. Gatz, J. Müller, T. Dullweber, and R. Brendel, "Analysis and optimization of the bulk and rear recombination of screen-printed PERC solar cells", *Energy Procedia*, vol. 27, pp. 95–102, 2012.

[104] E. Urrejola, K. Peter, H. Plagwitz, and G. Schubert, "Silicon diffusion in aluminum for rear passivated solar cells", *Appl. Phys. Lett.*, vol. 98, pp. 153508, 2011.

[105] J. Müller, K. Bothe, S. Gatz, and R. Brendel, "Modeling the formation of local highly aluminum-doped silicon regions by rapid thermal annealing of screen-printed aluminum", *Phys. Status Solidi RRL*, vol. 6, pp. 111–113, 2012.

[106] T. Lauermann, B. Fröhlich, G. Hahn, and B. Terheiden, "Diffusion-based model of local Al back surface field formation for industrial passivated emitter and rear cell solar cells", *Prog. Photovolt: Res. Appl.*, vol. 23, pp. 10–18, 2015.

[107] Y. Chen, P. P. Altermatt, J. Dong, *et al.*, "Al-alloyed local contacts for industrial PERC cells by local printing", in *Proc. 40th IEEE Photovoltaic Specialist Conf.*, Denver, USA, 2014, pp. 3322–3325.

[108] C. Kranz, B. Wolpensinger, R. Brendel, and T. Dullweber, "Analysis of local aluminum rear contacts of bifacial PERC+ solar cells", *IEEE J. Photovolt.*, vol. 6, pp. 830–836, 2016.

[109] A. D. Smigelskas and E. O. Kirkendall, "Zinc Diffusion in Alpha Brass", *Trans. AIME*, vol. 171, pp. 130–142, 1947.

[110] S. Großer, M. Werner, and C. Hagendorf, "Microstructure of void formation stages at local rear contacts," *Energy Procedia*, vol. 77, pp. 701–706, 2015.

[111] C. Kranz, U. Baumann, B. Wolpensinger, *et al.*, "Void formation in screen-printed local aluminum contacts modeled by surface energy minimization", *Sol. Ener. Mat. Sol. Cells*, vol. 158, pp. 11–18, 2016.

[112] B. Tjahjono, M. J. Yang, V. Wu, *et al.*, "Optimizing CELCO technology in one year of mass production", in *Proc. 28th European Photovoltaic Solar Energy Conf.*, Paris, France, 2013, pp. 775–779.

[113] Y. Gassenbauer, K. Ramspeck, B. Bethmann, *et al.*, "Rear-surface passivation technology for crystalline silicon solar cells: a versatile process for mass production", *IEEE J. Photovolt.*, vol. 3, pp. 125–130, 2012.

[114] V. Meemongkolkiat, K. Nakayashiki, D. S. Kim, *et al.*, "Investigation of modified screen-printing Al pastes for local back surface field formation", in *Proc. 4th World Conf. Photovoltaic Energy Conversion*, Hawaii, USA, 2006, pp. 1338–1341.

[115] M. Rauer, R. Woehl, K. Rühle, *et al.*, "Aluminum alloying in local contact areas on dielectrically passivated rear surfaces of silicon solar cells", *IEEE Electron Dev. Lett.*, vol. 32, pp. 916–918, 2011.

[116] M. Rauer, C. Schmiga, R. Woehl, *et al.*, "Investigation of aluminum-alloyed local contacts for rear surface-passivated silicon solar cells", *IEEE J. Photovolt.*, vol. 1, pp. 22–28, 2011.

[117] S. Gatz, K. Bothe, J. Müller, T. Dullweber, and R. Brendel, "Analysis of local Al-doped back surface fields for high efficiency screen-printed solar cells", *Energy Procedia*, vol. 8, pp. 318–323, 2011.

[118] F. Kiefer, T. Brendemühl, M. Berger, *et al.*, "Influence of solder pads to PERC solar cells for module integration", *Energy Procedia*, vol. 38, pp. 368–374, 2013.

[119] C. Kranz, J. H. Petermann, T. Dullweber, and R. Brendel, "Simulation-based efficiency gain analysis of 21.2%-efficient screen-printed PERC solar cells", *Energy Procedia*, vol. 92, pp. 109–115, 2016.

[120] R. Brendel, T. Dullweber, R. Peibst, C. Kranz, A. Merkle, and D. Walter, "Breakdown of the efficiency gap to 29% based on experimental input data and modeling", *Prog. Photovolt: Res. Appl.*, vol. 24, pp. 1475–1486 (2016).

[121] S. Braun, G. Hahn, R. Nissler, C. Pönisch, and D. Habermann, "Multi-busbar solar cells and modules: high efficiencies and low silver consumption", *Energy Procedia*, vol. 38, pp. 334–339, 2013.

[122] A. Faes, M. Despeisse, J. Levrat, *et al.*, "SmartWire Solar Cell Interconnection Technology", in *Proc. 29th Eur. Photovoltaic Solar Energy Conf.*, Amsterdam, The Netherlands, 2014, pp. 2555–2561.

[123] G. Hahn, "Status of selective emitter technology", in *Proc. 25th Eur. Photovoltaic Solar Energy Conf.*, Valencia, Spain, 2010, pp. 1091–1096.

[124] A. Wolf, A. Kimmerle, S. Werner, *et al.*, "Status and perspective of emitter formation by $POCl_3$-diffusion", in *Proc. 31st Eur. Photovoltaic Solar Energy Conf.*, Hamburg, Germany, 2015, pp. 414–419.

[125] A. Piechulla, S. Denzer, V. D. Mihailetchi, *et al.*, "Low pressure diffusions for high quality emitter formation in advanced p- and n-type solar cells", in *Proc. 31st Eur. Photovoltaic Solar Energy Conf.*, Hamburg, Germany, 2015, pp. 420–424.

Chapter 9

Al_2O_3 passivation in industrial solar cells: n-PERT

Emanuele Cornagliotti[1]

9.1 History of n-PERT solar cells

Due to historical and technological reasons, most crystalline silicon (c-Si) solar cells currently in production are fabricated using p-type substrates [1]. However, n-type silicon is, in general, less sensitive to impurity contamination compared to p-type silicon [2]. It is therefore easier to maintain high bulk lifetime during the ingot and cell fabrication process. In addition, n-type substrates do not suffer from light-induced degradation, which is caused mainly by boron–oxygen complexes [3]. Another important advantage is that n-type cells have a lower temperature coefficient than p-type cells [4,5], mainly driven by the higher open-circuit voltage (V_{oc}). Accordingly, photovoltaic modules based on n-type cells have the potential to outperform modules based on p-type cells, both in terms of efficiency and energy yield.

N-type cells can be classified depending on their passivation and contacting scheme. In interdigitated back-contact cells (IBC), both electrodes are placed on the rear (dark) side [6], enabling high photo-generated currents. Efficiency in excess of 25% was obtained on large area devices using this cell architecture [7]. Heterojunction cells (HJT) are another industrially relevant n-type technology [4]. HJT cells make use of amorphous silicon (a-Si) based passivation and, thanks to the displacement of the metal contact from the Si interface, can reach very high V_{OC} and efficiency. HJT cells can either feature electrodes on both sides of the cell or in a back-contact configuration. While in the former case a best cell efficiency of 25.1% was reached [8], a world-record efficiency of 26.6% could be achieved in the latter case [9].

N-type passivated emitter and rear, totally diffused (PERT) cells feature electrodes on both sides (Figure 9.1) and generally do not make use of passivated contact schemes, which results, in most cases, in lower efficiencies compared to IBC or HJT cells. However, n-PERT cells are currently considered as the natural

[1]Kaneka Belgium NV, Zaventem, Belgium

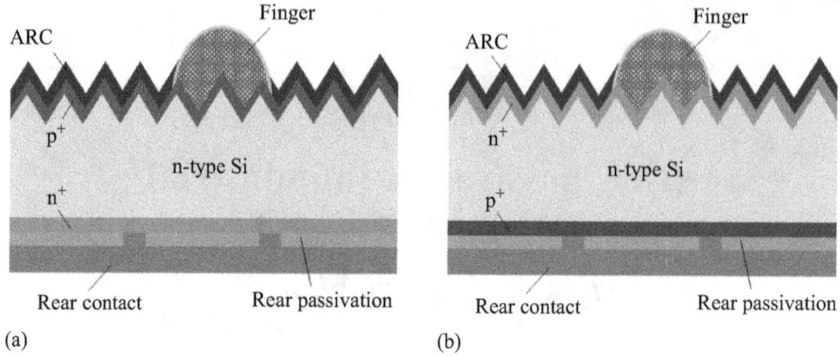

Figure 9.1 Typical cross section of a monofacial front junction (a) and rear-junction (b) n-PERT cell

industrial evolution of p-type passivated emitter and rear cells (PERC), as they can reach higher conversion efficiency than p-PERC, although at the expense of a slightly more complex process sequence (an additional doping step is required, e.g., boron diffusion). In addition, n-PERT cells are often inherently bifacial, thus opening the possibility for glass–glass module integration and higher energy yields due to bifacial gain [10].

N-PERT cells were first introduced by Dai *et al.* in 1993 [11]. In 2002, Zhao *et al.* reported efficiencies up to 21.1% and 21.9% using, respectively, Czochralski-grown (Cz) and float-zone (FZ) silicon [12]. In 2006, this result was improved to 22.7% [13] using a rear emitter configuration. Large area, rear junction, monofacial n-PERT solar cells were presented by Mertens *et al.* in 2013, reaching a conversion efficiency of 21.3% [14]. Improvement of this device structure was achieved in subsequent years by researchers at IMEC, leading to a top efficiency of 22.5% on large area Cz substrates [15–17].

Currently, bifacial n-PERT cell concepts are object of intense development, with several research groups developing different metallization solutions, using both screen printing [18–21] or co-plating [22]. Different approaches for junction and surface field formation can be implemented: in many cases a combination of conventional tube boron and phosphorus diffusion is employed, but these process steps can also be replaced by ion implantation [23,24] or diffusion from doped CVD layers [25].

A distinctive feature of n-PERT cells is the presence of a p^+-type doped surface (random-pyramid textured or planar) where Shockley–Read–Hall (SRH) recombination needs to be effectively suppressed in order to manufacture high efficiency devices. Although several passivation layers have been proposed in the past, amorphous aluminum oxide (Al_2O_3) has been quickly identified as the most promising solution [26] due to the excellent passivation quality, good thermal stability, and the industrially viable deposition processes.

In the following section, the properties of Al_2O_3 based p^+surface passivation will be described, focusing on the impact of deposition conditions, annealing sequence, capping layer material, Al_2O_3 thickness, and surface morphology.

9.2 Al_2O_3 passivation for n-PERT cells

9.2.1 Overview of passivation layers for p^+-type silicon surfaces

Several dielectric layers have been used in the past to passivate p^+-type silicon surfaces. Thermally grown silicon oxide (SiO_2) chemically passivates recombination centers by reducing the density of interface states (D_{it}). After a Forming Gas Anneal (FGA) [27] or aluminum anneal (the so-called alneal process [28]), peak D_{it} can be reduced to approximately 10^{10} cm^{-2} eV^{-1}. Thermal SiO_2 is also known to possess a relatively low amount ($<10^{12}$ cm^{-2}) of positive fixed charges (Q_f) [27,29]. The positive polarity is not optimal for p^+ passivation, as it could result in depletion or inversion of carriers in the near-surface. Nevertheless, due of the relatively low absolute Q_f value and low D_{it} (especially in the case of high growth temperatures [29]), thermal SiO_2 can be successfully employed to passivate simultaneously n^+ and p^+ silicon surfaces, as in the case of IBC cell structures [30].

Hydrogenated amorphous silicon nitride (a-SiN_x:H or, simply, SiN_x) deposited by plasma-enhanced chemical vapor deposition (PECVD) has also been used to passivate p^+ surfaces [31]. However, the Si–SiN_x interface can have lower electronic quality than the Si–SiO_2 interface, especially when a direct plasma reactor is used [32]. This can have an important impact on the SRH recombination rate at the surface, especially if the surface doping concentration is low. Therefore, in some cases, a thin chemical SiO_2 interlayer is grown before SiN_x deposition [33,34]. In addition, the high content of positive charges present in the SiN_x layer [35] can lead to passivation degradation mechanisms due to depletion or inversion of carriers (which creates a recombinative channel in the near-surface with strong injection level dependence) or even to parasitic shunting [36], when the inversion layer is not electrically isolated from the p^+ contact. Other materials that have been also used to passivate p^+-type surfaces include silicon carbide (SiC_x) [37,38] or silicon oxy-nitride (SiON) [39].

Al_2O_3 is of particular interest for n-PERT not only because of the low D_{it} at the Si–Al_2O_3 interface [40], but especially because of the high negative Q_f ($>10^{12}$ cm^{-2}) [41,42], which induces an accumulation layer at the Si surface, reducing thus the density of minority carriers (electrons) available for SRH recombination. In general, due to the high negative Q_f and low D_{it}, Al_2O_3 is suited for the passivation of a very wide range of doping levels from p-type base material ($N_A \sim 10^{15}$ cm^{-3}) to highly doped p^+ layers ($N_A > 10^{19}$ cm^{-3}). It has been also shown that, for boron diffused surfaces passivated either by atomic layer deposition (ALD), atmospheric pressure chemical vapor deposition (APCVD), or PECVD Al_2O_3, Q_f and the fundamental surface recombination velocity for electrons (S_{n_0}) are constant across a wide range of doping levels [43,44]. These properties, together with a good thermal stability, already made Al_2O_3 the material of choice for rear passivation of industrial p-type PERC cells [45].

9.2.2 Al_2O_3 for n-PERT: deposition, annealing sequence, and capping layers

The most common methods used by industry for the deposition of Al_2O_3 are ALD [46] and PECVD [47], but Al_2O_3 can also be deposited by APCVD [48], screen-printing [49], sputtering [50], or obtained by anodization of an aluminum layer [51].

In the case of ALD, the deposition temperature has an important impact on the amount of hydrogen present in the layer, with reduced hydrogen content for higher deposition temperatures, together with a higher mass density [52]. Since hydrogen is important for the passivation of unsaturated dangling bonds at the interface, a higher hydrogen content is normally desirable. On the other hand, hydrogen has been linked to the emergence of blistering upon thermal treatment, which is an undesired effect as it can dramatically reduce surface passivation [53–58]. Another crucial aspect influencing passivation quality regards the thermal treatment which is applied after the deposition, as different solar cell structures will require different thermal budgets, which have, in turn, a substantial impact on the passivation stability of the Al_2O_3 layer. It is well known that Al_2O_3 layers need to be treated at temperatures of at least 450 °C to fully activate the passivation [40]. This is particularly relevant for techniques such as spatial thermal ALD [59], since the wafers are maintained at the deposition temperature (~150–250 °C) for a very limited amount of time and there is no self-annealing effect during the deposition itself. A post-deposition anneal (PDA) is also functional to achieve some degree of gaseous desorption from the Al_2O_3 layer, with higher PDA temperatures reducing risk of subsequent blistering [56,60]. The PDA is, in production, normally combined with the capping layer deposition (e.g., SiN_x), if the latter requires deposition temperatures in the range of 450–600 °C.

Al_2O_3 layers are typically deposited in a thickness range comprised between 4 and 10 nm (for ALD) and 15–25 nm for PECVD [61], as thicker layers can result in an increase of blister density [55,56,58]. A higher Al_2O_3 thickness would also not be desirable from a cost perspective due to the high cost of the tri-methyl-aluminum (TMA) precursor. Moreover, Al_2O_3 features a refractive index (at 633 nm) comprised between 1.6 and 1.7 [62], which is not the optimal value for front side anti-reflection coating. Therefore, Al_2O_3 thickness needs to be reduced as much as possible (as far as the passivation properties are not hampered) and it needs to be capped by another layer to improve the photon collection capability of the cell. This optical requirement is of course especially important when Al_2O_3 is on the front (sunny) side of the cell, but can also be important when applied at the rear (dark) side [63]. In addition, Al_2O_3 needs to be capped to improve temperature stability [64], for long-term reliability purposes (in particular, to avoid humidity degradation [65]) and for protection from chemical reactions with screen-printed pastes (as at the rear side of PERC cells).

Typically, the material used to cap Al_2O_3 is SiN_x, but effective capping has also been reported by SiO_x [16], SiON [66], SiC_x [67] and stacks of SiO_x/SiN_x

[56,68]. Non-vacuum techniques such as liquid-phase deposited dielectric layers have been also tested [69]. In n-PERT cells, mostly SiN_x has been employed, with SiO_x used in case of a rear-side emitter configuration, due to better optical performance [63]. However, if SiO_x is capped by an Al metal electrode, it can only be used in combination with a low-temperature metallization process (e.g., Cu plating) due to the chemical reactions occurring between Al and SiO_x at high temperature [70].

The capping layer can affect the quality of the Al_2O_3 passivation in three ways. First, the deposition temperature of the capping layer can affect the amount of desorbed gaseous species from the layer and hence the amount of blisters [56] of the final stack. This is particularly relevant if a specific *ad-hoc* PDA is not carried out after the Al_2O_3 deposition. Second, the amount of hydrogen contained in the capping layer is a crucial factor for dangling bonds passivation at the Si/SiO_2 interface and can compensate for lack of hydrogen in Al_2O_3 itself (e.g., if the latter is deposited at particularly high temperatures). Third, the thermal expansion coefficient and the stress associated with the capping layer can play an important role in the delamination mechanisms, which lead to the formation of blisters.

In n-PERT solar cells with screen-printed contacts, a firing step is required at the end of the process, similarly to p-type PERC cells. This is normally the most challenging step for the thermal stability of Al_2O_3 as wafers can experience temperatures well above 800 °C for a few seconds. However, if the proper combination of Al_2O_3 thickness, post deposition anneal and capping layers are employed, passivation quality can even improve after the firing step, due, for instance, to hydrogen release from SiN_x [71]. If self-aligned plated contacts are employed and firing-through is not necessary, the firing step can instead be optimized to maximize surface passivation or to cure laser induced defects [17,22].

Summarizing, the quality of the final passivation depends on multiple factors: the hydrogen content of the Al_2O_3 layer, its thickness, the temperature profile of the PDA, of the capping layer deposition and of the annealing (or firing) step following the capping deposition.

9.2.3 Al_2O_3 for n-PERT: thickness dependence

In industrial manufacturing, a minimization of the deposited Al_2O_3 thickness is desired, as this reduces the consumption of TMA, which is a major contributor to the total cost of ownership of an Al_2O_3 deposition step. It is therefore important to study the dependence of passivation quality on the Al_2O_3 thickness.

In reference [16], the limiting recombination mechanism affecting a n-PERT p^+ emitter (passivated with an Al_2O_3/SiO_x stack on a planar <100> surface) was analyzed by determining the breakdown of the different recombination components to the total emitter saturation current (J_{0_e}), with the aid of the simulation software EDNA 2 [72,73]. For a 90 Ω/□ driven-in emitter, the Auger component of the emitter saturation current ($J_{0_e-\mathrm{Auger}}$) was ~17 fA/cm^2 (Figure 9.2). The radiative component being negligible (~0.06 fA/cm^2), the difference between the measured J_{0_e} and the J_{0_e} due to intrinsic recombination ($J_{0_e-\mathrm{Auger}} + J_{0_e-\mathrm{radiative}}$) has to be ascribed to SRH recombination mechanisms within the emitter bulk or at the emitter surface.

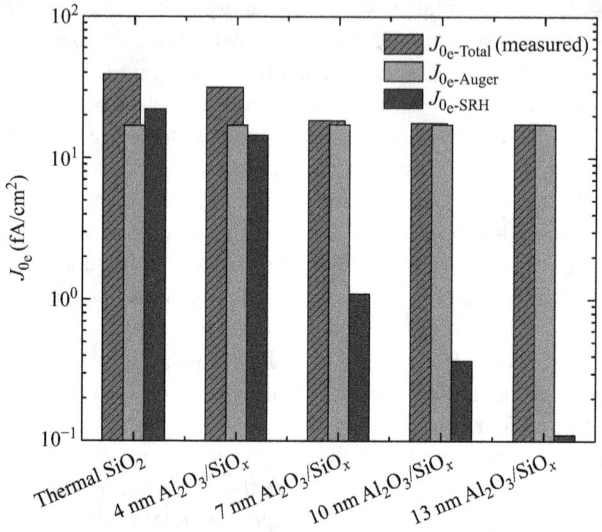

Figure 9.2 *J_{0_e} breakdown for a 90 Ω/□ boron emitter passivated by different ALD Al_2O_3 thicknesses (capped by PECVD SiO_x) as compared to a thermal SiO_2 passivation [16]. $J_{0_e-Total}$ is a measured value*

The J_{0_e}–SRH value thus obtained can therefore be considered as an upper limit for SRH recombination at the surface. Deposition of 7 nm of Al_2O_3 was sufficient to reduce the SRH recombination contribution to a small fraction (1.1 fA/cm^2) of the total J_{0_e}, with the remaining part being almost entirely ascribed to the Auger recombination in the emitter. With 13 nm of Al_2O_3, J_{0_e}–SRH approached 0.1 fA/cm^2, a marked reduction compared to the value of 22 fA/cm^2 obtained with thermal SiO_2.

Figure 9.3 collects the J_{0_e} of boron emitters as a function of Al_2O_3 thickness measured by different authors. In the work of Richter *et al.* [74], the impact of Al_2O_3 thickness on the passivation quality of a Al_2O_3/SiN_x stack was studied as a function of the thermal treatment applied (FGA or firing). A pronounced thickness dependency was found for the FGA case, while in the case of firing, very good passivation could be obtained even with very-thin layers (<5 nm). In the case of the 86 Ω/□ emitter, the total J_{0_e} is limited by Auger recombination, while it drops below 10 fA/cm^2 when a driven-in 133 Ω/□ emitter is used. In the case of PECVD Al_2O_3, thicknesses of around 15–25 nm are considered optimal, with passivation degradation which is observed for thicker layers [75,77]. However, excellent passivation could also be obtained with layers thinner than 10 nm. Duttagupta *et al.* [75] reported total J_{0_e} below 20 fA/cm^2 with only 5 nm of PECVD Al_2O_3 on a 80 Ω/□ emitter (Figure 9.3), while Saint-Cast *et al.* [76] showed constant J_{0_e} of 23 fA/cm^2 in the thickness range of 5–20 nm.

In general, as outlined by Figure 9.3, J_{0_e} tends to decrease as thickness is increased, with an optimum around 8–12 nm for ALD and 20–25 nm for PECVD. For such an optimal thickness range, and in the case of planar surfaces, J_{0_e} is mostly

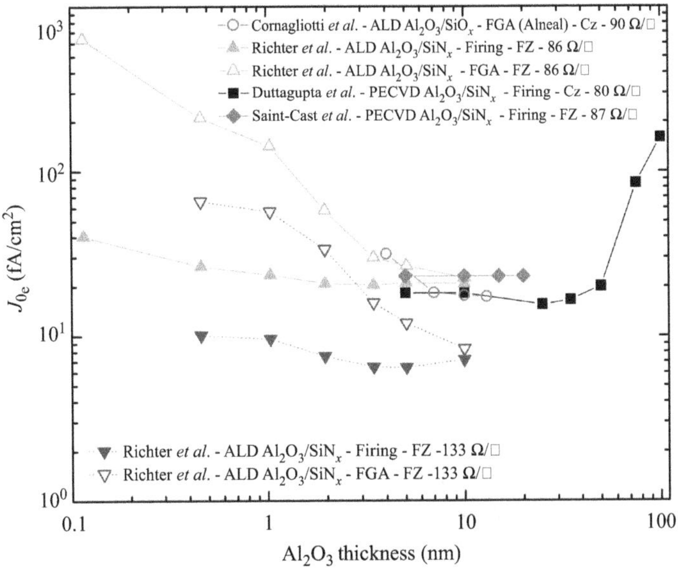

Figure 9.3 Boron emitter J_{0_e} measured on planar surfaces (FZ or Cz-Si) as a function of Al_2O_3 layer thickness for different deposition methods (ALD or PECVD), capping layer (SiN_x or SiO_x) and annealing type (FGA or Firing). Data from [16,74–76]

limited by the Auger recombination in the emitter. Above a critical thickness (higher in the case of PECVD layers), passivation can be hampered by increased layer blistering and J_{0_e} sharply increases, due to increased SRH recombination at the surface.

Note that the SRH recombination component tends to increase as surface doping concentration is decreased. However, thanks to the high negative Q_f and low D_{it}, Al_2O_3 has proven very effective also for the suppression of SRH surface recombination in lowly doped ($<10^{18}$ cm^{-3}) grown-in emitters in epitaxial wafers [78]. Thanks also to the low Auger recombination associated with these deep and lowly doped emitters, a V_{oc} of 700 mV could be obtained [79], a remarkable value for large area Cz-Si cells which do not make use of passivated contacts schemes.

9.2.4 Al_2O_3 for n-PERT: surface morphology dependence

In n-PERT cells, an important aspect is the dependence of p^+ emitter passivation on the silicon crystal orientation and on the morphological structure of the surface. Depending on the cell architecture (front junction or rear junction) and the processing sequence, the p^+ surface can consist of <100> planar or a <111> textured surface. It was observed by different research groups (see Table 9.1) that the J_{0_e} ratio between <111> textured and <100> planar surfaces ($J_{0_e,\text{text}<111>}/J_{0_e,\text{plan}<100>}$) is comprised, for Al_2O_3 based passivation, between 1.7 and 2.5, depending on the

Table 9.1 Ratio between J_{0_e} on a boron-diffused <111> textured surface and on a <100> planar surface as reported by different authors for different Al_2O_3-based passivation schemes. Data from reference [75] refer to PECVD Al_2O_3. Data from references [16, 74] refer to ALD Al_2O_3

	$J_{0_e,\text{text}<111>}/J_{0_e,\text{plan}<100>}$
Thermal SiO_2—Alneal	3.4 [16]
Al_2O_3—FGA [74]/N_2 Anneal [16]	2.1–2.3 [74], 2.4 [16]
Al_2O_3/SiN_x—FGA	2.2–2.5 [74]
Al_2O_3/SiN_x—Fired	1.7–1.9 [74], 1.5–2 [75]
Al_2O_3/SiO_x—Alneal	2.1 [16]

annealing and capping scheme. In contrast, $J_{0_e,\text{text}<111>}/J_{0_e,\text{plan}<100>}$ was found to be ~3.4 for thermal SiO_2 passivation [16].

In some cases, the value was found to be rather close to the geometric surface area ratio of 1.7 while in some other cases the value were reported to be slightly higher, up to 2.5. The difference between the geometric surface area ratio (~1.7) and the obtained $J_{0_e,\text{text}<111>}/J_{0_e,\text{plan}<100>}$ can be attributed to several factors. First of all, the different crystal orientation (<111> for textured versus <100> for planar) enhances recombination intrinsically because of the higher amount of dangling bonds in <111> surfaces compared to <100> surfaces. As a reference, the J_{0_e} ratio between planar <111> and <100> surfaces ($J_{0_e,\text{plan}<111>}/J_{0_e,\text{plan}<100>}$) has been estimated to be around 1–1.2 for both APCVD Al_2O_3 layers [80] and ALD Al_2O_3 layers [81]. For ALD Al_2O_3 it was also suggested that layers deposited on <111> surfaces may have a lower negative Q_f as compared to layers deposited on <100> surfaces [81]. Other potential factors, which are particularly difficult to quantify, are the existence of additional defects associated with enhanced stress at the edges and vertices of the pyramidal structure [82] or the difference in Auger contribution at textured surfaces compared to planar surfaces because of increased doping at pyramidal tips [83]. Also, blister density depends on surface morphology, due to the finite lateral diffusion length of hydrogen in Al_2O_3 [53], with textured surfaces presenting in general smaller but more numerous blisters and planar surface presenting instead fewer but larger blisters. These additional factors can be described by the J_{0_e} ratio between textured <111> and planar <111> surfaces ($J_{0_e,\text{text}<111>}/J_{0_e,\text{plan}<111>}$). This value can be above 2 (for APCVD Al_2O_3 layers, [80]), suggesting that the aforementioned effects have a particular relevance.

The low $J_{0_e,\text{text}<111>}/J_{0_e,\text{plan}<100>}$ ratio (<2.5) shows that Al_2O_3 is a material particularly suited for the passivation of textured p^+-type surfaces. Due to the intrinsic conformality of ALD deposition process, ALD Al_2O_3 is also regarded as a promising candidate for the passivation of black silicon–boron emitters [84].

9.3 Conclusions

Al_2O_3 is currently the material of choice for the passivation of p^+-type emitters in n-type silicon solar cells. Its outstanding passivation properties are able to suppress almost completely the SRH recombination at planar surfaces of such emitters, even in the case of lowly doped emitters. Al_2O_3 also provides excellent passivation of textured p^+ surfaces, making the Auger recombination in the emitter the limiting factor to a further decrease of the J_{0_e} of an n-PERT device. As a proof of its excellent passivation quality, large-area n-PERT devices with V_{oc} in excess of 700 mV and up to 22.5% efficiency have been demonstrated.

References

[1] "International Technology Roadmap for Photovoltaic," www.itrpv.net, 2016.

[2] D. MacDonald and L. J. Geerlings, "Recombination activity of interstitial iron and other transition metal point defects in p- and n-type crystalline silicon," *Applied Physics Letters*, vol. 85, no. 18, pp. 4061–4063, 2004.

[3] J. Schmidt, A. G. Aberle, and R. Hezel, "Investigation of carrier lifetime instabilities in Cz-grown silicon," *Conference Record of the 26th IEEE Photovoltaic Specialists Conference*, pp. 13–18, 1997.

[4] M. Taguchi, K. Kawamoto, S. Tsuge, *et al.*, "HIT cells – High efficiency crystalline Si cells with novel structure," *Progress in Photovoltaics: Research and Applications*, vol. 8, no. 5, pp. 503–513, 2000.

[5] J. Haschke, J. P. Seif, Y. Riesen, *et al.*, "The Impact of Silicon Solar Cell Architecture and Cell Interconnection on Energy Yield in Hot & Sunny Climates," *Energy & Environmental Science*, 2017.

[6] E. Van Kerschaver and G. Beaucarne, "Back-contact solar cells: a review," *Progress in Photovoltaics: Research and Applications*, vol. 14, no. 2, pp. 107–123, 2006.

[7] M. A. Green, K. Emery, Y. Hishikawa, W. Warta, and E. D. Dunlop, "Solar cells efficiency tables (version 48)," *Progress in Photovoltaics: Research and Applications*, vol. 24, no. 7, pp. 905–913, 2016.

[8] K. Yamamoto, "25.1% efficiency Cu metallized heterojunction crystalline Si solar cell," *Proceedings of 25th International Photovoltaic Science and Engineering Conference*, 2015.

[9] Kaneka, *Press Release*, 2017.

[10] R. Kopecek, I. Shoukry, and J. Libal, "Cost/kWh thinking and bifaciality: two allies for low-cost PV of the future," *Photovoltaics International*, vol. 30, pp. 88–97, December 2015.

[11] X. M. Dai, M. A. Green, and S. R. Wenham, "High efficiency n-silicon solar cells using rear junction structures," *Conference Record of the 23rd IEEE Photovoltaic Specialists Conference*, pp. 153–156, 1993.

[12] J. Zhao, A. Wang, P. P. Altermatt, and M. A. Green, "High efficiency PERT cells on n-type silicon substrates," *Conference Record of the 29th IEEE Photovoltaic Specialists Conference*, pp. 218–221, 2002.

[13] J. Zhao and A. Wang, "High efficiency rear emitter PERT cells on CZ and FZ n-type silicon substrates," *Conference Record of the 4th World Conference on Photovoltaic Energy Conversion (WCPEC)*, pp. 996–999, 2006.

[14] V. Mertens, T. Ballmann, J. Cieslak, *et al.*, "Large area n-type Cz double side contact back junction solar cell with 21.3% conversion efficiency," *Proceedings of 28th European Photovoltaic Solar Energy Conference and Exhibition*, pp. 714–717, 2013.

[15] M. Aleman, A. Uruena, P. Choulat, *et al.*, "Reducing front recombination losses to improve the efficiency of rear junction Cu-plated n-Si cells," *Proceedings of 29th European Photovoltaic Solar Energy Conference and Exhibition*, pp. 528–531, 2014.

[16] E. Cornagliotti, A. Uruena, M. Aleman, *et al.*, "Large-area n-type PERT solar cells featuring rear p+ emitter passivated by ALD Al_2O_3," *IEEE Journal of Photovoltaics*, vol. 5, no. 5, pp. 1366–1372, 2015.

[17] A. Uruena, M. Aleman, E. Cornagliotti, *et al.*, "Progress on large area n-type silicon solar cells with front laser doping and a rear emitter," *Progress in Photovoltaics: Research and Applications*, vol. 24, no. 8, p. 1149–1156, 2016.

[18] T. S. Boscke, D. Kania, A. Helbig, *et al.*, "Bifacial n-type cells with >20% front-side efficiency for industrial production," *IEEE Journal of Photovoltaics*, vol. 3, no. 2, pp. 674–677, 2013.

[19] I. Romijn, B. van Aken, J. Anker, *et al.*, "Industrial implementation of efficiency improvements in n-type solar cells and modules," *Proceedings of 27th European Photovoltaic Solar Energy Conference and Exhibition*, pp. 533–537, 2012.

[20] J. Lossen, F. Buchholz, C. Comparotto, *et al.*, "From lab to fab: bifacial n-type cells entering industrial production," *Proceedings of the 31st European Photovoltaic Solar Energy Conference and Exhibition*, pp. 965–968, 2015.

[21] B. Lim, T. Brendemuhl, T. Dullweber, and R. Brendel, "Loss analysis of n-type passivated emitter rear totally diffused back-junction silicon solar cells with efficiencies up to 21.2%," *IEEE Journal of Photovoltaics*, vol. 6, no. 2, pp. 447–453, 2016.

[22] E. Cornagliotti, R. Russell, L. Tous, *et al.*, "Bifacial n-PERT cells (Bi-PERT) with plated contacts for multi-wire interconnection," *Proceedings of the 32nd European Photovoltaic Solar Energy Conference and Exhibition*, pp. 420–425, 2016.

[23] H. Schulte-Huxel, F. Kiefer, S. Blankemayer, *et al.*, "Flip-flop cell interconnection enabled by an extremely high bifacial factor of screen-printed ion implanted n-PERT Si solar cells," *Proceedings of 32nd European Photovoltaic Solar Energy Conference and Exhibition*, pp. 407–412, 2016.

[24] J. Benick, R. Muller, N. Bateman, and M. Hermle, "Fully implanted n-type PERT solar cells," *Proceedings of 27th European Photovoltaic Solar Energy Conference and Exhibition*, pp. 676–679, 2012.

[25] N. Wehmeier, B. Lim, A. Merkle, *et al.*, "PECVD BSG diffusion sources for simplified high-efficiency N-PERT BJ and BJBC solar cells," *IEEE Journal of Photovoltaics*, vol. 6, no. 1, pp. 119–125, 2016.

[26] J. Benick, B. Hoex, M. C. M. Van de Sanden, W. M. M. Kessels, O. Schultz, and S. W. Glunz, "High efficiency n-type Si solar cells on Al_2O_3-passivated boron emitters," *Applied Physics Letters*, vol. 92, no. 25, p. 253504, 2008.

[27] A. G. Aberle, S. Glunz, and W. Warta, "Impact of illumination level and oxide parameters on Shockley Read Hall recombination at the Si-SiO_2 interface," *Journal of Applied Physics*, vol. 71, pp. 4422–4431, 1992.

[28] B. E. Deal, E. L. MacKenna, and P. L. Castro, "Characteristics of fast surface states associated with SiO_2-Si and Si_3N-SiO_2-Si structures," *Journal of the Electrochemical Society*, vol. 116, no. 7, pp. 997–1005, 1969.

[29] S. C. Vitkavage, E. A. Irene, and H. Z. Massoud, "An investigation of Si-SiO_2 interface charges in thermally oxidized (100), (110), (111), and (511) silicon," *Journal of Applied Physics*, vol. 68, p. 5262, 1990.

[30] M. Aleman, J. Das, T. Janssens, *et al.*, "Development and integration of a high efficiency baseline leading to 23% IBC cells," *Energy Procedia*, vol. 27, pp. 638–645, 2012.

[31] F. W. Chen, A. L. Tsu-Tsung, and J. E. Cotter, "PECVD silicon nitride surface passivation for high-efficiency n-type silicon solar cells," *Conference Record of the 4th World Conference on Photovoltaic Energy Conversion*, pp. 1020–1023, 2006.

[32] J. Schmidt and A. Aberle, "Carrier recombination at silicon–silicon nitride interfaces fabricated by plasma-enhanced chemical vapor deposition," *Journal of Applied Physics*, vol. 85, no. 7, pp. 3626–3633, 1999.

[33] R. C. Naber, N. Guillevin, A. R. Burgers, L. J. Geerlings, and A. W. Weeber, "ECN n-type silicon solar cell technology: an industrial process that yields 18.5%," *Conference Record of the 34th IEEE Photovoltaic Specialists Conference*, pp. 990–992, 2009.

[34] V. D. Mihailetchi, Y. Komatsu, and L. J. Geerlings, "Nitric acid pretreatment for the passivation of boron emitters for n-type base silicon solar cells," *Applied Physics Letters*, vol. 92, p. 063510, 2008.

[35] J. R. Elmiger and M. Kunst, "Investigation of charge carrier injection in silicon nitride/silicon junctions," *Applied Physics Letters*, vol. 69, p. 517, 1996.

[36] S. Dauwe, L. Mittelstadt, A. Metz, and R. Hezel, "Experimental evidence of parasitic shunting in silicon nitride rear surface passivated solar cells," *Progress in Photovoltaics: Research and Applications*, vol. 10, no. 4, pp. 271–278, 2002.

[37] R. Petres, J. Libal, T. Buck, *et al.*, "Improvements in the passivation of p+ Si surfaces by PECVD silicon carbide films," *Conference Record of the 4th World Conference on Photovoltaic Energy Conversion*, 2006.

[38] I. Martin, M. Vetter, M. Garin, *et al.*, "Crystalline silicon surface passivation with amorphous SiC_x:H films deposited by plasma enhanced chemical-vapour deposition," *Journal of Applied Physics*, vol. 98, no. 11, p. 114912, 2005.

[39] J. Seiffe, L. Gautero, M. Hofmann, *et al.*, "Surface passivation of crystalline silicon by plasma-enhanced chemical vapor deposition double layers of silicon-rich silicon oxynitride and silicon nitride," *Journal of Applied Physics*, vol. 109, no. 3, p. 034105, 2011.

[40] G. Dingemans, Nanolayersurface passivation schemes for silicon solar cells, University of Tilburg: PhD Thesis, 2011.

[41] G. Agostinelli, P. Delabie, P. Vitanov, *et al.*, "Very low surface recombination velocities on p-type silicon wafers passivated with a dielectric with fixed negative charge," *Solar Energy Materials and Solar Cells*, vol. 90, pp. 3438–3443, 2006.

[42] B. Hoex, J. J. H. Gielis, M. C. M. van de Sanden, and W. M. M. Kessels, "On c-Si surface passivation mechanism by the negative-charge-dielectric Al_2O_3," *Journal of Applied Physics*, vol. 104, p. 113703, 2008.

[43] L. E. Black, T. Allen, K. R. McIntosh, and A. Cuevas, "Effect of boron concentration on recombination at the p-Si–Al_2O_3 interface," *Journal of Applied Physics*, vol. 115, no. 9, p. 093707, 2014.

[44] F. J. Ma, S. Duttagupta, M. Peters, G. S. Samudra, A. G. Aberle, and B. Hoex, "Numerical modelling of silicon p+ emitters passivated by a PECVD AlO_x/SiN_x stack," *Energy Procedia*, vol. 33, pp. 104–109, 2013.

[45] T. Dullweber and J. Schmidt, "Industrial Silicon Solar Cells Applying the Passivated Emitter and Rear cell (PERC) Concept – A Review," *IEEE Journal of Photovoltaics*, vol. 6, no. 5, pp. 1366–1381, 2016.

[46] J. Schmidt, A. Merkle, R. Brendel, B. Hoex, M. C. M. Van de Sanden, and W. M. M. Kessels, "Surface passivation of high-efficiency silicon solar cells by atomic-layer-deposited Al_2O_3," *Progress in Photovoltaics: Research and Applications*, vol. 16, no. 6, pp. 461–466, 2008.

[47] P. Saint-Cast, J. Benick, D. Kania, *et al.*, "Very low surface recombination velocity on p-type c-Si by high-rate plasma-deposited aluminum oxide," *Applied Physics Letters*, vol. 95, p. 151502, 2009.

[48] L. Black and K. McIntosh, "Surface passivation of c-Si by atmospheric pressure chemical vapor deposition of Al_2O_3," *Applied Physics Letters*, vol. 100, p. 202107, 2012.

[49] T. Hayasaka, S. Kodama, M. Shimizu, M. Hamada, N. Tanaka, and T. Nojiri, "Field-effect surface passivation paste by screen-printing for high efficiency PERC," *Proceedings of the 32nd European Photovoltaic Solar Energy Conference and Exhibition*, pp. 677–679, 2016.

[50] G. Krugel, W. Wolke, F. Wagner, J. Rentsch, and R. Preu, "Sputtered aluminum oxide for rear side passivation of p-type silicon solar cells," *Proceedings of the 27th European Photovoltaic Solar Energy Conference and Exhibition*, pp. 1958–1962, 2012.

[51] P. H. Lu, K. Wang, Z. Lu, A. J. Lennon, and S. R. Wenham, "Anodic aluminum oxide passivation for silicon solar cells," *IEEE Journal of Photovoltaics*, vol. 3, no. 1, pp. 143–151, 2013.

[52] G. Dingemans, M. C. M. van de Sanden, and W. M. M. Kessels, "Influence of the deposition temperature on the c-Si surface passivation by Al_2O_3 films

synthesized by ALD and PECVD," *Electrochemical and Solid-State Letters*, vol. 13, no. 3, pp. H76–H79, 2010.

[53] L. Hennen, E. H. A. Grannemann, and W. M. M. Kessels, "Analysis of blister formation in spatial ALD Al_2O_3 for silicon surface passivation," *Conference Record of the 37th IEEE Photovoltaic Specialists Conference*, pp. 001049–001054, 2012.

[54] A. Richter, F. M. M. Souren, D. Schuldis, *et al.*, "Thermal stability of spatial ALD deposited Al_2O_3 capped by PECVD SiN_x for the passivation of lowly- and highly-doped p-type silicon surfaces," *Proceedings of the 27th European Photovoltaic Solar Energy Conference and Exhibition*, pp. 1133–1137, 2012.

[55] B. Vermang, H. Goverde, A. Lorenz, *et al.*, "On the blistering of atomic layer deposited Al_2O_3," *Conference Record of the 37th IEEE Photovoltaic Specialists Conference*, pp. 003562–003567, 2011.

[56] E. Cornagliotti, L. Tous, A. Uruena, *et al.*, "Integration of spatial ALD aluminum oxide for rear side passivation of p-type PERC/PERL solar cells," *Proceedings of 28th European Photovoltaic Solar Energy Conference and Exhibition*, pp. 976–981, 2013.

[57] T. Ludera, T. Lauermann, A. Zuschlag, G. Hahn, and B. Terheiden, "Al_2O_3/SiN_x-stacks at increased temperatures: avoiding blistering during contact firing," *Energy Procedia*, vol. 27, pp. 426–431, 2012.

[58] P. Saint-Cast, M. Hofmann, S. Kuhnhold, *et al.*, "A review of PECVD aluminium oxide for surface passivation," *Proceedings of the 27th European Photovoltaic Solar Energy Conference and Exhibition*, pp. 1797–1801, 2012.

[59] P. Poodt, A. Lankhorst, F. Roozeboom, K. Spee, D. Maas, and A. Vermeer, "High-speed spatial atomic-layer deposition of aluminum oxide layers for solar cell passivation," *Advanced Materials*, vol. 22, no. 32, pp. 3564–3567, 2010.

[60] B. Vermang, "Aluminium oxide as negatively charged surface passivation for industrial crystalline silicon solar cells," *PhD Thesis*, 2012.

[61] B. Veith, T. Dullweber, M. Siebert, *et al.*, "Comparison of ICP-AlO_x and ALD-Al_2O_3 layers for the rear surface passivation of c-Si solar cells," *Energy Procedia*, vol. 27, pp. 379–384, 2012.

[62] G. Dingemans and W. M. M. Kessels, "Status and prospects of Al_2O_3-based surface passivation schemes for silicon solar," *Journal of Vacuum Science & Technology A*, vol. 30, p. 040802, 2012.

[63] F. Duerinckx, M. Aleman, P. Choulat, I. Kuzma Filipek, E. Cornagliotti, and J. Szlufcik, "Quantifying internal optical losses for 21% n-Si rear junction cells," *Proceedings of the 29th European Photovoltaic Solar Energy Conference and Exhibition*, pp. 417–420, 2014.

[64] B. Veith, F. Werner, D. Zielke, R. Brendel, and J. Schmidt, "Comparison of the thermal stability of single Al_2O_3 layers and Al_2O_3/SiN_x stacks for the surface passivation of silicon," *Energy Procedia*, vol. 8, pp. 307–312, 2011.

[65] W. Liang, K. J. Weber, D. Suh, J. Yu, and J. Bullock, "Humidity degradation and repair of ALD Al_2O_3 passivated silicon," *Conference Record of the 39th IEEE Photovoltaic Specialists Conference (PVSC)*, pp. 38–44, 2013.

[66] H. T. T. Nguyen, N. Balaji, C. Park, *et al.*, "Al_2O_3/SiON stack layers for effective surface passivation and anti-reflection of high efficiency n-type c-Si solar cells," *Semiconductor Science and Technology*, vol. 32, no. 2, 2017.

[67] G. Lopez, P. R. Ortega, C. Voz, *et al.*, "Surface passivation and optical characterization of Al_2O_3/a-SiC_x stacks on c-Si substrates," *Beilstein Journal of Nanotechnology*, vol. 4, pp. 726–731, 2013.

[68] B. Vermang, H. Goverde, L. Tous, *et al.*, "Approach for Al_2O_3 rear surface passivation of industrial p-type Si PERC above 19%," *Progress in Photovoltaics: Research and Applications*, vol. 20, no. 3, pp. 269–273, 2012.

[69] B. Veith-Wolf, J. Wang, M. Hannu-Kuure, *et al.*, "Siloxane-based capping layers for Al_2O_3 as silicon-nitride replacement in industrial-type PERC solar cells," *Proceedings of the 29th European Photovoltaic Solar Energy Conference and Exhibition*, pp. 600–602, 2014.

[70] S. Roberts and P. J. Dobson, "Evidence for reaction at the Al-SiO_2 interface," *Journal of Physics D: Applied Physics*, vol. 14, no. 3, L17, 1981.

[71] A. Richter, S. Henneck, J. Benick, M. Horteis, M. Hermle, and S. W. Glunz, "Firing stable Al_2O_3/SiN_x layer stack passivation for the front side boron emitter of n-type Si solar cells," *Proceedings of the 25th European Photovoltaic Solar Energy Conference and Exhibition*, pp. 1453–1459, 2010.

[72] "EDNA 2," *www.pvlighthouse.com.*

[73] K. R. McIntosh and P. P. Altermatt, "A freeware 1D emitter model for silicon solar cells," *Conference Record of the 35th IEEE Photovoltaic Specialists Conference*, pp. 2188–2193, 2010.

[74] A. Richter, J. Benick, and M. Hermle, "Boron emitter passivation with Al_2O_3 and Al_2O_3/SiN_x stacks using ALD Al_2O_3," *IEEE Journal of Photovoltaics*, vol. 3, no. 1, pp. 236–245, 2013.

[75] S. Duttagupta, F. Lin, D. K. Shetty, A. G. Aberle, and B. Hoex, "Excellent boron emitter passivation for high-efficiency Si wafer solar cells using AlO_x/SiN_x dielectric stacks deposited in an industrial inline plasma reactor," *Progress in Photovoltaics: Research and Applications*, vol. 21, pp. 760–764, 2013.

[76] P. Saint-Cast, A. Richter, E. Billot, *et al.*, "Very low surface recombination velocity of boron doped emitter passivated with plasma-enhanced chemical-vapor-deposited AlO_x layers," *Thin Solid Films*, vol. 522, pp. 336–339, 2012.

[77] P. Saint-Cast, "Passivation of Si surfaces by PECVD aluminium oxide," *PhD Thesis*, 2012.

[78] R. Hao, T. S. Ravi, V. Siva, *et al.*, "Kerfless epitaxial mono crystalline Si wafers with built-in junction and from reused substrates for high-efficiency PERx cells," *IEEE Journal of Photovoltaics*, vol. 6, no. 6, pp. 1451–1455, 2016.

[79] I. Kuzma-Filipek, M. Recaman-Payo, F. Duerinckx, *et al.*, "22.5% n-PERT solar cells on epitaxially grown silicon wafers," *Proceedings of 32nd European Photovoltaic Solar Energy Conference and Exhibition*, pp. 560–564, 2016.

[80] L. Black, T. C. Kho, K. R. McIntosh, and A. Cuevas, "The influence of orientation and morphology on the passivation of crystalline silicon surfaces by Al_2O_3," *Energy Procedia*, vol. 55, pp. 750–756, 2014.

[81] W. Liang, K. J. Weber, D. Suh, *et al.*, "Surface Passivation of Boron-Diffused p-Type Silicon Surfaces With (1 0 0) and (1 1 1) Orientations by ALD Al_2O_3 Layers," *IEEE Journal of Photovoltaics*, vol. 3, no. 2, pp. 678–683, 2013.
[82] S. C. Baker-Finch and K. R. McIntosh, "The contribution of planes, vertices, and edges to recombination at pyramidally textured surfaces," *IEEE Journal of Photovoltaics*, vol. 1, no. 1, pp. 59–65, 2011.
[83] H. Wagner, S. Steingrube, B. Wolpensinger, *et al.*, "Analyzing emitter dopant inhomogeneities at textured Si surfaces by using 3D process and device simulations in combination with SEM imaging," *Conference Record of the 38th IEEE Photovoltaic Specialists Conference*, pp. 313–316, 2012.
[84] P. Repo, J. Benick, G. von Gastrow, *et al.*, "Passivation of black silicon boron emitters with atomic layer deposited aluminum oxide," *Physica Status Solidi – Rapid Research Letters*, vol. 7, no. 11, pp. 950–954, 2013.

Chapter 10

Double-layer dielectric stacks for advanced surface passivation of crystalline silicon solar cells

Shubham Duttagupta[1] *and Armin Gerhard Aberle*[1]

10.1 Introduction

Excellent surface passivation of the front and rear of crystalline silicon (c-Si) wafer solar cells is imperative for achieving higher efficiency, especially for cells with reduced thickness. In order to further reduce cost and improve the PV cell efficiency in mass-scale production, it is extremely important to evaluate, improve and develop efficient and cost-effective silicon surface passivation layers that are compatible with mass production. Due to the continuous decrease in wafer thickness and desire for higher energy conversion efficiency of c-Si wafer solar cells, electronic and optical losses at the rear surface are becoming increasingly important [1]. Multi-functional thin films deposited at low temperature (<450 °C) such as silicon oxide (SiO_x), silicon nitride (SiN_x), amorphous silicon (*a*-Si:H) and – more recently – aluminium oxide (AlO_x) and their stacks are continuously being evaluated for this purpose and have been successfully applied for cell efficiency improvements [2–11].

Rear surface passivation is especially important for high-efficiency (>20%) silicon solar cells where it is essential to reduce the effective surface recombination velocity at the rear to values below ~100 cm/s after metallisation [1,12]. High-efficiency solar cell structures based on *p*-type silicon substrates such as the 'aluminium local back surface field' (Al-LBSF) solar cell (or PERC cell, for 'passivated emitter and rear cell') [13] and 'passivated emitter rear (either totally or locally) diffused' (PERT/PERL) silicon wafer solar cells [14,15] are gradually becoming the next 'mainstream' industrial solar cell structures (see the International Technology Roadmap for Photovoltaics (ITRPV) [16]). These structures require the rear-side silicon surface to be passivated with a multifunctional thin film (generally a stack) that improves the performance of the solar cells both electronically and optically [see schematic in Figure 10.1(a)]. The polarity and density (i.e., number) of the fixed charge in the dielectric have a profound impact

[1]Solar Energy Research Institute of Singapore (SERIS), National University of Singapore (NUS), Singapore

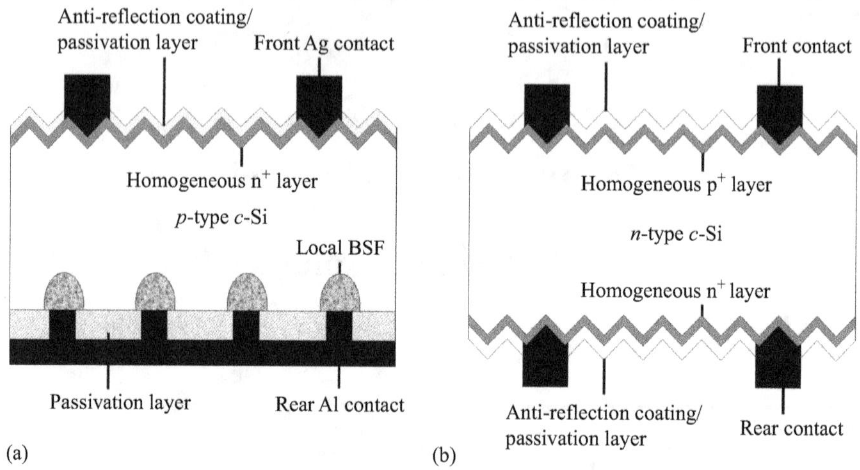

Figure 10.1 Schematic cross sections of (a) the industrial Al-LBSF silicon solar cell (or PERC cell) and (b) the industrial bifacial silicon solar cell with full-area phosphorus back surface field (BSF)

on the surface recombination velocity at the rear *p*-type moderately doped (undiffused) surface and consequently the solar cell's performance.

Although the majority of today's silicon solar cells are based on *p*-type Si wafers, the relative insensitivity of *n*-type c-Si material to various impurities and recombination active defects (such as the B-O complex) makes the use of *n*-type wafers very interesting [17,18]. These advantages are already exploited in today's highest-efficiency commercial silicon wafer solar cells produced by Panasonic and SunPower [19,20]. The emitter of an *n*-type silicon wafer solar cell is either a heterojunction or a heavily doped p^+-type c-Si layer formed by thermal diffusion or ion implantation. A simple high-efficiency bifacial *n*-type cell structure is shown in Figure 10.1(b). The electronic passivation of the surface of the p^+ emitter has proven to be very challenging in the past, especially before 2006 [21]. A cost-effective solution for the passivation of the p^+ emitter surface is key for *n*-type silicon cells to gain market share.

Indisputably, surface passivation of moderately doped and heavily doped *p*-type surfaces is of major importance to both *p*-type and *n*-type cells. Amorphous hydrogenated silicon nitride (*a*-SiN_x:H or, briefly, SiN_x) films (with either uniform or graded refractive index) deposited by plasma-enhanced chemical vapour deposition (PECVD) have become the state-of-the-art anti-reflection coating at the front of industrial silicon wafer solar cells. This coating not only reduces reflection losses but simultaneously provides excellent surface passivation for n^+ diffused emitters and hydrogen passivation of bulk defects [22–25]. Surface passivation of n^+ diffused emitters is well established [26–28]. Considering the practical importance and relevance of *p*-type Si surfaces, i.e. a moderately doped (undiffused) *p*-type Si surface and a heavily doped p^+-type Si surface, this chapter focuses on the investigation of surface passivation using double-layer dielectric stacks. We study a positively charged dielectric stack (i.e. SiO_x capped with SiN_x) as well as a negatively charged dielectric stack (i.e., AlO_x capped with SiN_x).

10.2 Deposition of PECVD SiO_x/SiN_x and AlO_x/SiN_x stacks

Silicon photovoltaic research at SERIS uses large-area (156 mm × 156 mm) *p*- and *n*-type monocrystalline Czochralski (Cz) Si wafers. Wafers with a bulk resistivity of 5–8 Ωcm, (100) surface orientation and a thickness of 180 μm were used for the experiments described in this book chapter. The wafers were saw damage etched in a KOH solution removing 15 μm of Si from each side (we refer to these as planar samples). Some wafers received alkaline texturing in KOH/IPA/DI water solution, resulting in a textured surface with upright random pyramids with a size of up to ~6 μm (we refer to these as textured samples). In this chapter, we report our surface passivation results on moderately doped and heavily doped *p*-type silicon surfaces. Undiffused *p*-type c-Si with a dopant density in the range of 1×10^{15}–1×10^{16} cm^{-3} is referred to as moderately doped silicon. The *n*-type wafers (textured or planar) were diffused in an industrial tube furnace (TS-81004, Tempress Systems B.V.). Symmetrical boron diffused p^+np^+ structures having planar and textured surfaces served as heavily doped *p*-type silicon surfaces. The active dopant profiles of the boron-diffused regions were measured with the electrochemical voltage (ECV) method, as shown in Figure 10.2. Prior to the diffusion and PECVD processes, unless otherwise mentioned, all samples received a Radio Corporation of America (RCA) clean with a final HF (hydrofluoric acid) dip followed by water rinsing and drying, to ensure H-terminated surfaces. All dielectric films were

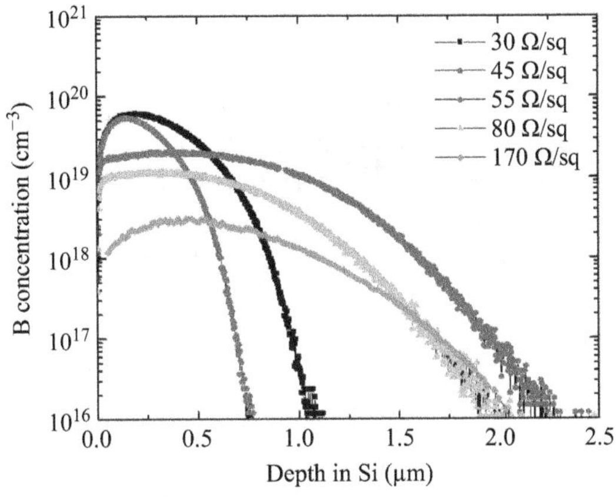

Figure 10.2 Boron dopant profile of p^+np^+ samples as measured by ECV profiling. The sheet resistance values determined by 4-point probe measurements are also shown in the graph

Table 10.1 Inline PECVD deposition parameters and film properties for the PECVD SiO_x, capping SiN_x and SiN_x

Dielectric	Process gas	Heater set temperature, T (°C)	Reactor pressure, p (mbar)	Plasma power, P (W)	Refractive index, n at 633 nm	Thickness, d (nm)
SiO_x	SiH_4+ N_2O	200	0.15	700	1.47	15
Capping SiN_x	$SiH_4 + NH_3$	300, 400	0.20	2,200	2.0	70
SiN_x	$SiH_4 + NH_3$	450	0.20	2,200	2.0	70

Table 10.2 Inline PECVD deposition parameters and film properties for the PECVD AlO_x and the capping layers – PECVD SiO_x or SiN_x

Dielectric	Process gas	T (°C)	p (mbar)	P (W)	n at 633 nm	d (nm)
AlO_x	TMA $[Al(CH_3)_3]$ + N_2O + Ar	350	0.10	1,500	1.61	25
Capping SiO_x	$SiH_4 + N_2O$	200	0.15	700	1.47	70
Capping SiN_x	$SiH_4 + NH_3$	300	0.20	2,200	2.05	70

deposited in a commercial inline microwave-powered remote PECVD reactor (SiNA-XS, Roth & Rau AG, Germany now Meyer Burger AG). The deposition conditions and film properties (such as refractive index and thickness) are summarised in Tables 10.1 and 10.2. Note: The c-Si wafer temperature during the PECVD process is estimated to be about 50 °C lower than the heater temperature. The transport speed of the sample carrier was adjusted to yield the desired thickness of the films. It is also worth mentioning here that during the time of this research (2010–2014), 20–25 nm thick AlO_x film was a standard, which today (in 2018) is significantly reduced to 7 (±2) nm allowing 3-fold reduction in cost. After deposition of the dielectric films onto both surfaces (in two separate runs), the samples received a post-deposition anneal in an industrial fast firing furnace (Ultraflex, Despatch, USA) for a few seconds at a heater set temperature of about 850 °C.

10.3 Characterisation of surface passivation

10.3.1 Evaluation of surface passivation of undiffused and diffused Si surfaces

Characterisation or evaluation of surface passivation requires measurement of S_{eff} (or J_{0e}). A direct measurement of S_{eff} or J_{0e} is not possible. It is, however, possible to measure the so-called effective minority carrier lifetime τ_{eff} of the sample, which takes into account the recombination mechanisms at the surfaces of the sample as well as within its bulk. The measurement of the effective lifetime τ_{eff} can, for example, be done using the contactless photoconductance method, which is a well-known and widely used method (see Refs. [29–32]). In the photoconductance method, τ_{eff} is measured as a function of the excess minority carrier concentration, which accounts for the recombination losses in the c-Si bulk (represented by τ_{bulk}) and the front and rear surfaces (represented by $\tau_{surface}$ for each of the two surfaces). The main assumption here is that both surfaces have a similar surface recombination velocity. It is widely accepted that this assumption is reasonable if the surface preparation is the same for both surfaces and a similar passivation scheme is applied on both sides, although some concern can be raised regarding the cleanliness of the surface to be passivated in the second PECVD run (when the passivation layer is deposited onto the surface). If both surfaces are passivated identically, the τ_{eff} of such a symmetrical structure is expressed as

$$\frac{1}{\tau_{eff}} = \frac{1}{\tau_{Bulk}} + \frac{1}{\tau_{Surface}} \tag{10.1}$$

$$\frac{1}{\tau_{Bulk}} = \left(\frac{1}{\tau_{Radiative}} + \frac{1}{\tau_{Auger}}\right)\text{intrinsic} + \frac{1}{\tau_{SRH}} \tag{10.2}$$

If S_{eff} is sufficiently small, it is expressed as [29]

$$S_{eff} = \frac{W}{2}\left(\frac{1}{\tau_{eff}} - \frac{1}{\tau_{Bulk}}\right) \tag{10.3}$$

For wafers with very good bulk lifetime, the total recombination losses can be assumed to be dominated by recombination at the wafer surfaces. Assuming $\tau_{bulk} = \infty$, one obtains an upper limit of the surface recombination velocity $S_{eff,max}$:

$$S_{eff,max} = \frac{W}{2\tau_{eff}} \tag{10.4}$$

In c-Si solar cells, surface passivation studies of heavily doped regions are very important. They are usually characterised by the emitter saturation current density J_{0e} ('e' stands for emitter, in case of absence of an emitter one can use J_{0s}, where 's' stands for surface, which can be a moderately doped c-Si surface or a high-low junction). This saturation current density is an important parameter which limits the open-circuit voltage and hence the efficiency of the solar cells. A better surface passivation will mean lower J_{0e}.

Kane and Swanson introduced a method [33] for the extraction of J_{0_e} from photoconductance measurements at high injection from specially prepared symmetrical lifetime samples [34,35]. This method is extensively used in our work to investigate the level of surface passivation on heavily doped c-Si surfaces. Photoconductance within the heavily doped regions can be neglected because they are usually very thin (0.2 to 5 μm) compared to the substrate (~180 μm) and the heavy doping in that region strongly reduces the minority carrier lifetime. At high-injection conditions, the injection level dependence of S_{eff} can be used to separate recombination in the heavily doped region and the bulk Si [33–35]. The effective

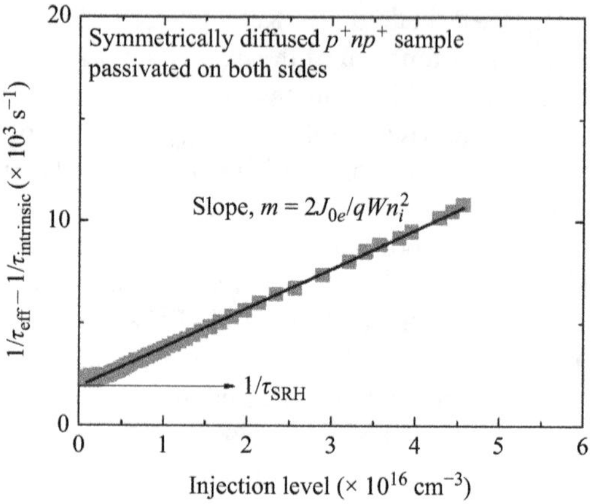

Figure 10.3 Intrinsic recombination (mainly Auger)-corrected inverse effective lifetime as a function of injection level of a symmetrically diffused and passivated c-Si sample

lifetime τ_{eff} of a symmetrically diffused wafer with or without passivation can be expressed as [36]:

$$\frac{1}{\tau_{eff}} = \frac{1}{\tau_{Bulk}} + \frac{2}{\tau_{Emitter}} \tag{10.5}$$

$$\frac{1}{\tau_{eff}} = \left(\frac{1}{\tau_{Radiative}} + \frac{1}{\tau_{Auger}}\right) + \frac{1}{\tau_{SRH}} + \frac{J_{0e}}{qWn_i^2}\left(2N_{doping} + \Delta n_{front} + \Delta n_{rear}\right) \tag{10.6}$$

Subtracting $1/\tau_{intrinsic}$ from $1/\tau_{eff}$ and assuming $\Delta n_{front} = \Delta n_{rear} = 2\Delta n$, (10.6) becomes

$$\frac{1}{\tau_{eff}} - \frac{1}{\tau_{Intrinsic}} = \frac{1}{\tau_{SRH}} + \frac{2J_{0e}}{qWn_i^2}\left(N_{doping} + \Delta n\right) \tag{10.7}$$

J_{0e} is thus obtained from the slope of the straight line when $\frac{1}{\tau_{eff}} - \frac{1}{\tau_{Intrinsic}}$ is plotted versus Δn (an example is shown in Figure 10.3) provided the measurement is taken at high injection where $\Delta n >> N_{doping}$ (excess carrier density being at least one order of magnitude higher than the bulk doping concentration). The analysis method also allows to extract the SRH bulk lifetime (τ_{SRH}) from the y-intercept of Figure 10.3, which can be a useful tool for the investigation of solar cell fabrication process steps. High-injection conditions can easily be satisfied when high-resistivity (>5 Ωcm) substrates are used. At high injection, τ_{SRH} is independent of Δn and thus J_{0_e} can be determined from (10.7).

10.3.2 Evaluation of electronic interface properties

The surface passivation of our samples is significantly influenced by the fixed charge density (Q_f) in the dielectric film (or stack), the density of interface defect states (D_{it}) and the capture cross section (σ) of holes and electrons at the c-Si/dielectric interface [26,37–41]. In order to determine the fixed charge density (Q_f) in the film, the density of interface defect states (D_{it}), contactless corona-voltage measurements were carried out using a PV-2000 metrology tool from Semilab Inc., USA [42–45]. The method uses corona charging in air to deposit an electric charge on a dielectric thus changing the electric field in the dielectric and in the semiconductor. The response is measured in a non-contact manner by using a contact potential difference (V_{CPD}) in the dark and under strong illumination. The contact potential difference (V_{CPD}) between the wafer and a vibrating reference electrode is determined by using Kelvin probe measurements in the dark and under illumination as a function of the deposited corona charge (Q_c). Corona charging changes the surface band bending in the crystalline silicon wafer. The change in surface potential due to corona charging at the Si-dielectric interface is defined as surface barrier (ΔV_{sb}). The flat-band voltage (V_{fb}) is obtained when there is no band bending observed upon corona charging (i.e. $\Delta V_{sb} = 0$).

This technique in principle measures the total charge (Q_{total}) in the dielectric. Total charge Q_{total} is defined as the amount of corona charge required to drive the initial condition (band bending due to the charge present in the PECVD layer) to a flat-band condition. It should be noted that the fixed insulator charge density Q_f is considered the main component in Q_{total} (for details, see Ref. [45]). It is assumed that the charges contributing to flat-band are all located near the silicon/dielectric interface. c-Si samples passivated with a leaky dielectric (such as PECVD SiN_x film) can suffer from hysteresis during a corona-charging measurement, which can potentially adversely influence the measurements. As a general procedure positive corona charge is used for negatively charged dielectrics (e.g. PECVD AlO_x or related stacks) and negative corona charge is used for positively charged dielectrics (e.g. PECVD SiN_x, SiO_x or related stacks).

The density of interface traps, D_{it}, be calculated as the ratio $\Delta Q_{IT}/\Delta V_{sb}$, where ΔV_{sb} is the change in the surface barrier due to the deposited quantum of corona charge and ΔQ_{it} is the interface trap charge density, which is the charge added to the interface traps when the surface barrier changes as ΔV_{sb}. Further details of the procedure of extraction of Q_{total} and D_{it} are discussed in detail by Wilson *et al.* [43,46]. V_{sb} corresponds to an energy position within the Si bandgap E_g. For *n*-type Si, $V_{sb} = 0$ corresponds to the Fermi energy level (E_f) below conduction band edge (E_C) (exact location is $E_C - E_f$). In this way, D_{it} can be extracted as a function of energetic position in the silicon bandgap. In this technique, the D_{it} spectra starts from flat-band condition ($V_{sb} = 0$ corresponding to E_f) and extends until deep inversion. This is the reason for very high D_{it} near E_f (as $V_{sb} = 0$ in this location as flat-band is achieved). This is due to the measurement limitation and D_{it} data under accumulation is inaccessible. Here, the D_{it} is valid from the onset of flat-band condition to deep inversion. The measurements of charge density and energy-dependent interface defect density D_{it} (E) were done on multiple passivated samples at different time intervals to validate repeatability and reliability of the measurement. As a uncertainty experiment few measurements were carried out comparing the contactless corona-voltage and traditional MOS (metal-oxide semiconductor) C–V (capacitance–voltage) technique confirms the $D_{it,midgap}$ values (which are indicative of interface quality) and charge density values are comparable.

10.4 Investigation of moderately doped *p*-type silicon surfaces

10.4.1 *Double-layer dielectric stack: PECVD SiO_x/SiN_x*

Thermally grown silicon dioxide (SiO_2) has been extensively studied in the past decades for the passivation of (un)diffused silicon surfaces [26,39,47,48]. A high-quality Si–SiO_2 interface was extremely crucial for the success in microelectronics application [47,48]. After an optimised anneal, the Si–SiO_2 interface can have a very low interface defect density, for example, $D_{it,midgap}$ in the order of 10^9 eV^{-1} cm^{-2} [47,49–51]. Such low $D_{it,midgap}$ is a prerequisite for excellent surface passivation of c-Si [1,12], which is evident by extremely low surface recombination velocity S_{eff} (2.4 cm/s) [52]. These thermal SiO_2 films were used for high-efficiency silicon solar cell devices that resulted in (then) world-record efficiencies of up to 24.7% [53,54]. However, higher processing temperature (800–1,100°C) for thermal SiO_2 can be detrimental for less pure Cz-grown or multicrystalline c-Si. Moreover, a high-temperature process can significantly increase the fabrication cost.

The most widely used alternative approach for low-temperature SiO_x is the PECVD method, which allows low-temperature (<400 °C) processing along with significantly higher deposition rates. Single-layer SiO_x generally does not provide a high-quality passivation, with $S_{eff,max}$ of about 500–1,000 cm/s in the as-deposited state [55–58]. A forming gas anneal (FGA) can improve passivation thereby lowering $S_{eff,max}$ to 100 cm/s, but these values are insufficient for high-performance silicon wafer solar cells [55,57]. Important to mention that some results in the literature indicated unstable surface passivation [55,59]. However, use of a capping layer (for example, SiN_x) with optimised annealing significantly improved the surface passivation capability with $S_{eff,max}$ of 2 and 11 cm/s on *n*- and *p*-type silicon [60,61]. The reason was attributed to effective hydrogen passivation (whereby the hydrogen comes from the capping SiN_x). This resulted in $D_{it,midgap}$ as low as 10^{11} eV^{-1} cm^{-2} and, in addition, a very low Q_f of 10^{11} cm^{-2} was measured [61]. Such low Q_f can be utilised for surface passivation of heavily doped p^+ c-Si, which otherwise seems not possible by dielectrics with high Q_f [21]. PECVD SiO_x with a capping layer was shown to provide very good surface passivation on moderately doped undiffused silicon (after optimised anneal). However, results after high-temperature annealing are rarely found in the literature. Given that the dielectric stack had shown potential on moderately doped *p*-type surface, it is important to investigate this further considering the relevance in industrial screen-printed solar cell applications that require high-temperature firing. For heavily doped n^+ and p^+ c-Si, thermal SiO_2 (with and without a SiN_x capping layer) after optimised annealing (or an anneal process) had demonstrated excellent passivation with J_{0e} of ~20 fA/cm^{-2} [27,28,36,62–66]. However, very limited data are available on PECVD SiO_x for surface passivation of heavily doped c-Si. Considering the

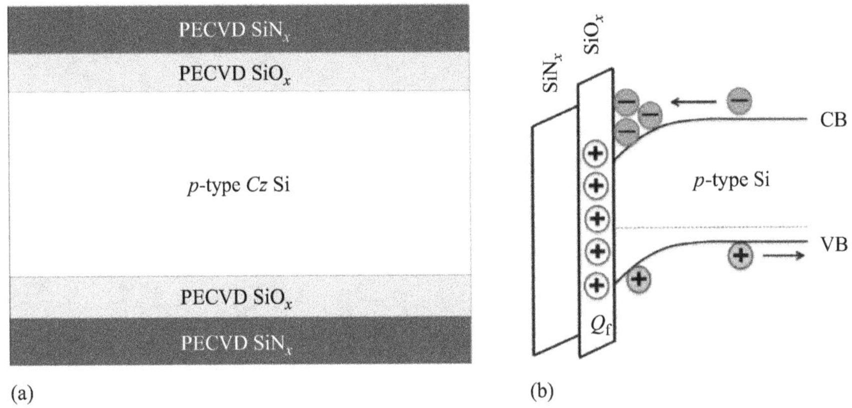

Figure 10.4 (a) Symmetrical test structure used for the evaluation of surface passivation and electrical interface properties and (b) equilibrium band diagram of c-Si/SiO$_x$/SiN$_x$ structure (bandgaps are not to scale and Q_f is assumed to be located near the interface)

economic and technological advantages, around ~2010, Duttagupta *et al.* re-visited the PECVD SiO$_x$ layers (with a capping layer) for effective surface passivation of c-Si surfaces for solar cells.

A schematic of the test structure evaluated for these surface passivation studies is shown in Figure 10.4(a). The corresponding equilibrium band diagram is shown in Figure 10.4(b).

Figure 10.5(a) presents S_{eff} values measured at an excess carrier concentration of 1×10^{15} cm^{-3}, using (10.3) (shown in the left y-axis and assuming a bulk lifetime of 1.9 ms [67]) and the measured 1-Sun iV_{oc} (open-circuit voltage) (shown in the right y-axis) as a function of the deposition temperature of the PECVD SiO$_x$ film. Note that the capping SiN$_x$ film was deposited at a fixed deposition temperature of 400 °C. We observe a strong dependence of the surface passivation quality on the SiO$_x$ deposition temperature. As is evident from Figure 10.5, the surface passivation quality varies by more than one order of magnitude when the SiO$_x$ deposition temperature changes from 100 to 400 °C. An optimum passivation with $S_{eff} < 8$ cm/s was obtained for a SiO$_x$ deposition temperature of 200 °C, corresponding to an iV_{oc} of 715 mV. It should be noted that the reported S_{eff} can be considered as an upper limit, as commercial-grade low-resistivity p-type Cz wafers were used; the bulk lifetimes of these wafers are known to be lower than the intrinsic lifetime that was used to calculate S_{eff}.

In order to better understand the impact of Q_{total} and $D_{it,midgap}$ on the surface passivation, Figure 10.5(b) also presents each of them as a function of the SiO$_x$ deposition temperature. It seems that $D_{it,midgap}$ is independent of the deposition temperature with a relatively low mid-gap value in the order of (5–6) $\times 10^{11}$ eV^{-1} cm^{-2} for deposition temperatures of $\geq$200 °C (Figure 10.5(b)). This suggests that, regardless of the SiO$_x$ deposition temperature ($\geq$200 °C), most of the defects at the Si/SiO$_x$ interface are well passivated. This can be due to effective hydrogenation from

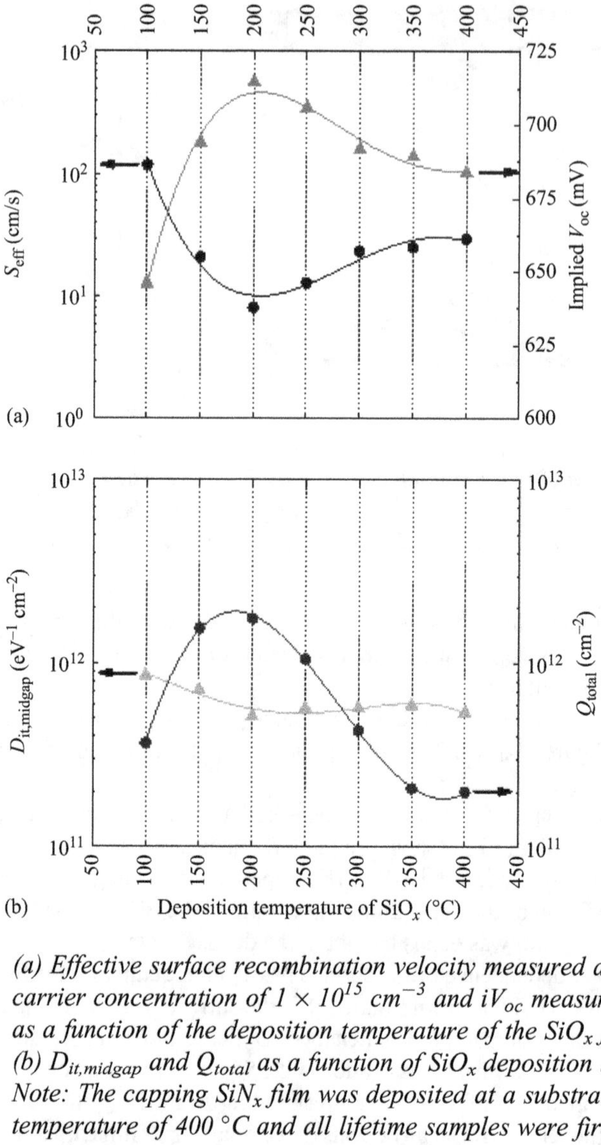

Figure 10.5 (a) Effective surface recombination velocity measured at an excess carrier concentration of 1×10^{15} cm^{-3} and iV_{oc} measured at 1-Sun as a function of the deposition temperature of the SiO_x film and (b) $D_{it,midgap}$ and Q_{total} as a function of SiO_x deposition temperature. Note: The capping SiN_x film was deposited at a substrate temperature of 400 °C and all lifetime samples were fired for a few seconds at a set peak temperature of 880 °C

the overlying SiN_x film during the industrial firing process (i.e. reducing the dangling bond density at the c-Si surface) similar as in Refs. [61,64,68–73]. It is interesting to observe the behaviour of Q_{total} as a function of deposition temperature (Figure 14.5(b)). This trend mimics the trend of S_{eff}. It thus seems that the surface passivation quality of this set of samples is mainly ruled by Q_{total}, i.e. by field-effect passivation [74]. The optimum surface passivation was achieved for a deposition temperature of 200 °C, with $Q_{total} = 1.75 \times 10^{12}$ cm^{-2} and $D_{it,midgap} = 5 \times 10^{11}$ eV^{-1} cm^{-2}.

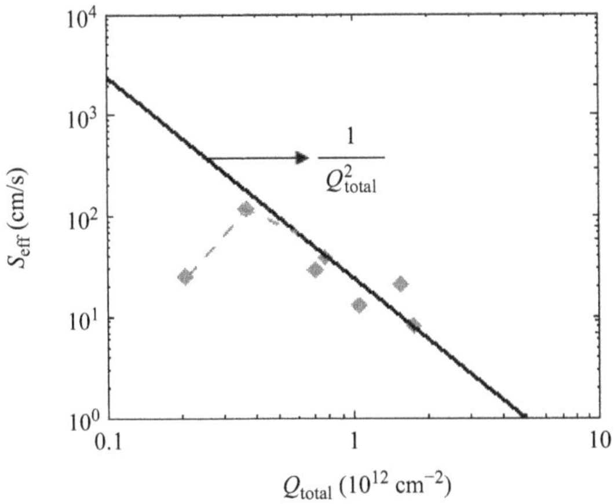

Figure 10.6 *Effective surface recombination velocity S_{eff} for p-type 1.5-Ωcm Si surfaces as a function of experimentally measured positive Q_{total}. The black straight line represents the theoretically predicted S_{eff} vs. $1/Q_{total}^2$, behaviour for inversion conditions at the surface of the p-type Si wafer [75]*

Figure 10.6 shows the relationship between measured S_{eff} and measured Q_{total}. For the investigated *p*-type Si wafers (1.5 Ωcm), the surface becomes inverted for an insulator charge density of about $+3 \times 10^{11}$ cm^{-2} [50]. As can be seen from Figure 10.6, for inversion conditions, S_{eff} is proportional to the inverse square of the total positive charge density, in accordance to theory [75]. On the other hand, for charge values up to 3×10^{11} cm^{-2}, the surface remains under depletion conditions. These are known from theory to cause a maximum in the surface recombination rate at a particular surface band bending (i.e. insulator charge) [50]. This is also observed in our experiments, see the dashed line in Figure 10.6. This set of data also allows investigation of the impact of Q_{total} on the injection-level dependence of τ_{eff} thereby showing difference in the 1-Sun V_{oc} and maximum power point (MPP) voltage conditions (V_{MPP}).

Figure 10.7 presents the injection-level dependent τ_{eff} for two SiO_x deposition temperatures: (a) 200 °C, (b) 400 °C. As the measured $D_{\mathrm{it,midgap}}$ values of these samples are similar (see Figure 10.5(b)), the difference in the passivation quality can be totally attributed to the Q_{total} within the film. This is a unique set of experimental data since it demonstrates the impact of a one order of magnitude variation in insulator charge (with a constant $D_{\mathrm{it,midgap}}$) on τ_{eff}. Normally, such experimental data and the modification of Q_{total} are studied by external corona charging, which often creates interface damage [76]. In Figure 10.7, both the curves demonstrate a similar injection level dependence. It is possible that this injection level dependence results from an asymmetric bulk lifetime τ_{bulk} (since commercial-grade *p*-type Cz Si wafers were used, which are known to contain impurities with asymmetric capture cross sections, causing an injection-level dependent τ_{bulk} [77]).

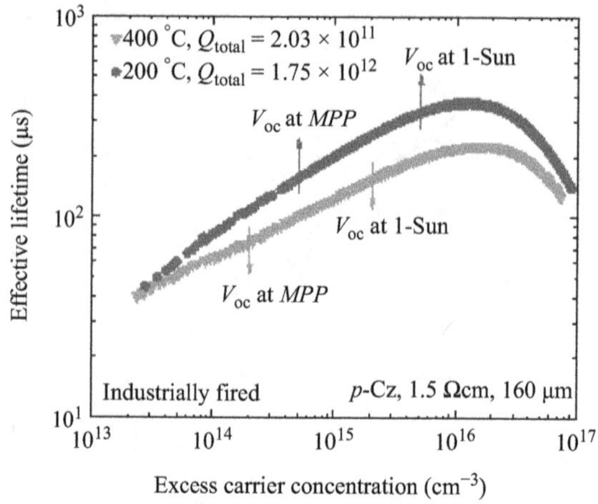

Figure 10.7 Measured τ_{eff} as a function of excess carrier concentration for symmetrically passivated p-type Cz Si wafers with PECVD SiO_x/SiN_x stacks, for two SiO_x deposition temperatures (200 and 400 °C)

Note that, based on the studies by Aberle *et al.* [50], the surface lifetime $\tau_{surface}$ will not suffer from an injection-level dependence since the Q_{total} is high enough ($Q > 5 \times 10^{11}$ cm^{-2}) to cause inversion. This is especially the case for the 200 °C sample where the charge is 1.75×10^{12} cm^{-2}. Based on the results from Figure 10.7, it appears that the τ_{eff} curve for the higher positive charge density shows higher 1-sun V_{oc} and V_{MPP} potential.

10.4.1.1 Evaluation of optical properties of the PECVD SiO_x films

The optical constants (refractive index *n* and extinction coefficient *k*) were measured by spectroscopic ellipsometry (GES5, Sopralab, France). Figure 10.8 shows the refractive index of the single PECVD SiO_x film used in this work. The refractive index at $\lambda = 633$ nm is measured as 1.47, while the photon absorption was below the detection limit. This indicates that a thin PECVD SiO_x layer is optically transparent and has a similar behaviour as thermal SiO_2, and thus can be used in combination with a SiN_x film to serve as an effective anti-reflection coating on silicon. Reflection measurements show less than 2% weighted average reflection (weighted over the 300–1,000 nm wavelength range of the solar spectrum) measured on random pyramid-textured Si surfaces passivated by SiO_x/SiN_x dielectric stacks. The hemispherical reflection (as a function of wavelength) of the samples was measured using spectrophotometry (Lambda 950, PerkinElmer, USA) in the 300–1,000 nm range. Excellent surface passivation quality for both dopant polarities enables the use of these layers for back-side applications (e.g. for all-back-contact cells), while the good anti-reflective properties add value as this stack can also be used for front-side applications in *n*-type Si wafer solar cells (with good optical and passivation properties).

Owing to the results discussed in the previous section, two deposition temperatures were selected (200 and 400 °C) for the SiO_x process and Al-LBSF solar

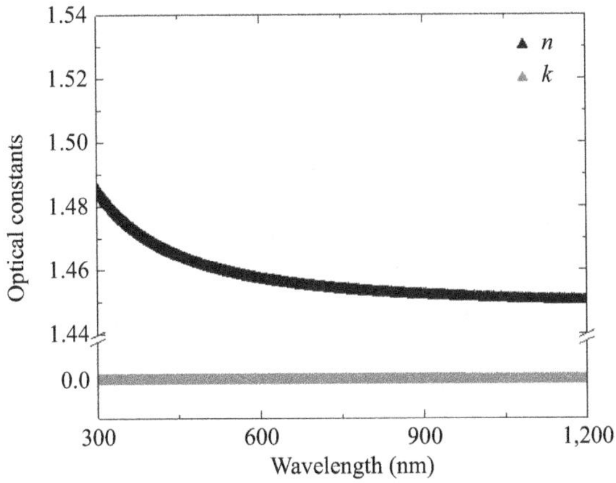

Figure 10.8 Refractive index (n) of PECVD SiO_x film as a function of wavelength, as obtained from spectroscopic ellipsometry measurements. The extinction coefficient (k) was found to be below the detection limit

cells were fabricated (more details in Ref. [78]). The SiO_x film deposited at 200 °C provided an almost 10 times higher Q_{total} compared to the one deposited at 400 °C. As expected from the results shown above, the Al-LBSF cells with SiO_x deposited at 200 °C showed an enhanced efficiency (0.22% average absolute improvement) compared to the batch of cells with SiO_x deposited at 400 °C. This is mainly due to improvements in the cell V_{oc} by 2.5 mV (average) and *FF* by 0.56% absolute (average). This is in agreement with Figure 10.7, as higher 1-Sun V_{oc} and V_{MPP} were measured on the lifetime test wafers. The reason for the *FF* improvement is still under investigation; however, a possible explanation to this increase is the impact of the injection dependence of ideality factor in the voltage range between MPP and open-circuit conditions. It was suggested in 2D simulations [79] that a higher positive insulator charge density (for example, using our 200 °C deposited SiO_x film) in combination with an LBSF (instead of a standard ohmic contact) can improve the cell J_{sc} by $\sim$1.7 mA/cm^2. However, this was not observed in this study, as no change in J_{sc} was observed. A better understanding of the impact of charge tailoring on the solar cell's operation is critical. More experimental work along with multidimensional modelling is needed to further investigate the impact of the polarity and the magnitude of Q_{total} on Al-LBSF or PERC cells. As a proof of concept, these initial results are quite promising, indicating the significant potential of charge tailoring on industrial Al-LBSF silicon solar cells.

10.4.2 Double-layer dielectric stack: PECVD AlO_x/SiN_x

Considering the practical importance of *p*-type Si surfaces, i.e. a moderately doped (undiffused) Si surface and a heavily doped p^+-type Si surface, it is important to study negatively charged dielectrics. This would fundamentally be a good choice because (i) negative charge repels the minority carrier electrons from the surface, thereby reducing the surface recombination rate, and (ii) accumulation conditions do not introduce widely speculated artefacts like increased sub-surface recombination in the space-charge region (SCR) that could arise at the inverted surface depending on the positive charge density as proposed in Refs. [40,80–82].

In recent years, amorphous aluminium oxide (AlO_x) has received tremendous interest in the photovoltaic community for its application as a surface passivation film for crystalline silicon wafer solar cells. Although the passivation properties were already reported by Hezel and Jaeger in 1989 [83], the technology was dormant for ~15 years until Agostinelli *et al.* used Al_2O_3 for rear surface passivation of *p*-type Si solar cells with spin-on sol-gel and atomic-layer-deposition (ALD) methods [84,85]. Soon after, a series of outstanding surface passivation results was reported by Hoex *et al.*, Dingemans *et al.* using the ALD method [6,72,86–90], and excellent solar cell results were demonstrated on *p*-type (>20%) and *n*-type silicon (~24%) [8,91]. Several other techniques were also evaluated like PECVD [92,93], sputtering [94] and spatial ALD [95].

AlO_x films provide excellent surface passivation on lightly doped *n*-type and *p*-type silicon as well as on heavily doped p^+ silicon [6,72,86,87,92]. In addition to a low interface defect density (~10^{11} eV^{-1} cm^{-2}), a distinguishing property of AlO_x is the presence of a high negative charge density (10^{12}–10^{13} cm^{-2}, in units of elementary charges) in the film. This is especially beneficial on *p*-type c-Si surfaces as the minority carriers (electrons) are effectively shielded from the c-Si surface. Due to its excellent surface passivation properties on *p*-type surfaces, AlO_x has been applied successfully in *p*-type Al-LBSF (or PERC) cells and *n*-type PERL cells [7,8].

Owing to the advantages of inline PECVD – i.e. high-rate deposition (~100 nm/min) and industrial applicability – this technology was adapted for commercial high-efficiency solar cells. As the inline PECVD process was a new topic (in 2009), a detailed understanding, process optimisation was required for optimum surface passivation results. In addition, more work was required for the development of high-quality surface passivation results especially after industrial high-temperature firing (instead of typical lab-type activation anneal, i.e. forming gas anneal).

A schematic of the test structure evaluated for surface passivation study is shown in Figure 10.9(a), while Figure 10.9(b) shows the corresponding equilibrium band diagram.

The optical constants (refractive index *n* and extinction coefficient *k*) of the deposited AlO_x were measured by spectroscopic ellipsometry (GES5, Sopralab, France). Figure 10.10 shows the refractive index of the single PECVD AlO_x film used in this work. The refractive index at $l = 633$ nm is measured as 1.6, while the photon absorption was below the detection limit. This indicates that a thin PECVD AlO_x layer is optically transparent and has a similar behaviour as thermal SiO_2, and thus can be used in combination with a SiN_x film to serve as an effective antireflection coating on silicon.

10.4.2.1 Importance of capping layers

Although a thin PECVD AlO_x layer (without a capping layer) after industrial firing is capable of providing excellent surface passivation results on both *p*-type and *n*-type silicon, there are several reasons why capping layers are beneficial: (i) cost: TMA (tetra-methyl aluminium) is an expensive chemical. Reducing its

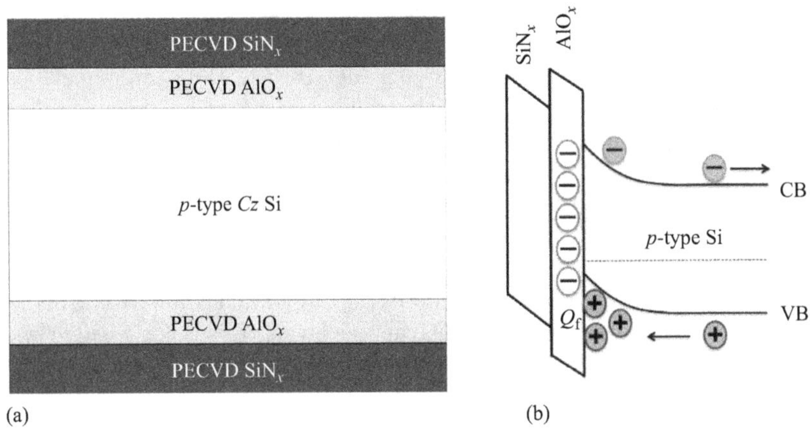

Figure 10.9 *(a) Symmetrical test structure used for the evaluation of surface passivation and electrical interface properties and (b) equilibrium band diagram of c-Si/AlO$_x$/SiN$_x$ structure (bandgaps are not to scale and Q_f is assumed to be located near the interface)*

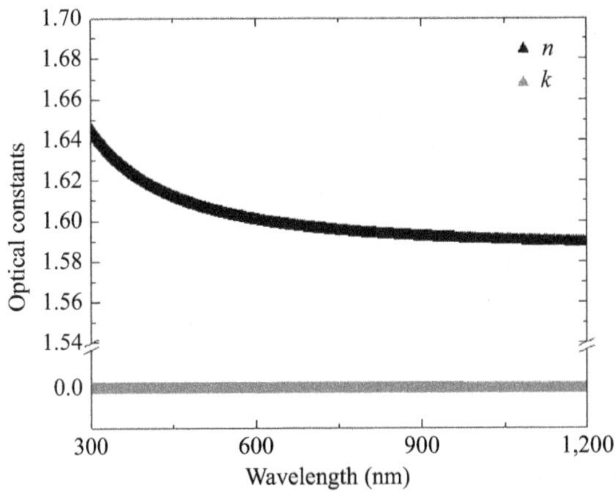

Figure 10.10 *Refractive index (n) of a PECVD AlO$_x$ film as a function of wavelength, as obtained from spectroscopic ellipsometry measurements. The extinction coefficient (k) was found to be below the detection limit*

consumption via a thinner layer will ensure lower cost in fabrication, (ii) application perspective: when AlO$_x$ is used for front-side applications, e.g. passivating the front p^+ layer of *n*-type solar cells, an effective anti-reflection coating is required. A SiN$_x$ capping layer is a good choice for this. Such a capping layer is also beneficial for rear surface applications. A SiO$_x$ capping layer is optically more suitable for rear surface applications due to its lower refractive index than AlO$_x$, (iii) improved stability: application of a capping layer provides a wider process window and, in the case of SiN_x, also gives improved chemical stability. Both AlO_x or/and SiO_x require a capping layer. In addition, when applied at the rear of industrial screen-printed cells, a capping SiN_x protects the underlying AlO_x (or AlO_x/SiO_x) from metal paste interaction that can potentially destroy the stability of the AlO_x. Moreover, the use of a SiN_x capping layer provides an effective hydrogenation of surface defect states, significantly enhancing the surface passivation quality.

The surface passivation results of a 25-nm AlO_x film without a capping layer, a 25-nm AlO_x film with a 70-nm SiN_x capping layer and a 25-nm AlO_x film with 70 nm SiO_x as capping layer are shown in Figure 10.11. In the as-deposited state, as expected, the surface passivation is inferior, although minor improvements were observed for as-deposited SiN_x and SiO_x capping layers. This is attributed to *in-situ* annealing of the PECVD AlO_x layer during deposition of the capping layers.

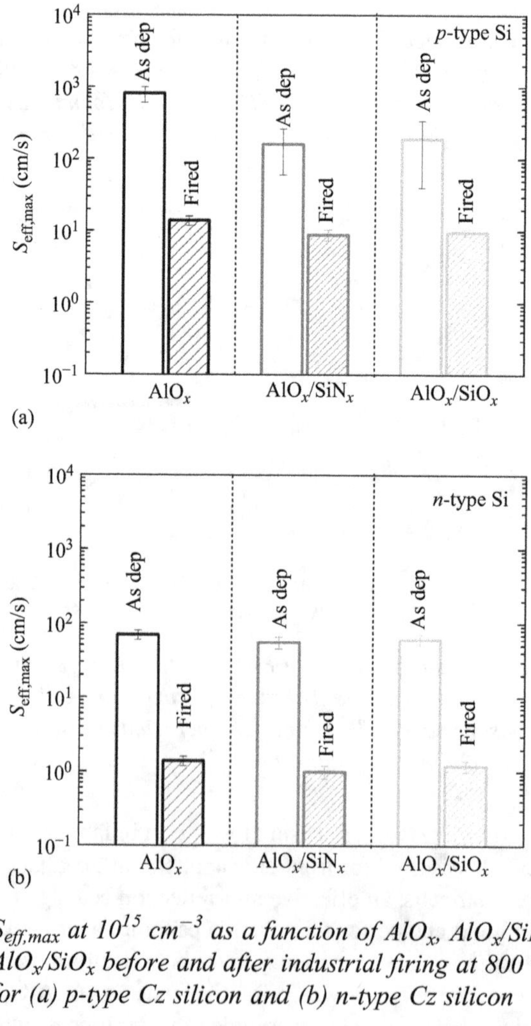

Figure 10.11 $S_{eff,max}$ *at* 10^{15} cm^{-3} *as a function of* AlO_x, AlO_x/SiN_x *or* AlO_x/SiO_x *before and after industrial firing at 800 °C for (a) p-type Cz silicon and (b) n-type Cz silicon*

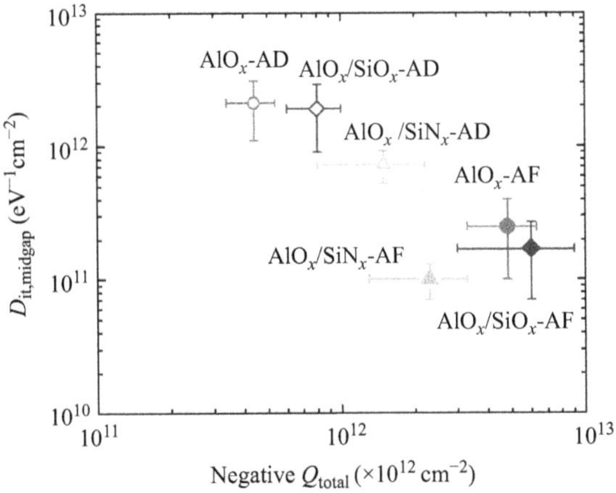

Figure 10.12 Measured $D_{it,midgap}$ as a function of negative Q_{total} for the investigated AlO_x films with and without capping layer (SiN_x or SiO_x), before and after industrial firing

After firing, the surface passivation quality is drastically improved, regardless of whether or not a capping layer is used. It should be noted that the SiN_x capping layer resulted in the best values with $S_{eff,max}$ of ~8.8 cm/s and <2 cm/s, respectively, for *p*-type and *n*-type silicon. A SiO_x capping layer resulted in $S_{eff,max}$ of ~9.7 cm/s and <2 cm/s, respectively, for *p*-type and *n*-type silicon. Thus, the application of each capping layer yielded remarkable surface passivation.

10.4.2.2 Surface passivation mechanism

Contactless corona-voltage measurements were used for measuring total charge density (Q_{total}) and energy-dependent $D_{it}(E)$ required to investigate the electronic interface properties of these films. Figure 10.12 shows $D_{it,midgap}$ as a function of Q_{total} for the investigated AlO_x with and without capping layer (SiN_x or SiO_x). As can be seen, the as-deposited films show very high $D_{it,midgap}$, which explains the inferior passivation by the as-deposited films. The excellent passivation results after industrial firing are explained by a drastic reduction in $D_{it,midgap}$ (9×10^{10}–2.5×10^{11} eV^{-1} cm^{-2}) combined with a significant increase in the negative fixed charge density (2.5×10^{12}–7×10^{12} cm^{-2}). Based on these results, it follows that the surface passivation mechanism for the investigated films is ruled by both good chemical and good field-effect passivation.

10.5 Investigation on heavily doped p^+ silicon surface

Excellent emitter passivation is essential for high-efficiency Si wafer solar cells. Low recombination in the emitter is required to attain high open-circuit voltage (V_{oc}), which is a prerequisite for highly efficient solar cells. The electronic passivation of heavily doped p^+ c-Si surfaces has proven to be very challenging in the past, especially with positively charged dielectrics. The most commonly used positively charged dielectrics in the PV community – thermal oxide (SiO_2) and amorphous hydrogenated silicon nitride (*a*-SiN_x:H or, briefly, SiN_x) – are typically found to be an unsuitable choice for p^+ silicon surface passivation due to their intrinsic positive fixed charge (Q_f) and asymmetric capture cross sections for electrons and holes ($\sigma_{n/p}$) [21,36,86,96,97]. During the 1980s, King and Swanson performed a detailed study on p^+ silicon surface passivation by thermal SiO_2 [98]. In 2002, Kerr *et al.* reported one of the best results for SiO_2 passivated (after *anneal* or forming gas anneal) p^+ silicon surfaces, with emitter saturation current density (J_{0e}) values of ~20 fA/cm^2 [36]. In fact, Kerr *et al.* were also one of the first to study surface passivation of p^+ silicon by PECVD SiN_x (in the as-deposited state). It was found that SiN_x can only provide a modest level of surface passivation on p^+ silicon surfaces, which was later confirmed by other studies [21,36,96]. On the other hand, Chen *et al.* reported an excellent p^+ silicon surface passivation (with $J_{0_e} < 13$ fA/cm^2) by Si-rich SiN_x films after a prolonged anneal at 450 °C, and also after an industrial firing process [99,100]. Recently, it was also shown that SiO_2 and SiO_2/SiN_x stacks provide a modest level of surface passivation on p^+ silicon, whereas negatively charged Al_2O_3 films grown by ALD give outstanding surface passivation on similar p^+ silicon surfaces [97]. The latter was also demonstrated at the solar cell device level, with efficiencies of up to 23.9% for *n*-type PERL c-Si wafer solar cells [8,86,101]. In 2008, Mihailetchi *et al.* reported excellent J_{0e} values of 23 fA/cm^2 for p^+ silicon surfaces passivated by a stack consisting of an ultrathin oxide grown wet-chemically by HNO_3 and then capped with a SiN_x film [102]. Thus in general (except for the work of Chen *et al.*), it was observed that a single PECVD SiN_x film is not suitable for p^+ silicon surface passivation [21,36,96]. One of the reasons is the very high fixed positive charge density of $>10^{12}$ cm^{-2} in PECVD SiN_x films on c-Si. Very recently it was observed that the fixed charge density in the SiN_x film can be reduced by adding a thin silicon oxide layer between the c-Si wafer and the SiN_x film, and this resulted in good surface passivation on low-resistivity undiffused *p*-type c-Si wafers, with effective surface recombination velocities (S_{eff}) of <11 cm/s [61]. Although passivation results on *p*-type silicon using inline PECVD AlO_x was already demonstrated in 2009 [92], it was required to obtain such high-quality passivation results on p^+-type Si, which is vital for *n*-type solar cells. Furthermore, ALD Al_2O_3 films provided excellent surface passivation on p^+ silicon for an arbitrary sheet resistance [86]. However, results are compromised for n^+-doped Si passivated by Al_2O_3 films grown by plasma-assisted and O_3-based atomic layer deposition (ALD) [103]. This can be a problem for application in solar cells where simultaneous passivation of n^+ and p^+ surface is required, for example, certain solar cell structures such as interdigitated back-contact (IBC) solar cells require simultaneous passivation of both dopant polarities [104]. In SERIS, Duttagupta *et al.* have successfully optimised and implemented PECVD AlO_x films in a commercial inline tool that enabled excellent passivation of both n^+ and p^+ c-Si surfaces using a single AlO_x film or a stack – and, importantly, that are high-temperature firing stable. Given this promising result on undiffused c-Si wafers, the passivation performance of SiO_x/SiN_x and AlO_x/SiN_x stack on p^+ diffused *n*-type c-Si wafers will be shown in this chapter.

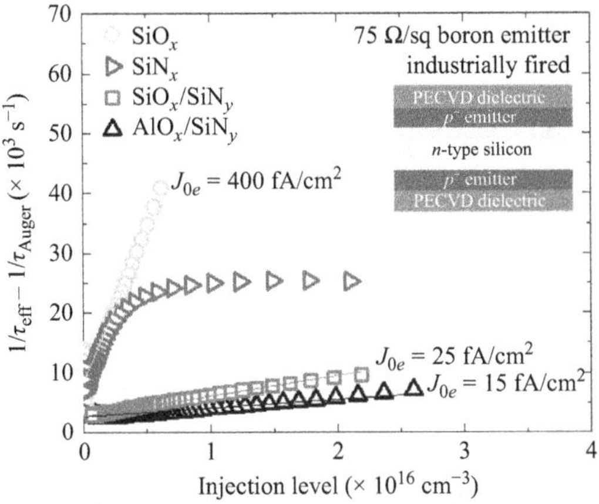

Figure 10.13 Measured injection level dependence of the Auger-corrected inverse effective lifetime of symmetrical p^+np^+ samples symmetrically passivated by SiO_x, SiN_x, SiO_x/SiN_x and AlO_x/SiN_x films. All layers were deposited by PECVD and the samples received a post-deposition anneal in an industrial firing furnace

Figure 10.13 shows the measured Auger-corrected inverse effective carrier lifetimes as a function of injection level for 75-Ω/sq planar p^+ silicon layers passivated by a single-layer PECVD SiO_x, a single-layer SiN_x, a SiO_x/SiN_x dielectric stack and an AlO_x/SiN_x dielectric stack. A single PECVD SiO_x film provides poor surface passivation, both in the as-deposited state (J_{0e} ~ 500–600 fA/cm^2) and after an industrial firing process (~350–400 fA/cm^2), as is evident from the higher slope. However, it is expected that these results can be improved by a FGA, but this is beyond the scope of this work. In the as-deposited state, the SiO_x/SiN_x stack resulted in a J_{0e} value of ~450 fA/cm^2. A significant improvement was observed after industrial firing with set temperatures of >800 °C, resulting in an excellent level of p^+ surface passivation with J_{0e} values of ~25 fA/cm^2. This can most likely be attributed to a significantly improved chemical passivation by a reduction of the interface defect density between at the SiO_x/c-Si interface by hydrogen released from the SiN_x films during the high-temperature step. Experiments on pyramid-textured samples showed an increase in J_{0e} by a factor of ~1.8 compared to planar samples, which can be understood by the increased surface area (factor 1.73). Assuming a short-circuit current density of 40 mA/cm^2 and using the ideal one-diode model, it can be calculated that a J_{0e} of 25 fA/cm^2 would limit the one-Sun open-circuit voltage to 726 mV at 25 °C for an *n*-type c-Si solar cell. One of the key observations here is that we observe a linear relationship of the Auger-corrected effective lifetime curves with positively charged PECVD SiO_x/SiN_x dielectric stacks (allowing the extraction of J_{0_e} values using (10.7)), which is not the case for other positively charged dielectrics such as PECVD SiN_x [36,96,105]. A similar non-linearity has also been reported by other authors [36,96,103]. This non-linearity is a deviation from the standard theory, which can be attributed to increased surface recombination at the interface either due to (i) high positive fixed charge density ($>10^{12}$ cm^{-2}) or (ii) higher asymmetry of the capture cross sections (σ) for electrons and holes at the c-Si/SiN_x interface. Therefore, in order to quantify the level of surface passivation by PECVD SiN_x, $V_{oc,implied}$ of these symmetrically passivated p^+np^+ samples was reported as also mentioned by others [31,36,96,106]. A $V_{oc,implied}$ of about 630 ± 10 mV was measured, which implies an inferior surface passivation compared to the p^+np^+ samples passivated by PECVD SiO_x/SiN_x stacks, which showed a $V_{oc,implied}$ of ~690 ± 5 mV after an industrial firing process. Please note that $V_{oc,implied}$ also depends on factors such as the bulk doping level and the bulk carrier lifetime of the used c-Si wafers, as well as the optical properties of the passivation coating; hence, care must be taken when comparing experimental values between different works.

Using the negatively charged PECVD AlO_x/SiN_x films provided outstanding surface passivation on heavily doped *p*-type silicon surfaces, with a similar level of surface passivation on such surfaces as Al_2O_3 films grown by ALD. In fact, PECVD AlO_x films with a thickness of 5 nm are shown to be sufficient to obtain an excellent surface passivation [105,107,108]. It is emphasised that in the present work the thermal activation of the AlO_x film was done using a process that is widely used in the PV industry, whereas in earlier works with ALD-grown Al_2O_3 films the thermal activation was done by a 30 min anneal at 425 °C in a N_2 environment. The present work thus demonstrates that a standard industrial fast firing process is sufficient for obtaining excellent passivation of p^+ silicon emitters with an AlO_x/SiN_x dielectric stack. This is an important result with respect to the industrial application of such stacks on silicon wafer solar cells. It should also be noted that 25 nm PECVD AlO_x films without a capping SiN_x could also effectively be activated by a standard industrial firing process and resulted in similarly low J_{0_e} values as obtained for the AlO_x/SiN_x stacks. A single AlO_x film, however, cannot simultaneously act as an efficient anti-reflection coating on the front surface of a high-efficiency *n*-type solar cell due to its low refractive index; thus, the focus in this work is on AlO_x/SiN_x stacks. Assuming a short-circuit current density of 40 mA/cm^2 and an ideal diode law, the J_{0e} result for the 80 Ω/sq emitter represents a 1-sun open-circuit voltage limit of the solar cell of 736 mV at 25 °C.

Figure 10.14 shows J_{0e} results as a function of the sheet resistance. PECVD SiO_x/SiN_x and PECVD AlO_x/SiN_x stacks show excellent passivation results comparable to the best published results on p^+ silicon [21,36,86,100,102,109,110]. It is important to emphasise that such high-quality passivation for a wide range of the p^+ sheet resistance was achieved using a low-temperature dielectric deposited in an industrial reactor and fired in an industrial firing furnace.

10.5.1 *Role of fixed charge density and extraction of S_{n0} at p^+ c-Si surfaces*

The electron surface recombination velocity parameter S_{n_0} at the p^+ type c-Si surface is a critical parameter for the interface quality, as it quantifies the recombination activity of that interface. From the Shockley–Read–Hall (SRH) theory [37,38],

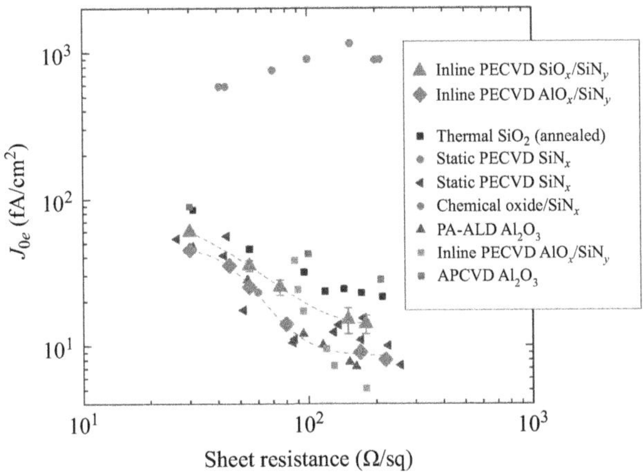

Figure 10.14 *J_{0e} values of p^+ emitters passivated by industrially fired PECVD SiO_x/SiN_x (large triangles) and AlO_x/SiN_x stacks (large diamonds) as a function of the p^+ sheet resistance. The dashed line is a guide to the eye. For comparison, other published results for passivated p^+ Si are also included from the literature [21,36,86,100,102,109,110]*

extended by Girish *et al.* and Aberle *et al.* to include the effects of interface charges and illumination [50,111], S_{n0} is proportional to the product of the N_{st} (the number of surface states per unit area) and the electron capture cross section (σ_n):

$$S_{n0} \equiv v_{th}\sigma_n N_{st} \tag{10.8}$$

where v_{th} is the thermal velocity of the charge carriers. Taking both the measured boron diffusion profile and the Q_f (as measured above) into account, the S_{n_0} values were extracted for the passivated p^+ silicon surface by matching the experimental and simulated effective lifetime curves using Sentaurus TCAD [112] and using the physical models from Refs. [113,114]. Figure 10.14 shows the simulated J_{0_e} of a symmetrically passivated 75 Ω/sq p^+np^+ sample as a function of S_{n0}. This figure also shows the role of considering the amount and polarity of charge for the extraction of S_{n_0}. A J_{0_e} of 25 fA/cm^2 (from Figure 10.15) and Q_f of $+10^{11}$ cm^{-2} as determined in this work for industrially fired PECVD SiO_x/SiN_x stack corresponds to an S_{n_0} of about 400–500 cm/s. However, taking the measured J_{0_e} value of 15 fA/cm^2 for the PECVD AlO_x/SiN_x stack and the measured Q_f of -10^{12} cm^{-2} we can see that this results in an S_{n0} value of $\sim 10^4$ cm/s. Such low surface recombination velocity is comparable to values reported earlier for ALD grown Al_2O_3 films [86]. The J_{0_e} vs. S_{n_0} curve for an insulator charge of $+10^{11}$ cm^{-2} is almost the same as that for zero charge. It is also clear that the surface passivation mechanism for the SiO_x/SiN_x stack completely relies on 'chemical passivation', while the AlO_x/SiN_x stack relies on 'field-effect passivation'. The role of charge (both in terms of polarity and in terms of magnitude) in surface passivation of p^+ silicon is illustrated.

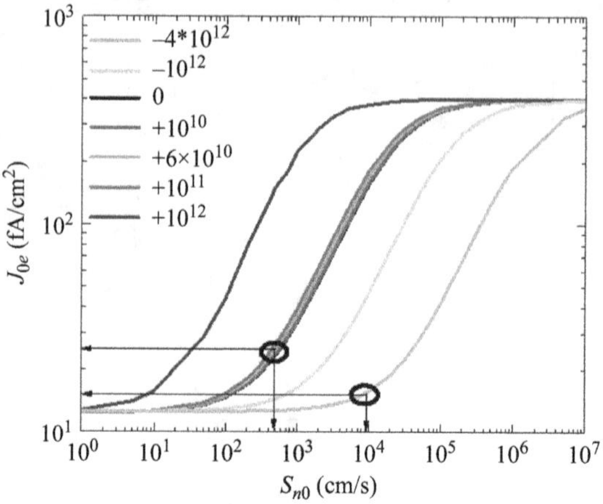

Figure 10.15 Simulated J_{0e} values of a 75 Ω/sq p^+np^+ sample symmetrically passivated with a PECVD SiO_x/SiN_x stack, as a function of the S_{n0} and the fixed insulator charge density

Obviously, a negative fixed charge is more beneficial as it repels the minority carrier electrons from the interface. As shown in Figure 10.4, a film with negative insulator charge of -1×10^{12} cm^{-2} with a similar S_{n0} value of 400 cm/s results to lower J_{0_e} of <18 fA/cm^2. On the other hand, films with Q_f of >+1 × 12 cm^{-2} (such as for SiN_x) show poor surface passivation with J_{0e} >100 fA/cm^2 (considering a similar S_{n0}). This explains (at least partly) the detrimental effect of a very high positive fixed charge in the case of passivating p^+ silicon surfaces. However, experimentally it is often found that the J_{0e} values are much higher than 100 fA/cm^2, which is likely to be due to higher S_{n0} at these c-Si/SiN_x interfaces. This could either be due to asymmetric capture cross sections or higher interface defect densities at these interfaces. Interestingly, it was observed that J_{0e} can be below 150 fA/cm^2 when D_{it} is low (10^{11} eV^{-1} cm^{-2}) and Q_f is below 10^{12} cm^{-2}. In addition to this simulation study, it is further corroborated from field effect measurements that a positively charged dielectric can effectively passivate p^+ silicon surfaces, provided the fixed charge Q_f is <+1 × 10^{12} cm^{-2} and D_{it} remains low.

10.6 Industrial Al-LBSF (or PERC) solar cells

The operation of high-efficiency Si solar cells is significantly influenced by the surface recombination behaviour of rear-passivating dielectrics, for example, an injection-level dependent surface recombination velocity (giving a 'non-linear' cell behaviour) [26,115–118]. The polarity and magnitude of the fixed charge in the rear dielectric may have a significant impact on the efficiency of *p*-type Si solar cells. For example, Dauwe *et al.* showed that SiN_x can lead to reduced solar cell efficiency when applied on the rear of *p*-type Si wafer solar cells with local rear contacts [119]. It was explained that this is the case for the PERC solar cell structure, which features a standard ohmic contact between the metal and the silicon wafer. This is attributed to the presence of an inversion layer induced by the high positive charge density in nearly-stoichiometric SiN_x that allows a current flow ('short circuit') between the inversion layer and the rear metal contacts. This leads to an effect called 'parasitic shunting' and particularly reduces the short-circuit current density J_{sc} [119]. Later it was shown that the use of Si-rich SiN_x reduced the effect of parasitic shunting [4]. Si-rich SiN_x generally has a lower positive charge density compared to nearly-stoichiometric SiN_x films [120]. Consequently, Si-rich SiN_x leads to a poorly conductive inversion layer (or none at all), as shown in Ref. [4]. For Al-LBSF solar cells, however, parasitic shunting is typically not observed as the inversion layer is interrupted by the local back surface field [4,12,79,121–126]. However, Schultz *et al.* evaluated the impact of different rear dielectric films on Al-LBSF cells (Al-LBSF cells were prepared by laser-fired contacts) [10,127]. It was demonstrated that application of a thin thermal SiO_2 beneath the SiN_x improved their cell performance significantly (compared to the cells with SiN_x rear passivation only) showing higher J_{sc} and improved IQE performance at longer wavelengths. Due to increased industrial applicability of Al-LBSF cells, it is important to understand the performance of *p*-type Al-LBSF Si wafer solar cells with different rear surface passivation layer.

Three low-temperature deposited dielectrics (AlO_x/SiN_x, SiO_x/SiN_x and SiN_x) are investigated for rear surface passivation of *p*-type Al-LBSF Si wafer solar cells. From one-Sun current-voltage (I-V), external quantum efficiency, reflection measurements and device simulation, the impacts of the rear-surface passivating dielectrics on the performance of the Al-LBSF solar cells are analysed. Furthermore, the polarity and magnitude of the fixed charge in the rear passivating dielectric is correlated to the fill factor (FF) loss for Al-LBSF solar cells. The process sequence can be found in detail in Ref. [82]. The measured one-Sun I-V parameters of the four groups of cells are summarised in Table 10.3. A clear increase in efficiency in the range of 0.8%–1.5% absolute for the Al-LBSF structure (with AlO_x/SiN_x or SiO_x/SiN_x as rear dielectrics) is observed compared to the Al-BSF reference cells. This is mainly because of improvements in the open-circuit voltage (V_{oc}) and, particularly, the short-circuit current density (J_{sc}). The FF of the Al-BSF cells is slightly higher, mainly because of lower series resistance due to full-area metal coverage at the rear. The V_{oc} increase of about 20–30 mV was observed for the Al-LBSF cells with AlO_x/SiN_x or SiO_x/SiN_x rear dielectric indicating reduced rear surface recombination losses due to the dielectrics. J_{sc} improvement of up to 1–2 mA/cm^2 was obtained for the same Al-LBSF cells compared to the Al-BSF counterparts, indicating increased rear internal reflection in combination with reduced recombination losses due to the rear dielectrics.

Cell V_{oc} is indicative of recombination at the surfaces and in the bulk of the Si at the measured light intensity (one-Sun in this case). It should be noted that all three dielectrics showed very good surface passivation results and similar injection level dependence on the undiffused *p*-Cz wafers used in this work. The results are published elsewhere, see Refs. [78,128]. Contactless corona-voltage measurement reveals high negative Q_{total} of 2×10^{12} cm^{-2} for AlO_x/SiN_x and positive Q_{total} of 7×10^{11} cm^{-2} and 4×10^{12} cm^{-2} for SiO_x/SiN_x and SiN_x, respectively, as

Table 10.3 One-sun I-V parameters measured under standard testing conditions (25 °C cell temperature, 100 mW/cm², AM 1.5 G) for the Al-LBSF silicon solar cells with three different rear passivating dielectrics (AlO_x/SiN_x, SiO_x/SiN_x and SiN_x). For comparison, results of standard Al-BSF cells are also included. Average values of 16 cells processed under each batch are also mentioned

	Group 1 AlO_x/SiN_x		Group 2 SiO_x/SiN_x		Group 3 SiN_x		Group 4 Al-BSF	
	Best	**Ave.**	**Best**	**Ave.**	**Best**	**Ave.**	**Best**	**Ave.**
V_{oc} (mV)	661.6	659.0	649.7	648.7	643.8	642.7	632.8	633.0
J_{sc} (mA/cm^2)	39.2	39.1	39.1	39.0	38.6	38.5	37.6	37.5
FF (%)	79.7	79.9	78.8	78.4	77.7	77.6	80.6	80.4
Eff* (%)	20.7	20.6	20.0	19.8	19.3	19.2	19.2	19.14

*Without stabilisation.

measured in this work. However, the difference lies in mid-gap interface defect density $D_{it,midgap}$. For the AlO_x/SiN_x and SiO_x/SiN_x stacks $D_{it,midgap}$ is similar (in the range of 10^{11} eV^{-1} cm^{-2}), while the SiN_x exhibits a higher interface defect density with $D_{it,midgap}$ values of $\sim 8 \times 10^{11}$ eV^{-1} cm^{-2}, which is related to the differences in surface passivation among these dielectrics in this case. If in the finished solar cells the surface recombination at the rear is ruled by $D_{it,midgap}$, then the difference in surface passivation results among these dielectrics can be partially correlated to the differences in the cells' V_{oc}.

The analysis showed large differences in J_{sc} extracted under 1 Sun and 0.3 Suns for AlLBSF cells having SiO_x/SiN_x (difference of ~1.5 mA/cm²) and SiN_x (difference of ~2.5 mA/cm²) as rear surface passivation schemes, whereas almost no difference was observed for the Al-LBSF cells with AlO_x/SiN_x rear passivation. This indicates that a severe non-linearity is present in the Al-LBSF cells having a SiO_x/SiN_x and, particularly, a SiN_x rear surface passivation. Schultz *et al.* [10,127] also showed that Al-LBSF cells (LBSF from laser fired contacts) with SiN_x (having high positive charge in the range of 10^{12} cm^{-2})* as rear passivation had reduced J_{sc} compared to cells with thin thermal SiO_2/SiN_x (having low positive charge in the range of 10^{11} cm^{-2})* as rear passivation. The observed J_{sc} effect was ascribed to parasitic shunting. On the other hand, in a recent study by Dullweber *et al.* on screen-printed Al-LBSF cells, AlO_x/SiN_x (having high negative charge in the range of 10^{12} cm^{-2})* and thin thermal SiO_2/SiN_x (having low positive charge in the range of 10^{11} cm^{-2})* did not show any difference in J_{sc}. This confirms that Al-LBSF cells having a rear dielectric with a small positive fixed charge or a negative fixed charge can both result in very good cell efficiency.

Significant differences in the FF of the Al-LBSF cells were observed for different rear-passivating dielectrics (AlO_x/SiN_x, SiO_x/SiN_x and SiN_x), see Table 10.3.

*Value estimated by the present author (no value was mentioned in the publications).

A FF loss analysis (based on the method described in Ref. [129]) revealed a striking correlation in the FF loss, especially arising from recombination processes with ideality factors not equal to one, which are referred to as 'non-ideal recombination' ($J_{0,\text{non-ideal}}$) in Al-LBSF cells (more details in Ref. [82]). As the cell processing was identical apart from the rear surface dielectrics, the differences in the FF loss due to $J_{0,\text{non-ideal}}$ can be attributed to the different rear surfaces. AlO_x/SiN_x with its negative fixed charge density generates accumulation conditions at the rear Si surface. On the other hand, SiO_x/SiN_x and SiN_x have low and high positive fixed charge density, respectively, which depletes or even inverts the rear surface. The inverted condition can result in additional bulk recombination, due to the formation of an SCR beneath the Si surface (more explanations in Refs. [40,80,81,130]). This SCR may well be the source of $J_{0,\text{non-ideal}}$. From Ref. [82], this effect is more pronounced for the dielectric with the higher positive charge density (SiN_x), leading to a higher FF loss due to $J_{0,\text{non-ideal}}$. Additionally, as mentioned by Aberle *et al.* [26,116], if an injection level dependent recombination mechanism is present in a solar cell, then the FF loss due to $J_{0,\text{non-ideal}}$ determined by the FF loss analysis include the influence of this injection level dependence (or 'non-linearity' of the cell). Thus, a detailed loss analysis of the Al-LBSF cells fabricated in this work needs to take their non-linearity into account.

Three different rear surface passivation schemes such as AlO_x/SiN_x, SiO_x/SiN_x and SiN_x were investigated in this section for application on large-area (239.0 cm^2) *p*-type Si Al local back surface field (Al-LBSF) silicon wafer solar cells. The Al-LBSF solar cells were fabricated and analysed by one-Sun current–voltage (I–V), external quantum efficiency and reflection measurements in combination with PC1D device simulations. A champion solar cell efficiency of 20.7% was obtained for Al-LBSF solar cells with an AlO_x/SiN_x rear surface passivation, while the solar cells with SiO_x/SiN_x and SiN_x rear surface passivation showed lower efficiencies. The analysis revealed a severe non-linearity in the Al-LBSF cells fabricated with SiO_x/SiN_x and SiN_x as rear surface passivation, which was not the case for Al-LBSF cells fabricated with AlO_x/SiN_x rear passivation. Despite the fact that these dielectrics showed a very similar level of surface passivation on dedicated lifetime samples, significant differences in the quantum efficiency and short-circuit current density were observed. It was hypothesised that this difference originates either from the non-linearity of the cells or/and from differences in polarity and amount of fixed charge density in the rear surface passivating dielectric. A detailed analysis revealed that the FF of Al-LBSF solar cells with a positively charged dielectric is strongly reduced due to a significant increase in non-ideal recombination. This non-ideal recombination is found to increase for higher positive charge densities and could be due to additional recombination in the SCR beneath the Si surface.

10.7 Summary and conclusions

This chapter presented an investigation of surface passivation at the moderately and heavily doped *p*-type surfaces using dielectric stacks of either PECVD SiO_x/SiN_x or AlO_x/SiN_x layers.

High-quality SiO_x layers were developed and optimised for c-Si surface passivation. SiO_x layers were deposited using a low-temperature (<300 °C) PECVD method. It was shown that inserting a thin PECVD SiO_x (10–30 nm) between the c-Si and the SiN_x reduced positive fixed charge density significantly (by almost one order of magnitude) with a drastic reduction of the interface defect density. The stack yields very low $S_{eff,max}$ of 7 and 8 cm/s on *n* and *p*-type c-Si, respectively, and shows remarkably low J_{0e} of 15 fA/cm^2 on p^+ type c-Si surfaces.

Experimentally the charge density in SiO_x/SiN_x dielectric stacks deposited by an industrial low-temperature PECVD method was varied by up to one order of magnitude (10^{11}–10^{12} cm^{-2}), without any impact on $D_{it,midgap}$. These films feature good interface quality ($D_{it,midgap}$ ~5×10^{11} eV^{-1} cm^{-2} after industrial firing), which is independent of the SiO_x deposition temperature. The fundamental relationship between Q_{total} and S_{eff} was verified, which previously was investigated only by simulations or external corona charging. Injection-level dependent effective carrier lifetime curves and measured charge densities showed higher cell V_{oc} and *FF* potential achievable with higher positive fixed insulator charge (under the assumption that the $D_{it,midgap}$ remains constant in the finished solar cells). As a proof-of-concept, Al-LBSF cells were fabricated with films having either relatively high or low positive insulator charge density. The films with higher positive charge density used as the rear passivating layer showed improved conversion efficiency by 0.2% (absolute), which came mainly from an improvement in V_{oc} and FF. The results indicate that dielectric charge tailoring is a very effective and promising approach for silicon solar cells, and has both fundamental and technological relevance.

In order to further reduce surface recombination at *p*-type c-Si surfaces and allow a better process window, the use of high negatively charged films was shown to be beneficial, whereby both field-effect and chemical passivation can be utilised. In this work, significantly improved surface passivation results using *dynamically* deposited plasma AlO_x (with or without a SiN_x or SiO_x capping layer) were demonstrated with ultra-low $S_{eff,max}$ of 1 and 8 cm/s on *n* and *p* type c-Si, respectively, after industrial firing. PECVD AlO_x (with or without a SiN_x or SiO_x capping layer) was demonstrated to provide exceptional passivation of p^+ type c-Si surfaces with J_{0e} of 9 fA/cm^2, for a large range of sheet resistances. Importantly, the PECVD AlO_x technology developed in SERIS is able to passivate both n^+ type and p^+ type c-Si surfaces simultaneously, which has specific significance for devices that need a single dielectric film for passivation, for example, IBC cells.

The polarity and amount of fixed charge have a profound impact on the surface recombination velocity and the solar cell's operation. All three dielectric films (SiN_x, SiO_x/SiN_x, AlO_x/SiN_x) developed in SERIS were applied at the rear surface of full-area (239.0 cm^2) *p*-type Si Al-LBSF solar cells and their performance was investigated using quantum efficiency and 1-Sun I-V measurements. Screen-printed Al-LBSF solar cells with AlO_x/SiN_x rear surface passivation gave the best PV efficiency (up to 20.7%). Today (2018), PERC cell efficiencies have reached above 21.5% in our lab jointly with our industrial partners, specifically with very thin (~7 nm) PECVD AlO_x capped with SiN_x and there are reports of above 23% cell efficiencies achieved by leading solar cell companies in their advanced R&D pilot-lines.

Noteworthy to mention, that AlO_x/SiN_x technology today is an industry standard for fabricating PERC cells with an installed capacity of >35 GW, which is projected to grow >50% in 2018 alone and is expected to dominate for the next decade.

References

[1] Aberle, A., 'Surface Passivation of Crystalline Silicon Solar Cells: A Review', *Progress in Photovoltaics: Research and Applications*, 2000, 8, pp. 473–487.

[2] Hofmann, M., Janz, S., Schmidt, C., *et al.*, 'Recent Developments in Rear-Surface Passivation at Fraunhofer Ise', *Solar Energy Materials and Solar Cells*, 2009, 93, (6–7), pp. 1074–1078.

[3] Mueller, T., Schwertheim, S., Scherff, M., and Fahrner, W., 'High Quality Passivation for Heterojunction Solar Cells by Hydrogenated Amorphous Silicon Suboxide Films', *Applied Physics Letters*, 2008, 92, (3), p. 033504.

[4] Dauwe, S., 'Low-Temperature Surface Passivation of Crystalline Silicon and Its Application to the Rear Side of Solar Cells', Doctor of Philosophy, University of Hannover, 2004.

[5] Taguchi, M., Yano, A., Tohoda, S., *et al.*, '24.7% Record Efficiency Hit Solar Cell on Thin Silicon Wafer', *IEEE Journal of Photovoltaics*, 2014, 4, (1), pp. 96–99.

[6] Hoex, B., Heil, S.B.S., Langereis, E., van de Sanden, M.C.M., and Kessels, W.M.M., 'Ultralow Surface Recombination of C-Si Substrates Passivated by Plasma-Assisted Atomic Layer Deposited Al_2O_3', *Applied Physics Letters*, 2006, 89, (4), pp. 042112–042113.

[7] Schmidt, J., Merkle, A., Brendel, R., Hoex, B., van de Sanden, M.C.M., and Kessels, W.M.M., 'Surface Passivation of High-Efficiency Silicon Solar Cells by Atomic-Layer-Deposited Al_2O_3', *Progress in Photovoltaics: Research and Applications*, 2008, 16, pp. 461–466.

[8] Benick, J., Hoex, B., van de Sanden, M.C.M., Kessels, W.M.M., Schultz, O., and Glunz, S., 'High Efficiency N-Type Si Solar Cells on Al_2O_3-Passivated Boron Emitters', *Applied Physics Letters*, 2008, 92, (25), p. 253504.

[9] Glunz, S.W., 'High-Efficiency Crystalline Silicon Solar Cells', *Advances in OptoElectronics*, 2007.

[10] Schultz, O., Rentsch, J., Grohe, A., Glunz, S.W., and Willeke, G.P., Dielectric Rear Surface Passivation for Industrial Multicrystalline Silicon Solar Cells', *4th World Conference on Photovoltaic Energy Conversion*, (2006).

[11] Saint-Cast, P., Benick, J., Kania, D., *et al.*, 'High-Efficiency C-Si Solar Cells Passivated with Ald and Pecvd Aluminum Oxide', *IEEE Electron Device Letters*, 2010, 31, (7), pp. 695–697.

[12] Aberle, A., Warta, W., Knobloch, J., and Voss, B., Surface Passivation of High Efficiency Silicon Solar Cells', *21st IEEE Photovoltaic Specialists Conference*, (1990).

[13] Blakers, A., Wang, A., Milne, A., Zhao, J., and Green, M., '22.8% Efficient Silicon Solar Cell', *Applied Physics Letters*, 1989, 55, (13), pp. 1363–1365.
[14] Zhao, J., Wang, A., Altermatt, P., Wenham, S., and Green, M., '24% Efficient Perl Silicon Solar Cell: Recent Improvements in High Efficiency Silicon Cell Research', *Solar Energy Materials and Solar Cells*, 1996, 41–42, pp. 87–99.
[15] Wang, A., Zhao, J., Green, M.A., *et al.*, '24% Efficient Silicon Solar Cells', *Applied Physics Letters*, 1990, 57, (6), pp. 602–604.
[16] International Technology Roadmap for Photovoltaics (ITRPV)', (2016, Seventh edition edn.).
[17] Macdonald, D. and Geerligs, L.J., 'Recombination Activity of Interstitial Iron and Other Transition Metal Point Defects in P- and N-Type Crystalline Silicon', *Applied Physics Letters*, 2004, 85, (18), pp. 4061–4063.
[18] Cotter, J.E., Guo, J.H., Cousins, P.J., Abbott, M.D., Chen, F.W., and Fisher, K.C., 'P-Type Versus N-Type Silicon Wafers: Prospects for High-Efficiency Commercial Silicon Solar Cells', *Electron Devices, IEEE Transactions on*, 2006, 53, (8), pp. 1893–1901.
[19] World's Highest Conversion Efficiency of 26.33% Achieved in a Crystalline Silicon Solar Cell. Available At: http://Www.Nedo.Go.Jp/English/News/Aa5en_100109.Html. (2016).
[20] Smith, D.D., Reich, G., Baldrias, M., Reich, M., Boitnott, N., and Bunea, G., 'Silicon Solar Cells with Total Area Efficiency above 25 %', in, *2016 IEEE 43rd Photovoltaic Specialists Conference (PVSC)*, (2016).
[21] Altermatt, P.P., Plagwitz, H., Bock, R., *et al.*, 'The Surface Recombination Velocity at Boron-Doped Emitters: Comparison between Various Passivation Techniques', in, *21th European Photovoltaic Solar Energy Conference* (WIP Renewable Energies, Munich, 2006).
[22] Aberle, A.G., 'Surface Passivation of Crystalline Silicon Solar Cells: A Review', *Progress in Photovoltaics*, 2000, 8, (5), pp. 473–487.
[23] Nagel, H., Aberle, A.G., and Hezel, R., 'Optimised Antireflection Coatings for Planar Silicon Solar Cells Using Remote PECVD Silicon Nitride and Porous Silicon Dioxide', *Progress in Photovoltaics*, 1999, 7, (4), pp. 245–260.
[24] Cuevas, A., Kerr, M.J., and Schmidt, J., Passivation of Crystalline Silicon Using Silicon Nitride', *3rd World Conference on Photovoltaic Energy Conversion*, (2003).
[25] Duerinckx, F. and Szlufcik, J., 'Defect Passivation of Industrial Multicrystalline Solar Cells Based on PECVD Silicon Nitride', *Solar Energy Materials and Solar Cells*, 2002, 72, (1–4), pp. 231–246.
[26] Aberle, A.G., *Crystalline Silicon Solar Cells: Advanced Surface Passivation and Analysis*, (UNSW Publishing and Printing Services, 1999).
[27] Kerr, M.J., Schmidt, J., and Cuevas, A., 'Comparison of High Quality Surface Passivation Schemes for Phosphorus Diffused Emitters', in, *16th European Photovoltaic Solar Energy Conference*, (2000).
[28] Kerr, M.J., Schmidt, J., Cuevas, A., and Bultman, J.H., 'Surface Recombination Velocity of Phosphorus-Diffused Silicon Solar Cell Emitters

Passivated with Plasma Enhanced Chemical Vapor Deposited Silicon Nitride and Thermal Silicon Oxide', *Journal of Applied Physics*, 2001, 89, (7), pp. 3821–3826.

[29] Sproul, A.B., Green, M.A., and Stephens, A.W., 'Accurate Determination of Minority Carrier- and Lattice Scattering-Mobility in Silicon from Photoconductance Decay', *Journal of Applied Physics*, 1992, 72, (9), pp. 4161–4171.

[30] Cuevas, A., Sinton, R.A., and Stuckings, M., Determination of Recombination Parameters in Semiconductors from Photoconductance', *IEEE Conference on Optoelectronic and Microelectronic Materials and Devices proceedings*, (1996).

[31] Sinton, R. and Cuevas, A., 'Contactless Determination of Current–Voltage Characteristics and Minority-Carrier Lifetimes in Semiconductors from Quasi-Steady-State Photoconductance Data', *Applied Physics Letters*, 1996, 69, (17), pp. 2510–2512.

[32] Nagel, H., Berge, C., and Aberle, A., 'Generalized Analysis of Quasi-Steady-State and Quasi-Transient Measurements of Carrier Lifetimes in Semiconductors', *Journal of Applied Physics*, 1999, 86, (11), pp. 6218–6221.

[33] Kane, D.E. and Swanson, R.M., Measurement of the Emitter Saturation Current by a Contactless Photoconductivity Decay Method', *18th IEEE Photovoltaic Specialists Conference*, (1985).

[34] Cuevas, A., 'The Effect of Emitter Recombination on the Effective Lifetime of Silicon Wafers', *Solar Energy Materials and Solar Cells*, 1999, 57, pp. 277–290.

[35] Cuevas, A., Stocks, M., Macdonald, D., and Sinton, R., Applications of the Quasi-Steady-State Photoconductance Technique', *2nd World Conference on Photovoltaic Energy Conversion*, (1998).

[36] Kerr, M.J.: 'Surface, Emitter and Bulk Recombination in Silicon and Development of Silicon Nitride Passivated Solar Cells', Doctor of Philosophy, Australian National University, 2002.

[37] Shockley, W. and Read, W.T., 'Statistics of the Recombinations of Holes and Electrons', *Physical Review*, 1952, 87, (5), p. 835.

[38] Hall, R.N., 'Electron-Hole Recombination in Germanium', *Physical Review*, 1952, 87, (2), p. 387.

[39] Grove, A.S. and Fitzgerald, D.J., 'Surface Effects on P-N Junctions: Characteristics of Surface Space-Charge Regions under Non-Equilibrium Conditions', *Solid-State Electronics*, 1966, 9, (8), pp. 783–806.

[40] Steingrube, S., Altermatt, P., Steingrube, D., Schmidt, J., and Brendel, R., 'Interpretation of Recombination at C-Si/SiNx Interfaces by Surface Damage', *Journal of Applied Physics*, 2010, 108, (1), p. 14506.

[41] Ma, F.J., Samudra, G.G., Peters, M., *et al.*, 'Advanced Modeling of the Effective Minority Carrier Lifetime of Passivated Crystalline Silicon Wafers', *Journal of Applied Physics*, 2012, 112, (5), p. 054508.

[42] Wilson, M., D'Amico, J., Savtchouk, A., *et al.*, Multifunction Metrology Platform for Photovoltaics', *37th IEEE Photovoltaic Specialists Conference* (2011).

[43] Wilson, M., Marinskiy, D., Byelyayev, A., *et al.*, 'The Present Status and Recent Advancements in Corona-Kelvin Non-Contact Electrical Metrology of Dielectrics for Ic-Manufacturing', *ECS Transactions*, 2006, 3, (3), pp. 3–24.

[44] Edelman, P., Savtchouk, A., Wilson, M., *et al.*, 'Non-Contact C-V Measurements of Ultra Thin Dielectrics', *The European Physical Journal – Applied Physics*, 2004, 27, pp. 495–498.

[45] Wilson, M., Lagowski, J., Savtchouk, A., Jastrzebski, L., and D'Amico, J., 'Cocos (Corona Oxide Characterization of Semiconductor) Metrology: Physical Principles and Applications', in Gupta, D.C. and Brown, G.A. (eds.), *ASTM Conference on Gate Dielectric Oxide Integrity*, (ASTM, 1999).

[46] Wilson, M., Lagowski, J., Jastrzebski, L., Savtchouk, A., and Faifer, V., 'Cocos (Corona Oxide Characterization of Semiconductor) Non-Contact Metrology for Gate Dielectrics', *AIP Conference Proceedings*, 2001, 550, (1), pp. 220–225.

[47] Reed, M. and Plummer, J., 'Chemistry of Si-SiO_2 Interface Trap Annealing', *Journal of Applied Physics*, 1988, 63, (12), pp. 5776–5793.

[48] Green, M.L., Gusev, E.P., Degraeve, R., and Garfunkel, E.L., 'Ultrathin (<4 Nm) Sio(2) and Si-O-N Gate Dielectric Layers for Silicon Microelectronics: Understanding the Processing, Structure, and Physical and Electrical Limits', *Journal of Applied Physics*, 2001, 90, (5), pp. 2057–2121.

[49] Eades, W. and Swanson, R., 'Calculation of Surface Generation and Recombination Velocities at the Si-SiO_2 Interface', *Journal of Applied Physics*, 1985, 58, (11), pp. 4267–4276.

[50] Aberle, A., Glunz, S., and Warta, W., 'Impact of Illumination Level and Oxide Parameters on Shockley-Read-Hall Recombination at the Si-SiO_2 Interface', *Journal of Applied Physics*, 1992, 71, (9), pp. 4422–4431.

[51] Jin, H., Weber, K.J., Dang, N.C., and Jellett, W.E., 'Defect Generation at the Si–SiO_2 Interface Following Corona Charging', *Applied Physics Letters*, 2007, 90, (26), pp. 262109.

[52] Kerr, M. and Cuevas, A., 'Very Low Bulk and Surface Recombination in Oxidized Silicon Wafers', *Semiconductor Science and Technology*, 2002, 17, (1), p. 35.

[53] Wang, A., Zhao, J., and Green, M.A., '24% Efficient Silicon Solar Cells', *Applied Physics Letters*, 1990, 57, (6), pp. 602–604.

[54] Zhao, J., Wang, A., and Green, M.A., '24.5% Efficiency Silicon Pert Cells on Mcz Substrates and 24.7% Efficiency Perl Cells on Fz Substrates', *Progress in Photovoltaics: Research and Applications*, 1999, 7, pp. 471–474.

[55] Leguijt, C., Lölgen, P., Eikelboom, J.A., *et al.*, 'Very Low Surface Recombination Velocities on 2.5 Ohm.Cm Si Wafers, Obtained with Low-Temperature PECVD of Si-Oxide and Si-Nitride', *Solar Energy Materials and Solar Cells*, 1994, 34, (1–4), pp. 177–181.

[56] Leguijt, C., Lölgen, P., Eikelboom, J.A., *et al.*, 'Low Temperature Surface Passivation for Silicon Solar Cells', *Solar Energy Materials and Solar Cells*, 1996, 40, (4), pp. 297–345.

[57] Dingemans, G., van de Sanden, M.C.M., and Kessels, W.M.M., 'Excellent Si Surface Passivation by Low Temperature SiO_2 Using an Ultrathin Al_2O_3 Capping Film', *Physica Status Solidi – Rapid Research Letters*, 2011, 5, (1), pp. 22–24.

[58] Duttagupta, S., Ma, F.-J., Hoex, B., and Aberle, A.G., 'Excellent Surface Passivation of Heavily Doped P+ Silicon by Low-Temperature Plasma-Deposited SiOx/SiNy Dielectric Stacks with Optimised Antireflective Performance for Solar Cell Application', *Solar Energy Materials and Solar Cells*, 2014, 120, Part A, (0), pp. 204–208.

[59] Zhao, J., Wang, A., and Green, M., 'Performance Degradation in Cz(B) Cells and Improved Stability High Efficiency Pert and Perl Silicon Cells on a Variety of Seh Mcz(B), Fz(B) and Cz(Ga) Substrates', *Progress in Photovoltaics: Research and Applications*, 2000, 8, pp. 549–558.

[60] Chen, Z., Pang, S.K., Yasutake, K., and Rohatgi, A., 'Plasma-Enhanced Chemical-Vapor-Deposited Oxide for Low Surface Recombination Velocity and High Effective Lifetime in Silicon', *Journal of Applied Physics*, 1993, 74, (4), pp. 2856–2859.

[61] Dingemans, G., Mandoc, M.M., Bordihn, S., van de Sanden, M.C.M., and Kessels, W.M.M., 'Effective Passivation of Si Surfaces by Plasma Deposited SiOx/a-SiNx:H Stacks', *Applied Physics Letters*, 2011, 98, (22), pp. 222102-222102-222103.

[62] Szlufcik, J., Sivoththaman, S., Nlis, J.F., Mertens, R.P., and Van Overstraeten, R., 'Low-Cost Industrial Technologies of Crystalline Silicon Solar Cells', *Proceedings of the IEEE*, 1997, 85, (5), pp. 711–730.

[63] Rohatgi, A., Doshi, P., Moschner, J., Lauinger, T., Aberle, A.G., and Ruby, D.S., 'Comprehensive Study of Rapid, Low-Cost Silicon Surface Passivation Technologies', *IEEE Transactions on Electron Devices*, 2000, 47, (5), pp. 987–993.

[64] Schmidt, J., Kerr, M., and Cuevas, A., 'Surface Passivation of Silicon Solar Cells Using Plasma-Enhanced Chemical-Vapour-Deposited Sin Films and Thin Thermal SiO_2/Plasma Sin Stacks', *Semiconductor Science and Technology*, 2001, 16, (3), pp. 164–170.

[65] Van Erven, A.J.M., Bosch, R.C.M., Weeber, A.W., and Bijker, M.D., 'Effects of Different Firing Profiles on Layer Characteristics and Passivation Properties of Industrial Etp Deposited Silicon Nitride Films', in, *19th European Photovoltaic Solar Energy Conference*, (2004).

[66] Van Erven, A.J.M., Bosch, R.C.M., and Bijker, M.D., 'Textured Silicon Surface Passivation by High-Rate Expanding Thermal Plasma Deposited Sin and Thermal SiO_2/Sin Stacks for Crystalline Silicon Solar Cells', *Progress in Photovoltaics: Research and Applications*, 2008, 16, (7), pp. 615–627.

[67] Richter, A., Glunz, S., Werner, F., Schmidt, J., and Cuevas, A., 'Improved Quantitative Description of Auger Recombination in Crystalline Silicon', *Physical Review B*, 2012, 86, (16), p. 165202.

[68] Nickel, N.H., 'Hydrogen Diffusion through Silicon/Silicon Dioxide Interfaces', *Journal of Vacuum Science & Technology B*, 2000, 18, (3), pp. 1770–1772.

[69] Nickel, N.H., Jackson, W.B., and Walker, J., 'Hydrogen Migration in Polycrystalline Silicon', *Physical Review B*, 1996, 53, (12), pp. 7750–7761.

[70] Larionova, Y., Mertens, V., Harder, N., and Brendel, R., 'Surface Passivation of N-Type Czochralski Silicon Substrates by Thermal-SiO_2/Plasma-Enhanced Chemical Vapor Deposition Sin Stacks', *Applied Physics Letters*, 2010, 96, (3), p. 032105.

[71] Dingemans, G., Beyer, W., van de Sanden, M.C.M., and Kessels, W.M.M., 'Hydrogen Induced Passivation of Si Interfaces by Al_2O_3 Films and SiO_2/Al_2O_3 Stacks', *Applied Physics Letters*, 2010, 97, (15), p. 152106.

[72] Dingemans, G., Einsele, F., Beyer, W., van de Sanden, M.C.M., and Kessels, W.M.M., 'Influence of Annealing and Al_2O_3 Properties on the Hydrogen-Induced Passivation of the Si/SiO_2 Interface', *Journal of Applied Physics*, 2012, 111, (9), p. 093713.

[73] Dekkers, H.F.W., Beaucarne, G., Hiller, M., Charifi, H., and Slaoui, A., 'Molecular Hydrogen Formation in Hydrogenated Silicon Nitride', *Applied Physics Letters*, 2006, 89, (21), p. 211914.

[74] Aberle, A., Glunz, S., and Warta, W., 'Field Effect Passivation of High Efficiency Silicon Solar Cells', *Solar Energy Materials and Solar Cells*, 1993, 29, (2), pp. 175–182.

[75] Kuhlmann, B., Aberle, A.G., and Hezel, R., 'Two-Dimensional Numerical Optimisation Study of Inversion Layer Emitters of Silicon Solar Cells', in, *13th European Photovoltaic Solar Energy Conference*, (1995).

[76] Glunz, S.W., Biro, D., Rein, S., and Warta, W., 'Field-Effect Passivation of the SiO_2-Si Interface', *Journal of Applied Physics*, 1999, 86, (1), pp. 683–691.

[77] Altermatt, P.P., Steingrube, S., Yang, Y., *et al.*, 'Higly Predictive Modeling of Entire Si Solar Cells for Industrial Applications', in, *24th European Photovoltaic Solar Energy Conference*, 2014, 5, (4), pp. 1014–1019.

[78] Duttagupta, S., Hameiri, Z., Hoex, B., and Aberle, A.G., 'Dielectric Charge Tailoring in PECVD SiOx/SiNx Stacks and Its Impact on Industrial P-Type Silicon Wafer Solar Cell Efficiency', *IEEE Journal of Photovoltaics*, 2014, 5, (4), pp. 1014–1019.

[79] Mittelstädt, L., Dauwe, S., Metz, A., and Hezel, R., 'Highly Efficient All Silicon Nitride Passivated Multicrystalline Silicon Solar Cells', in, *17th European Photovoltaic Solar Energy Conference*, (2001).

[80] Steingrube, S., Altermatt, P., Schmidt, J., and Brendel, R., 'Modelling C-Si/SiNx Interface Recombination by Surface Damage', *Physica Status Solidi – Rapid Research Letters*, 2010, 4, (3–4), pp. 91–93.

[81] Ma, F.-J., Samudra, G., Peters, M., *et al.*, 'Advanced Modeling of the Effective Minority Carrier Lifetime of Passivated Crystalline Silicon Wafers', *Journal of Applied Physics*, 2012, 112, (5), p. 054508.

[82] Duttagupta, S., 'Advanced Surface Passivation of Crystalline Silicon for Solar Cell Applications', Doctor of Philosophy, National University of Singapore, 2014.

[83] Hezel, R. and Jaeger, K., 'Low-Temperature Surface Passivation of Silicon for Solar Cells', *Journal of The Electrochemical Society*, 1989, 136, (2), pp. 518–523.

[84] Agostinelli, G., Vitanov, P., Alexieva, Z., *et al.*, 'Surface Passivation of Silicon by Means of Negative Charge Dielectric', in, *19th European Photovoltaic Solar Energy Conference*, (2004).

[85] Agostinelli, G., Delabie, A., Vitanov, P., *et al.*, 'Very Low Surface Recombination Velocities on P-Type Silicon Wafers Passivated with a Dielectric with Fixed Negative Charge', *Solar Energy Materials and Solar Cells*, 2006, 90, (18–19), pp. 3438–3443.

[86] Hoex, B., Schmidt, J., Bock, R., Altermatt, P.P., van de Sanden, M.C.M., and Kessels, W.M.M., 'Excellent Passivation of Highly Doped P-Type Si Surfaces by the Negative-Charge-Dielectric Al_2O_3', *Applied Physics Letters*, 2007, 91, (11), pp. 112107–112103.

[87] Hoex, B., Gielis, J.J.H., van de Sanden, M.C.M., and Kessels, W.M.M., 'On the C-Si Surface Passivation Mechanism by the Negative-Charge-Dielectric Al_2O_3', *Journal of Applied Physics*, 2008, 104, (11), pp. 113703–113707.

[88] Dingemans, G., Engelhart, P., Seguin, R., *et al.*, 'Stability of Al_2O_3 and Al_2O_3/a-SiNx:H Stacks for Surface Passivation of Crystalline Silicon', *Journal of Applied Physics*, 2009, 106, (11), p. 114907.

[89] Dingemans, G., Seguin, R., Engelhart, P., van de Sanden, M.C.M., and Kessels, W.M.M., 'Silicon Surface Passivation by Ultrathin Al_2O_3 Films Synthesized by Thermal and Plasma Atomic Layer Deposition', *Physica Status Solidi – Rapid Research Letters*, 2010, 4, (1–2), pp. 10–12.

[90] Dingemans, G., Terlinden, N.M., Pierreux, D., Profijt, H.B., Van De Sanden, M.C.M., and Kessels, W.M.M., 'Influence of the Oxidant on the Chemical and Field-Effect Passivation of Si by Ald Al_2O_3', *Electrochemical and Solid-State Letters*, 2011, 14, (1), pp. H1–H4.

[91] Schmidt, J., Merkle, A., Bock, R., *et al.*, 'Progress in the Surface Passivation of Silicon Solar Cells', in, *23rd European Photovoltaic Solar Energy Conference*, (2008).

[92] Saint-Cast, P., Kania, D., Hofmann, M., Benick, J., Rentsch, J., and Preu, R., 'Very Low Surface Recombination Velocity on P-Type C-Si by High-Rate Plasma-Deposited Aluminum Oxide', *Applied Physics Letters*, 2009, 95, (15), p. 151502.

[93] Miyajima, S., Irikawa, J., Yamada, A., and Konagai, M., 'High Quality Aluminum Oxide Passivation Layer for Crystalline Silicon Solar Cells Deposited by Parallel-Plate Plasma-Enhanced Chemical Vapor Deposition', *Applied Physics Express*, 2010, 3, p. 3.

[94] Li, T.-T. and Cuevas, A., 'Effective Surface Passivation of Crystalline Silicon by Rf Sputtered Aluminum Oxide', *Physica Status Solidi (RRL) – Rapid Research Letters*, 2009, 3, (5), pp. 160–162.

[95] Werner, F., Veith, B., Tiba, V., *et al.*, 'Very Low Surface Recombination Velocities on P- and N-Type C-Si by Ultrafast Spatial Atomic Layer Deposition of Aluminum Oxide', *Applied Physics Letters*, 2010, 97, (16), pp. -.

[96] Libal, J., Petres, R., Buck, T., *et al.*, 'N-Type Multicrystalline Silicon Solar Cells: Bbr3 -Diffusion and Passivation of P+ -Diffused Silicon Surfaces', *20th European Photovoltaic Solar Energy Conference*, (2005).

[97] Benick, J.: 'High-Efficiency N-Type Solar Cells with a Front Side Boron Emitter', Doctor of Philosophy, Albert-Ludwigs-University of Freiburg, 2010.

[98] King, R.R. and Swanson, R.M., 'Studies of Diffused Boron Emitters – Saturation Current, Bandgap Narrowing, and Surface Recombination Velocity', *IEEE Transactions on Electron Devices*, 1991, 38, (6), pp. 1399–1409.

[99] Chen, F., Tsu, and Cotter, J., 'Passivation of Boron Emitters on N-Type Silicon by Plasma-Enhanced Chemical Vapor Deposited Silicon Nitride', *Applied Physics Letters*, 2006, 88, (26), pp. 263514–263513.

[100] Chen, F.W., Romijn, I., Weeber, A., Tan, J., Hallam, B., and Cotter, J.E., 'Relationship between PECVD Silicon Nitride Film Composition and Surface and Edge Passivation', in, *22nd European Photovoltaic Solar Energy Conference*, (2007).

[101] Glunz, S.W., Benick, J., Biro, D., *et al.*, 'N-Type Silicon-Enabling Efficiencies >20% in Industrial Production', *35th IEEE Photovoltaic Specialists Conference*, (2010).

[102] Mihailetchi, V.D., Komatsu, Y., and Geerligs, L.J., 'Nitric Acid Pretreatment for the Passivation of Boron Emitters for N-Type Base Silicon Solar Cells', *Applied Physics Letters*, 2008, 92, (6), pp. 063510–063513.

[103] Hoex, B., van de Sanden, M.C.M., Schmidt, J., Brendel, R., and Kessels, W.M.M., 'Surface Passivation of Phosphorus-Diffused N+ Type Emitters by Plasma-Assisted Atomic-Layer Deposited Al_2O_3', *Physica Status Solidi – Rapid Research Letters*, 2011.

[104] Cousins, P.J., Smith, D.D., Luan, H.-C., *et al.*, 'Generation 3: Improved Performance at Lower Cost', *35th IEEE Photovoltaic Specialists Conference* (2010).

[105] Duttagupta, S., Lin, F., Shetty, K.D., *et al.*, 'State-of-the-Art Surface Passivation of Boron Emitters Using Inline PECVD AlOx/SiNx Stacks for Industrial High-Efficiency Silicon Wafer Solar Cells', in, *38th IEEE Photovoltaic Specialists Conference (PVSC)*, (2012).

[106] Cuevas, A. and Sinton, R.A., 'Prediction of the Open-Circuit Voltage of Solar Cells from the Steady-State Photoconductance', *Progress in Photovoltaics: Research and Applications*, 1997, 5, (2), pp. 79–90.

[107] Duttagupta, S., Lin, F., Shetty, K.D., Aberle, A.G., and Hoex, B., 'Excellent Boron Emitter Passivation for High-Efficiency Si Wafer Solar Cells Using AlOx/SiNx Dielectric Stacks Deposited in an Industrial Inline Plasma Reactor', *Progress in Photovoltaics: Research and Applications*, 2013, 21, (4), pp. 760–764.

[108] Duttagupta, S., Hoex, B., and Aberle, A.G., 'Progress with Industrially-Feasible Surface Passivation of Heavily-Doped P-Type and N-Type Crystalline Silicon by PECVD SiOx/SiNx with Optimised Antireflective Performance', in, *28th European Photovoltaic Solar Energy Conference*, (2013).

[109] Saint-Cast, P., Richter, A., Billot, E., *et al.*, 'Very Low Surface Recombination Velocity of Boron Doped Emitter Passivated with Plasma-Enhanced Chemical-Vapor-Deposited AlOx Layers', *Thin Solid Films*, 2012, 522, (0), pp. 336–339.
[110] Black, L. and McIntosh, K., 'Surface Passivation of C-Si by Atmospheric Pressure Chemical Vapor Deposition of Al_2O_3', *Applied Physics Letters*, 2012, 100, (20), p. 202107.
[111] Girisch, R.B.M., Mertens, R.P., and De Keersmaecker, R.F., 'Determination of Si-SiO_2 Interface Recombination Parameters Using a Gate-Controlled Point-Junction Diode under Illumination', *IEEE Transactions on Electron Devices*, 1988, 35, (2), pp. 203–222.
[112] Sentaurus Tcad, Version F-2011.09, Synopsys Inc., Mountain View, Ca', (2011).
[113] Altermatt, P.P., Schmidt, J., Heiser, G., and Aberle, A.G., 'Assessment and Parameterisation of Coulomb-Enhanced Auger Recombination Coefficients in Lowly Injected Crystalline Silicon', *Journal of Applied Physics*, 1997, 82, (10), pp. 4938–4944.
[114] Altermatt, P., Schumacher, J., Cuevas, A., *et al.*, 'Numerical Modeling of Highly Doped Si:P Emitters Based on Fermi-Dirac Statistics and Self-Consistent Material Parameters', *Journal of Applied Physics*, 2002, 92, (6), pp. 3187–3197.
[115] Aberle, A., Lauinger, T., Schmidt, J., and Hezel, R., 'Injection-Level Dependent Surface Recombination Velocities at the Silicon-Plasma Silicon Nitride Interface', *Applied Physics Letters*, 1995, 66, (21), pp. 2828–2830.
[116] Aberle, A., Robinson, S., Wang, A., Zhao, J., Wenham, S., and Green, M., 'High-Efficiency Silicon Solar Cells: Fill Factor Limitations and Non-Ideal Diode Behaviour Due to Voltage-Dependent Rear Surface Recombination Velocity', *Progress in Photovoltaics: Research and Applications*, 1993, 1, pp. 133–143.
[117] Aberle, A.G., Glunz, S.W., Stephens, A.W., and Green, M.A., 'High-Efficiency Silicon Solar Cells: Si/SiO_2, Interface Parameters and Their Impact on Device Performance', *Progress in Photovoltaics: Research and Applications*, 1994, 2, (4), pp. 265–273.
[118] Aberle, A.G., Altermatt, P.P., Heiser, G., *et al.*, 'Limiting Loss Mechanisms in 23-Percent Efficient Silcion Solar Cells', *Journal of Applied Physics*, 1995, 77, (7), pp. 3491–3504.
[119] Dauwe, S., Mittelstädt, L., Metz, A., and Hezel, R., 'Experimental Evidence of Parasitic Shunting in Silicon Nitride Rear Surface Passivated Solar Cells', *Progress in Photovoltaics: Research and Applications*, 2002, 10, (4), pp. 271–278.
[120] Mäckel, H. and Lüdemann, R., 'Detailed Study of the Composition of Hydrogenated SiNx Layers for High-Quality Silicon Surface Passivation', *Journal of Applied Physics*, 2002, 92, (5), pp. 2602–2609.
[121] Knobloch, J., Aberle, A.G., and Voss, B., 'Cost Effective Processes for Silicon Solar Cells with High Performance', in, *9th European Photovoltaic Solar Energy Conference*, (1989).

[122] Knobloch, J., Aberle, A.G., Warta, W., and Voss, B., 'Dependence of Surface Recombination Velocities at Silicon Solar Cell Surfaces on Incident Light Intensity', in, *8th European Photovoltaic Solar Energy Conference*, (1988).

[123] Aberle, A.G.: 'Investigations on Surface Passivation of High-Efficiency Silicon Solar Cells', Doctor of Philosophy, University of Freiburg, 1991.

[124] Dauwe, S., Mittelstadt, L., Metz, A., Schmidt, J., and Hezel, R., Low-Temperature Rear Surface Passivation Schemes for >20% Efficient Silicon Solar Cells', *3rd World Conference on Photovoltaic Energy Conversion*, (2003).

[125] Mittelstädt, L., Dauwe, S., Metz, A., Hezel, R., and Häßler, C., 'Front and Rear Silicon-Nitride-Passivated Multicrystalline Silicon Solar Cells with an Efficiency of 18.1%', *Progress in Photovoltaics: Research and Applications*, 2002, 10, (1), pp. 35–39.

[126] Ramanathan, S., Meemongkolkiat, V., Das, A., Rohatgi, A., and Koehler, I., 'Understanding and Fabrication of 20% Efficient Cells Using Spin-on-Based Simultaneous Diffusion and Dielectric Passivation', *IEEE Journal of Photovoltaics*, 2012, 2, (1), pp. 22–26.

[127] Schultz, O., Hofmann, M., Glunz, S.W., and Willeke, G.P., Silicon Oxide/Silicon Nitride Stack System for 20% Efficient Silicon Solar Cells', *31st IEEE Photovoltaic Specialists Conference*, (2005).

[128] Duttagupta, S., Hameiri, Z., Wilson, M., Hoex, B., and Aberle, A., 'Crystalline Silicon Surface Passivation with Industrial Plasma Silicon Nitride and Aluminium Oxide', in, *22nd International Photovoltaic Science and Engineering Conference*, (2012).

[129] Khanna, A., Mueller, T., Stangl, R.A., Hoex, B., Basu, P.K., and Aberle, A.G., 'A Fill Factor Loss Analysis Method for Silicon Wafer Solar Cells', *IEEE Journal of Photovoltaics*, 2013, 3, (4), pp. 1170–1177.

[130] Ma, F.-J., Hoex, B., Samudra, G., and Aberle, A., 'Modelling and Simulation of Field-Effect Surface Passivation of Crystalline Silicon-Based Solar Cells', *Energy Procedia*, 2012, 15, pp. 155–161.

Chapter 11
Hydrogenated amorphous silicon nitride ($a\text{-}SiN_x$:H) as surface passivation layer

Machteld Lamers[1] and Arthur Weeber[1,2]

11.1 Introduction

Hydrogenated amorphous silicon nitride (a-SiN$_x$:H) coatings have three key functionalities for application to wafer-based crystalline silicon PV technology:

- It is the standard antireflection coating because of its high transparency and tunable refractive index n;
- It acts as hydrogen (H) source for bulk passivation; and
- It passivates the surface of the cell.

For crystalline silicon solar cell applications, a balance between these three properties needs to be found and should be tuned to optimize the module power output. In general, layers with refractive indexes n between 2.0 and 2.1 are deposited by plasma-enhanced chemical vapor deposition (PECVD). A higher n could result in better surface passivation and light coupling on module level but will result in too high parasitic absorption. Therefore, coatings with lower n (about 2.05) are generally applied.

The passivating properties of the layer were previously correlated to cell output properties like the open-circuit voltage V_{OC} [1,2]. During the last decade, more details about the passivating properties and mechanisms behind it have been elucidated.

The bulk-passivating properties of a-SiN$_x$:H are related to the release of H from the layer, which diffuses into the Si wafer where it can passivate defects and impurities. The bulk-passivating quality is related to the amount of H present in a-SiN$_x$:H, the (thermal) energy available to break H from (mostly) N–H bonds in a-SiN$_x$:H and the density of the a-SiN$_x$:H layer itself. During the high-temperature firing step for contact formation, H is released from a-SiN$_x$:H and increasingly so for higher temperatures and longer times [3–5]. The atomic H diffuses through the a-SiN$_x$:H layer and Si wafer, and bonds to defects and most common transition

[1]ECN Part of TNO – Solar Energy, the Netherlands
[2]Delft University of Technology, Photovoltaic Materials and Devices Group, the Netherlands

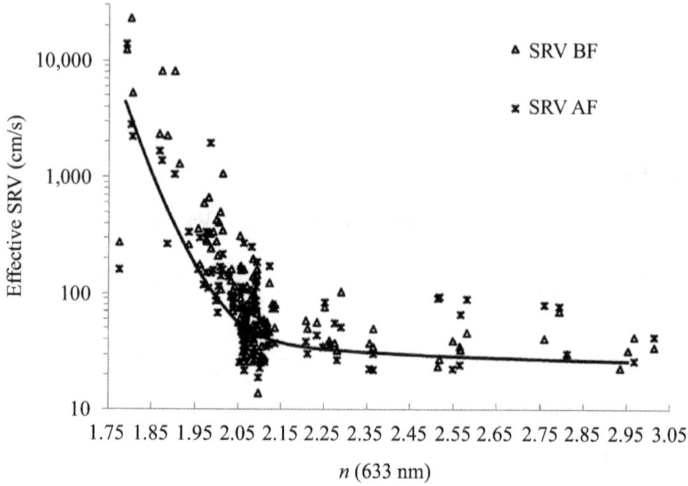

Figure 11.1 Relation between the SRV of a-SiN$_x$:H layers and its refractive index n at a wavelength of 633 nm [6]. Both surface recombination velocity (SRV) of a-SiNx:H before (BF) and after (AF) a short thermal treatment similar to the so-called contact firing are shown

metal impurities (for example, Fe) it encounters, thereby passivating them. Logically, layers with a high N–H density will release more H than layers with hardly any N–H. Atomic H present in a-SiN$_x$:H can easily form H_2 in case these layers also have a low total atomic density, thus an open structure, thereby effectively reducing the concentration of atomic H in the layer. Due to the open structure, the formed H_2 can easily effuse to the ambient [3]. The resultant low atomic H concentration in the Si wafer will cause in a much smaller improvement of the V_{OC} of the solar cell. For high Si–N bond density (and low N–H bond density), also corresponding to a high total atomic density, the number of atomic H released from the layer is too small to passivate most of the defect, thereby hardly improving V_{OC} [1]. An optimum was found for layers with a Si–N bond density of 1.2–1.3 × 10^{23} cm^{-3} [1]. The overall result on solar cell's efficiency significantly depends on the bulk quality of the Si wafer, i.e. the number of defects and impurities to be passivated [6].

First studies on the surface-passivating properties of a-SiN$_x$:H showed a trend that layers with higher *n* will result in better surface passivation (see Figure 11.1) [6–10]. Furthermore, it can be seen that the surface recombination can withstand a thermal anneal similar to contact firing. The refractive index *n* can easily be adjusted by varying the composition of the layers, which is directly related to gas flow ratio of the precursor gases used (silane (SiH_4) and ammonia (NH_3)). The dependence of *n* and extinction coefficient *k* on composition is depicted in Figure 11.2 [6]. However, a more recent study carried out by Duttagupta *et al.* [11] showed that also for layers with lower *n* an extremely surface recombination velocity (SRV) can be reached. Using both a batch (static deposition) and inline system (dynamically deposited a-SiN$_x$:H), SRV values below 10 cm/s were found

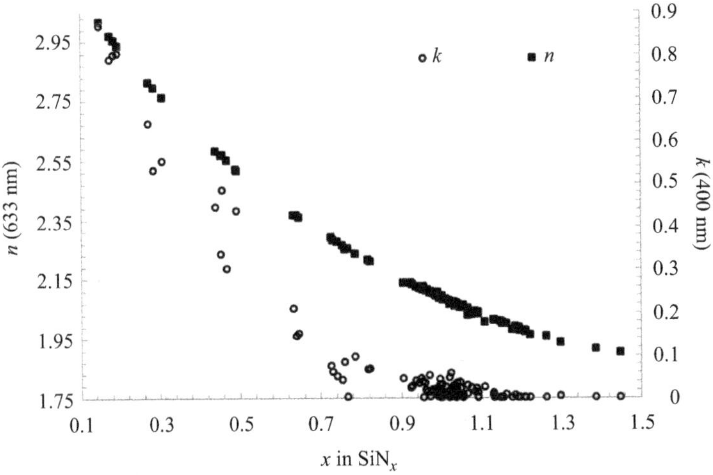

Figure 11.2 Relation of n and k on composition of the a-SiN$_x$:H layer [6]

for layers with n in the range of 2.0. Of course, the mechanisms to reduce recombination losses at the surface are more complicated.

The surface-passivating quality is determined by the combined effect of two mechanisms: field-effect passivation as a result of the number of fixed charges (Q_f) at the interface with Si and chemical passivation determined by the density of states in the bandgap at the interface (D_{it}). For D_{it}, it holds that a low value will result in a low recombination rate. Doping type and density, besides other factors, determine the effect of Q_f on the surface passivation [7]. A high Q_f on lightly doped material will result in a strong field-effect contribution to the surface passivation. Inversion can occur if the doped material is of the same polarity, while accumulation occurs for doped material with the opposite polarity. The situation changes for highly doped materials with the same polarity to the fixed charge as now inversion can no longer occur, and depletion occurs. More details are given by Aberle [7,12]. As Q_f and D_{it} are properties of the interface of a-SiN$_x$:H/Si, this indicates that they are determined by the initial growth of the a-SiN$_x$:H layer [13].

11.2 Surface-passivating properties related to bond structure

Both D_{it} and Q_f determine the final surface passivation quality obtained for a-SiN$_x$:H layers. The number of fixed chargers Q_f results in a local field effect and is caused by the atomic structure at the interface between the materials (a-SiN$_x$:H and Si) [14]. The created local field depends on the polarizabilities of the different materials and the variation in volume density of these polarizabilities over a certain distance at the interface. There are different geometrical bonding topologies in a-SiN$_x$:H which can be charged: the so-called K- and N-centers, see Figure 11.3 [15]. These centers are, respectively, *Si-N3 and *N = Si2, * indicates that these centers can be neutral, positively charged, negatively charged or bonded to H [16]. Assuming that K- and

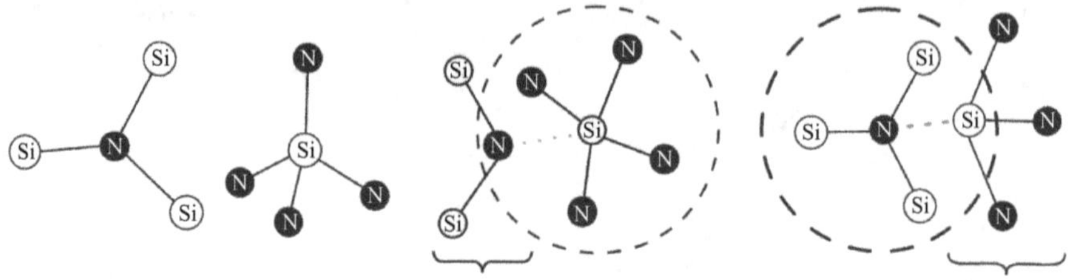

Figure 11.3 Standard bonding topologies of Si and N atoms in a-SiN_x:H. (a) Ideally coordinated Si and N atoms have three and four neighbors, respectively, (b) the N-center can be seen as a twofold coordinated N- or a fivefold coordinated Si-center; and (c) the K-center can be seen as a fourfold coordinated N- or threefold coordinated Si-center [16]

N-centers also occur at the interface, this may lead to the conclusion that Q_f is determined by the variation and volume fractions of K- and N-centers at the interface region of the a-SiN_x:H/Si interface [13].

During a-SiN_x:H deposition, the gas ratio between SiH_4 and NH_3 and temperature determine both the initial growth conditions and bulk properties [1,17,18]. Hence, a clear correlation exists between the properties of the layer, Q_f and D_{it}. To demonstrate these, we prepared samples using 275 μm thick p-type Float Zone (FZ) (100) wafers, which were double-side mirror polished with a base resistivity of 2.5 Ωcm. With remote PECVD, we deposited 80–100 nm thick a-SiN_x:H layers with varying composition on one side of the wafer. From Fourier-transformed infrared (FTIR) spectroscopy data, deconvoluted bond densities were calculated. Using ellipsometry, n was determined. We determined Q_f and D_{it} using capacitance–voltage (CV) metal-insulator-semiconductor MIS analysis. We produced CV-MIS samples by evaporating 300 nm of aluminum on the full area of the uncoated rear side and circular dots with varying sizes (0.1–1.91 mm^2) on the coated front side [6].

In literature, the data regarding the exact conversion values needed to convert the area of the FTIR spectrum per bond to a density is inconclusive, therefore the data are shown in peak area of the bond density [12] (see Figure 11.4(a)).

A clear and positive correlation between Q_f and D_{it} is found which can be seen in Figure 11.4(b). The H-Si-N3 bond (H-bonded K-defect) and the Si–N bond in locally distorted configurations follow the same trend, with a maximum around $n = 2.3$. This indicates that the material with $n = 2.3$ has relatively more K-centers (assuming also an increase in neutral and charged K-centers) and is more distorted than other a-SiN_x:H configurations. For high N-content ($n < 2.0$–2.1) the layer distortion decreases as the Si–N bond density peak at 790 cm^{-1} drops. Additionally, the layer shows a sharp increase in N–H bond densities while the Si–H peak is very small or no longer observed. For these layers, the number of K-centers goes down as the distortion lowers, just as Q_f reduces. For higher n and low SRV, Q_f and D_{it}

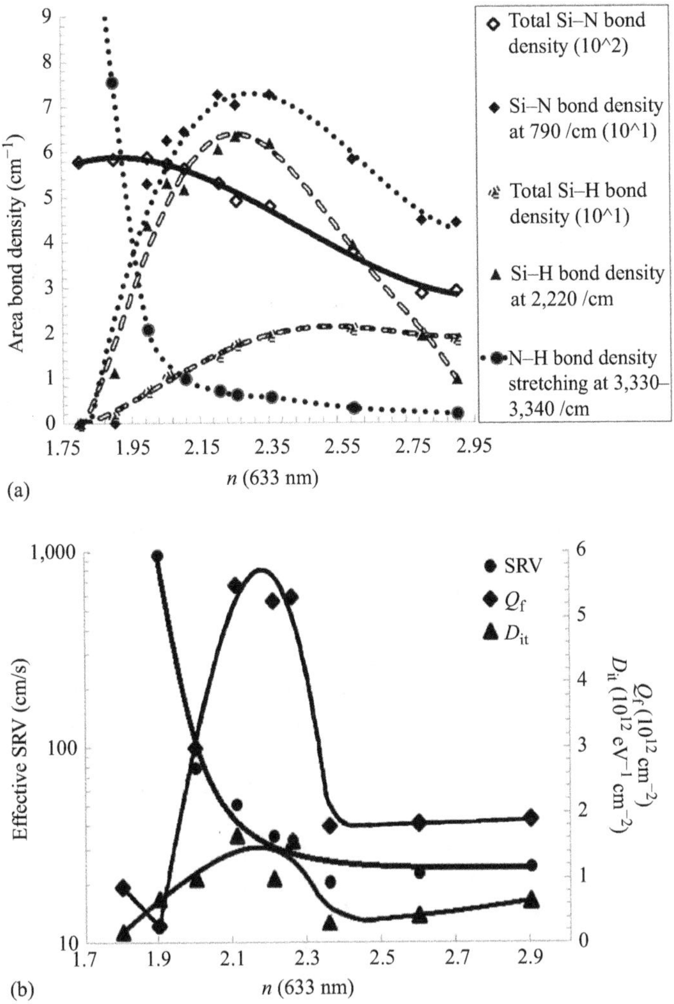

Figure 11.4 Relation between the bond densities, effective SRV, Q_f, D_{it} and n at 633 nm wavelength [12]: (a) interdependence of (the peak areas of) the Si–H bond density, the deconvoluted Si-H bond density at 2,220 cm^{-1}, the Si–N bond density, the deconvoluted Si–N bond density at 790 cm^{-1} (distorted), N–H bond density and n; and (b) relation between the effective SRV, Q_f, D_{it}, and n

decrease again to a certain level, simultaneously with a drop in H-bonded K-centers and distortion in the layer, and $D_{\rm it}$ becomes the more limiting factor in passivation quality and is regulated by the high amount of H available.

Demichelos *et al.* found a comparable interdependence between *n*, the Si–H bond density in the H-Si-N3 configuration and the Si–N bond density in the locally distorted configuration [19], and Wolf *et al.* found a similar trend between *n* and $Q_{\rm f}$ [20].

11.3 Effect of nitridation of the interface

During the initial growth of a-SiN_x:H, the surface is nitridated, which can be described as N-insertion into the Si surface region. The interface region itself is found to be approximately 1–3 nm wide. Depending on the deposition temperature and the NH_3 gas flow, different bonds and nitridation characteristics emerge as observed by Dai *et al.* [21]. At temperatures below 300 °C, no surface alteration can be observed. Above 350 °C, –NH_2 starts adhering to the Si surface. Beyond 400 °C, nitrogen begins to be inserted into the Si-Si bonds at the surface of the Si crystal; the degree of insertion increases with temperature. This nitridation displays a direct relation with Q_f and was demonstrated by Takakura *et al.*, though no change in D_{it} could be observed [22]. To study the effect of nitridation on passivation in more detail, we employed various NH_3 plasma treatments prior to a-SiN_x:H layer deposition. To determine a possible dependence on dopant type as well as crystal orientation, we used two different materials, both double-side mirror polished: 275 μm thick p-type FZ (100) wafers, with a base resistivity of 2.5 Ω.cm, and 275 μm thick n-type FZ (111) wafers, with a 3.5 Ω.cm base resistivity. Shortly before nitridation, the wafers received an HF dip to remove any surface oxide. As mentioned above the total process consisted of a NH_3 plasma pretreatment to vary nitridation, and an ensuing a-SiN_x:H layer deposition, the latter being the same for all pretreatments. We performed both process steps using a remote PECVD system. We measured the effective lifetime of the minority carriers by means of symmetric test structures with nitridation and a-SiN_x:H coating on both sides. Furthermore, we used samples with nitridation and coating on one side to determine Q_f and D_{it} using CV-MIS. We conducted the NH_3 plasma pretreatments at temperatures of between 300 and 500 °C [12].

Figures 11.5(a) and 11.5(b) show the effect of the temperature of a NH_3 plasma on Q_f and D_{it}. We can see a positive correlation between Q_f and D_{it}. A small effect starts being observable already at 300 °C, suggesting that the onset of nitridation starts at somewhat lower temperatures than suggested by Dai *et al.* [21]. The nitridation effect is indicated by the increase in Q_f. Figure 11.5(b) shows the effect on the effective SRV. A lower effective SRV, thus better passivation, is obtained when the surface is slightly nitridated; this way we obtained Q_f values up to about 8×10^{12} cm^{-2}, corresponding to pretreatment temperatures of 400 °C. Under these circumstances, and confirmed by Dai *et al.* [21], NH_2 starts adhering to the surface, but no or very little insertion of nitrogen in the Si–Si bond structure occurs. The reduction in effective SRV compared to samples without plasma pretreatment (nitridation) at 300 °C could be related to the onset of nitridation of the thin native oxide present on the wafer surface. Above 450 °C, N insertion into Si–Si lattice takes place, corresponding well to a higher effective SRV, and being related to a further increase in Q_f and a dominating high D_{it}. A maximum was found at 400 °C for n-type (111) and at 300 °C for p-type (100). The results clearly demonstrate that tuning of the effective SRV is possible. This possibility does not depend on the bulk properties of the a-SiN_x:H layer [12].

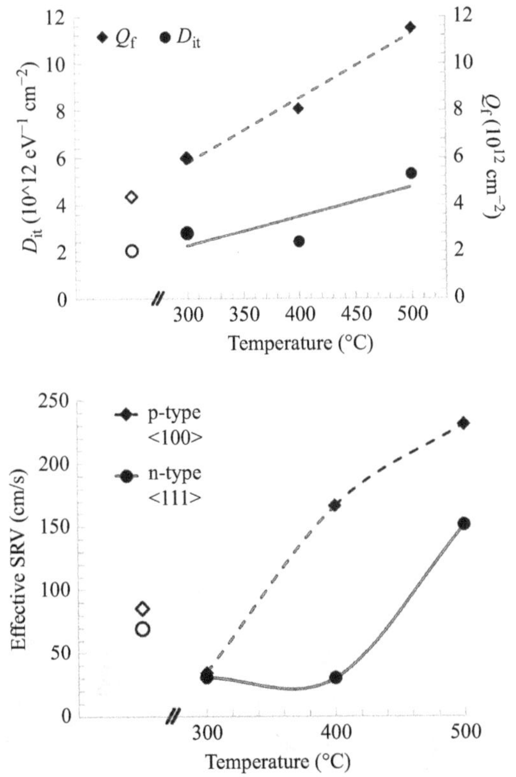

Figure 11.5 The effect of the temperature of a NH_3 plasma treatment on Q_f and D_{it} (a) and effective SRV (b) on lightly doped material. Closed and open symbols are, respectively, samples with and without pretreatment. The lines are guides to the eye [12]

We produced similar samples for cross-sectional TEM (transmission electron microscopy) specimens by mechanical polishing, then dimpling and Ar^+-ion sputtering; we used liquid nitrogen to cool the samples. The acceleration voltage was progressively reduced to 1.0 keV during the final stage of the Ar^+-ion sputtering to minimize damages to the samples. We performed TEM with a JEOL 2010F operated at 200 kV. Figure 11.6 shows the HRTEM (high-resolution TEM) image of the (100) sample [13].

A white line is drawn over the high-amplitude periodic signal. We observed that the first few atomic layers of c-Si at the surface display a distorted expanded crystal structure; furthermore, we see that the high-amplitude signal is shifted left as well as lowered in intensity. This indicates the presence of large strain and the possible insertion of other smaller atoms. A dark band is visible at the interface, possibly indicating the presence of light atoms, like N and/or H. Such a dark band is not visible at a SiO_2/c-Si interface [23]. Large strain was also present in the (111) oriented sample, depicted in the inverse fast Fourier-transformed (FT) image

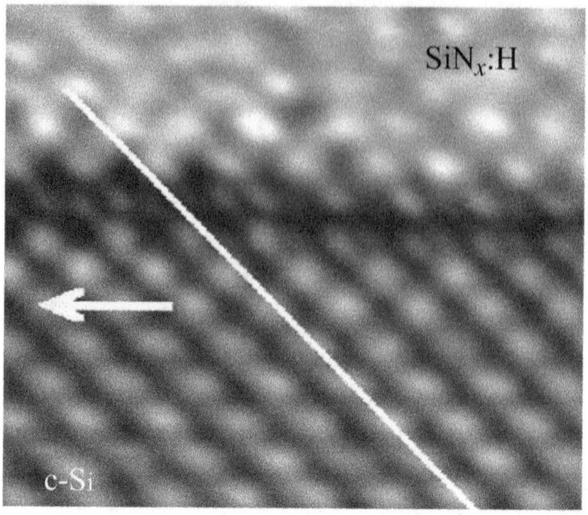

Figure 11.6 HRTEM image of a-SiN$_x$:H/c-Si interface of the (100) sample [13]

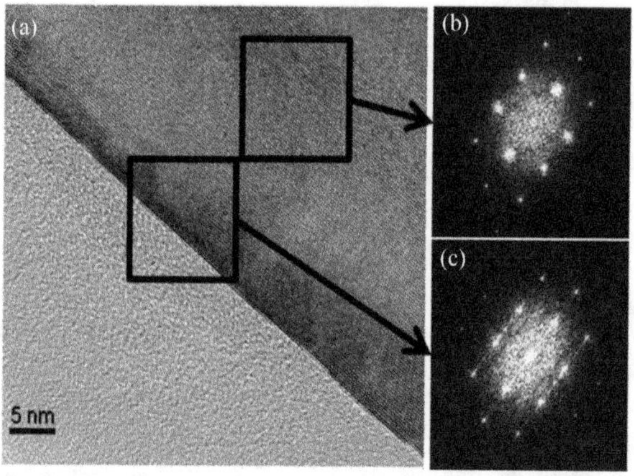

Figure 11.7 (a) HRTEM image of a-SiN$_x$:H/c-Si interface of the (111) sample; (b) FT image of c-Si; and (c) FT image of the interface of a-SiN$_x$:H and c-Si [13]

(Figure 11.7(c)), and corresponding to the dark band in Figure 11.7(a). No such strain, however, was found in the c-Si bulk (Figure 11.7(b)) [13].

On the a-SiN$_x$:H side of the interface, we found a 2 nm wide gradient in the plasmon loss energy, measured with electron energy loss spectroscopy (EELS); it is shown in Figure 11.8. One can see that the plasmon loss energy increases from 17.0 eV to 19.8 eV over 3 nm (from −1 nm to 2 nm). From the high-loss region of the EELS spectrum, we can also determine the chemical composition, as shown in Figure 11.9. Park *et al.* found a relation between the plasmon loss energy and the composition of a-SiN$_x$:H [24]. We found that this relation is likewise valid for the interface. The relation between plasmon loss energy and composition in conjunction with the increase in plasmon loss energy occurring at the Si side (1 nm)

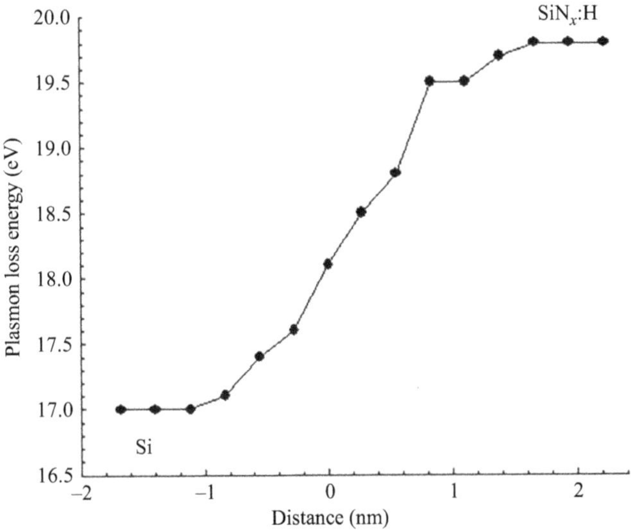

Figure 11.8 The plasmon loss energy over the interface of the (100) sample. The interface is at 0 nm [13]

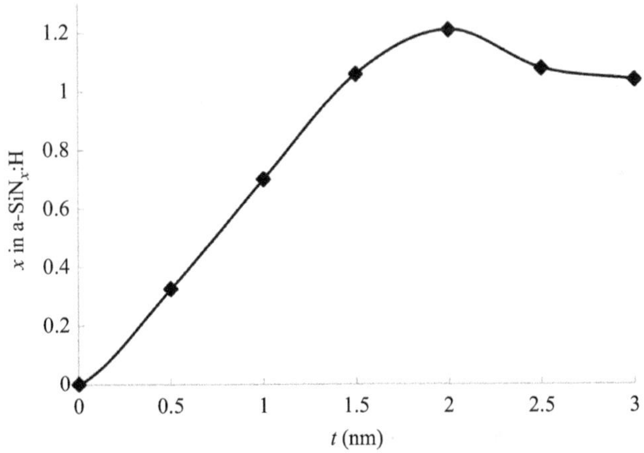

Figure 11.9 Experimentally determined change in composition at the interface of a-SiN$_x$:H (x = 1) and a (100) sample. The interface is at 0 nm [13]

indicates the presence of nitrogen in the first few nanometers of the Si substrate, besides hydrogen. Nitrogen was also located in the first few nanometers of a Si substrate as measured by Ikarashi *et al.* [25]. The composition is determined from the $L_{2,3}$ peak onset in the high-loss region of EELS (around 99 eV) [13].

The nitridation effect is determined by analyzing the plasmon loss energy. We measured this with an EELS line scan on the samples (see Figure 11.10) [13].

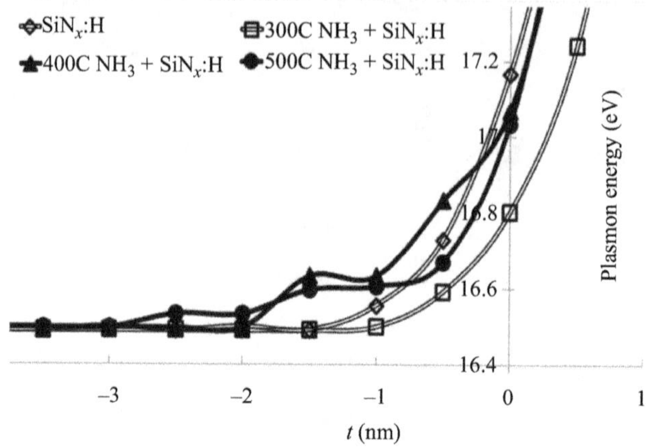

Figure 11.10 Plasmon loss energy over the interface for different treatments of NH_3 prior to a-SiN_x:H deposition. Each line represents the averaged value of three line scans [13]

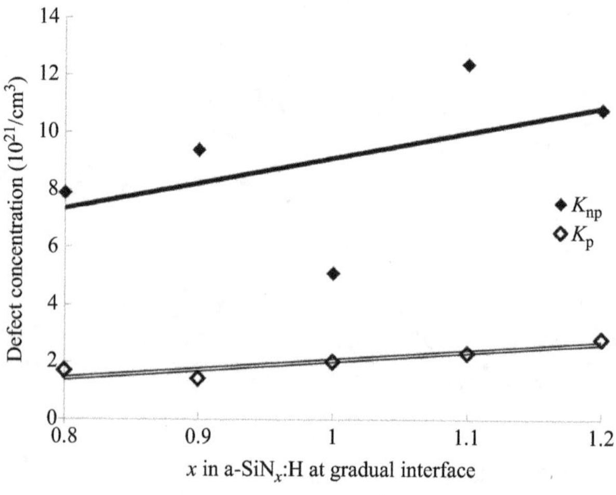

Figure 11.11 MD simulations of defect concentrations at gradual interfaces for increasing N concentration at the interface (higher x value); p refers to passivated meaning the defect has an H atom bonded, np indicates nonpassivated [13]

Molecular dynamics (MD) simulation shows that gradual interfaces show significantly fewer defects than sharp interfaces [13]. It also shows that a higher amount of N in a-SiN_x:H at the interface with Si leads to a higher defect concentration as is seen in Figure 11.11. This higher defect concentration results in a higher Q_f. The variation in volume density of the K-centers at the interface creates a local field. We presume that this is the cause of the macroscopically detected Q_f.

The location of Q_f is dispersed over the interface, starting a few nm in the c-Si to about 2 nm in the a-SiN_x:H [13].

11.4 *Ab initio* studies

To deepen our understanding of the interface, Hintzsche *et al.* [13,26] conducted *ab initio* density functional theory (DFT) studies for the interface of a-$Si_3N_{3.5}H_{0.8}$ and c-Si. They used 100 atoms of a-$Si_3N_{3.5}H_{0.8}$ and 56 c-Si atoms. a-$Si_3N_{3.5}H_{0.8}$ corresponds to the configuration of a standard a-SiN_x:H layer on the front side of solar cells. The orientation of the c-Si part is (111). The amorphous structures were obtained by heating the samples to about 2,500 K and then cooling them again by using two different cooling strategies. For the first strategy, the samples were cooled steadily until the systems were entirely frozen. This was typically the case at a temperature of about 1,500 K. Afterwards, the structure was relaxed at 0 K. About 30 interface structures were generated using this strategy. Under the second strategy, about 500 samples were relaxed immediately to ground state. To prevent an intermixture of the c-Si and a-SiN_x:H part, the degree of freedom of the c-Si atoms in the accompanying MD simulations was adjusted: while the c-Si atoms in the middle of the c-Si part always stayed fixed, the c-Si atoms of the two outermost c-Si layers were allowed to either move in all directions or only parallel to the interfaces. This led to systems with either gradual or sharp interfaces. Figure 11.12 displays the averaged concentration of Si, N, and H of quenched samples with sharp interfaces (sharp, quenched), cooled ones with sharp interfaces (sharp,

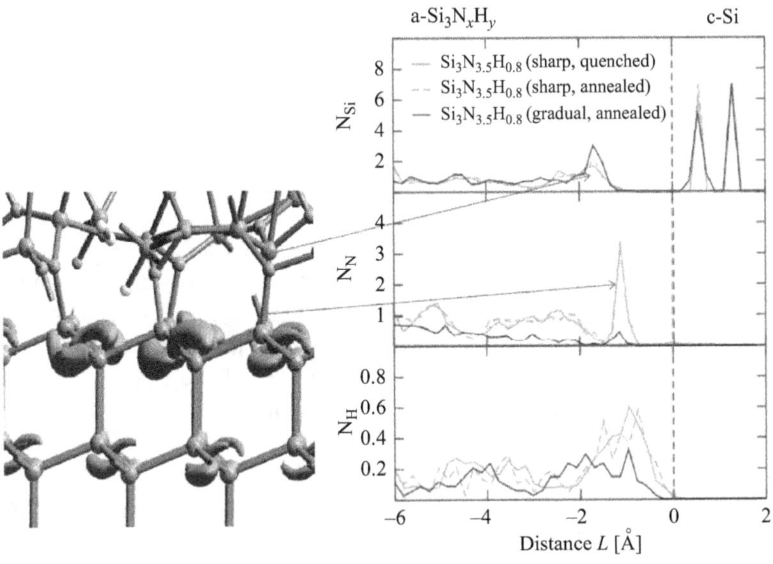

Figure 11.12 The structure and concentrations at the a-$Si_3N_{3.5}H_{0.8}$/c-Si interface [13] from ab initio theory studies

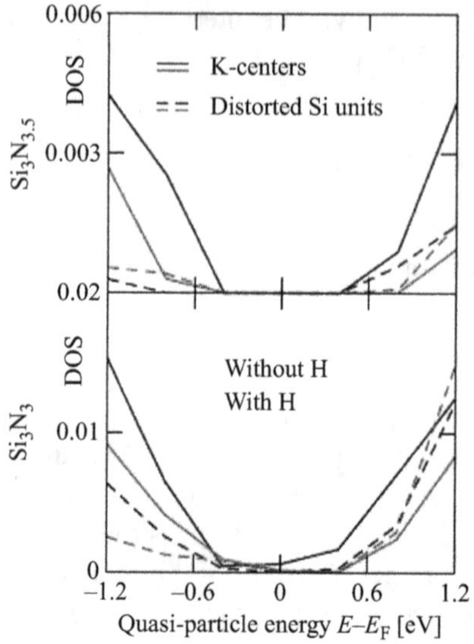

Figure 11.13 The density of states (DOS) of a-$Si_3N_{3.5}$ and a-Si_3N_3 close to Fermi level. The total DOS is related to K-centers and distorted Si units (full and broken lines). Additional hydrogen (gray lines) reduces the concentration of electronic defect states in the gap. Note that the y-axes are differently scaled [13]

annealed), and cooled ones with gradual interfaces (gradual, annealed). The figure also depicts the typical bonding structures of a c-Si/aSiN_x:H interface [13].

We can see that for a-$Si_3N_{3.5}H_{0.8}$ with constant material properties, a higher concentration of N and H is found at the interface. For a gradient interface however, the interface mostly consists of hydrogen. This could explain the dark band found in the HRTEM results presented in the previous section (Figure 11.6). Contrary to experimental measurements, no N is seen in the c-Si due to the described restrictions on the atom movement in the simulations. In addition, the dimensions of *ab initio* model structures are still small compared to real systems. While the models contain about 150 atoms and cover the length scale of a few Angstrom, real systems contain about 10^{23} atoms/cm^3 and require a description at nanometer-scale. In addition, initial growth of a material in plasma cannot be modeled using *ab initio* methods [13].

Figure 11.13 shows the density of states (DOS) of a-$Si_3N_{3.5}$ ($x = 1.17$) and a-Si_3N_3 ($x = 1$). With decreasing x, the band gap shrinks, and the DOS increases due to K-centers (solid lines) and distorted Si units (broken lines). The inclusion of hydrogen yields a further decrease of the density of states in the gap (black and gray lines). The states caused by N-centers were found outside the bandgap. We can also see that the concentration of N-centers does not influence D_{it}. We can interpret these observations in several ways. For example, it is well known that the optical absorption of a system increases when the band gap shrinks. Therefore, the absorption for a-Si_3N_3 must be higher than for a-$Si_3N_{3.5}$. We furthermore suggest that the number of defect states, which are mainly K-defects, are in direct relation to D_{it} and Q_f. D_{it} is further increased by distorted Si units [13].

11.5 Surface-passivating effect of a-SiN$_x$:H layers onto different solar cell architectures

A clear effect on solar cells can be expected, depending on surface doping, following the effect of the NH_3 plasma treatments prior to a-SiN$_x$:H deposition. As attested in Figure 11.8, also surfaces without extra treatment by NH_3 plasma show an increase in plasma loss energy in the cSi, indicating N insertion. Therefore, a significant amount of nitridation may occur during standard a-SiN$_x$:H deposition. Thus, both the NH_3 flow and temperature used in a-SiN$_x$:H deposition will determine the final Q_f and D_{it}. However, we did not observe a difference in cell performance for p-type cells with highly doped n-type emitter when varying the NH_3 plasma treatment temperature (300–500 °C) prior to front side a-SiN$_x$:H deposition [12]. For this cell design, these variations in surface passivation thus do not change the effective passivating quality of the front side for this cell design. When altering the cell design to n-type cells with a p-type emitter, the NH_3 plasma treatment temperature (300–500 °C) prior to front side a-SiN$_x$:H deposition realized a variation of 20 mV in cell voltage. No cell experiments regarding the rear side of p-type cells were done as an increase in Q_f and D_{it} was shown to have an unfavorable effect on cell performance by others [13,27].

For the above-described cell experiments, we fabricated solar cells and semi-fabrics using both n-type monocrystalline (mono) Cz Si and p-type multicrystalline (multi) Si and mono Cz Si wafers. Semi-fabricated Si wafers were textured with an alkaline or acidic procedure for respectively mono and multi Si wafers. The 180 μm thick and square p-type Cz mono Si had a base resistivity of 1.3 Ω.cm. The square p-type multi Si wafers were 210 μm thick with the same base resistivity. The semi-square n-type Cz mono Si were 180 μm thick and had a base resistivity of 2.5 Ω.cm. All wafers were sized 156 × 156 mm^2 [12]. The emitter (phosphorus for p-type base substrate material and boron for n-type) was made with a high-temperature diffusion step. On the rear of the n-type wafers, a back surface field layer was realized by phosphorus diffusion. After the succeeding glass removal, the wafers were cleaned and passivated with a-SiN$_x$:H. The wafers were subjected to a plasma NH_3 pretreatment to alter Q_f and D_{it} before a-SiN$_x$:H layer deposition on the front side [12]. On the rear side of the n-type wafer, a-SiN$_x$:H was also deposited, but without pretreatments. Metallization was applied by screen printing. For the p-type wafers, the rear side was fully covered by aluminum, while on the front side a silver H-pattern metallization scheme was applied. For the n-type wafers, both on the front and rear side, a silver H-pattern was realized. Contacting was established by a co-firing procedure [28,29]. The structures of the fabricated n-type and p-type cells are shown in Figure 11.14.

The effect on the surface passivation due to the presence of the positive fixed charge at the interface with a-SiN$_x$:H is different for the two types of emitters. For n-type emitters (on p-type cells), the positive fixed charge of the a-SiN$_x$:H will result

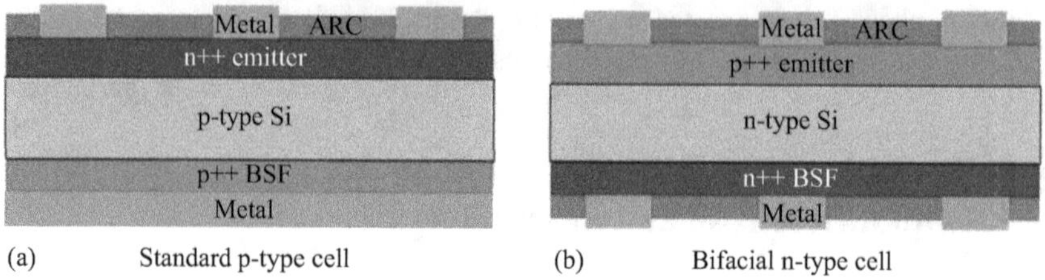

Figure 11.14 Structure of a solar cell [12]

in accumulation. For p-type emitters (on n-type cells), the positive fixed charge will result in depletion, and in case of low doping, in inversion which in final solar cells will induce inversion layer shunting as shown by Dauwe [27]. The quality of the emitter determines to a large extent the effect of the surface passivation. No effect of the surface passivation can be seen if recombination rates in the emitter are high. This is also indicated in the Shockley–Read–Hall formalism in which the recombination rate significantly depends on the doping density [30]. Thus, to examine the effect of Q_f on the cell output, the effect of doping must be included.

The cell voltage (V_{OC}) is directly related to surface passivation. In Figure 11.15(a), the V_{OC} is shown as a function of Q_f for n-type cells, both modeled with PC1D [12,31] and experimentally found. The trend in V_{OC} is due to the combined effect of increased D_{it} and increased depletion caused by higher Q_f as both increase simultaneously with increased nitridation. It is found that V_{OC} reduces up to 20 mV with increasing positive fixed charge as more depletion occurs at the interface. Both J_{SC} and FF were also affected as a drop of 2 mA/cm^2 and 1% were found, respectively. This strong effect, independent of the a-SiN$_x$:H deposited, shows the degree of influence of the surface passivation. PC1D simulations of the cell agreed with the cell results and showed that for negative fixed charges the recombination in the emitter is limiting the V_{OC} [12].

As shown in Figure 11.15(b), a similar observation can be made for p-type solar cells. The emitter of these cells had a high surface doping of 8×10^{20} atoms/cm^3 as determined with secondary ion mass spectroscopy. Implied V_{OC} as determined (using p-type mono semi-fabricates) showed no influence by Q_f due to different nitridations. PC1D modeling [31] was done to illustrate the effect of the maximal doping of the emitter and shown here as a function of the fixed charge. A p-type cell was modeled for which the doping of the phosphorus emitter (depth 300 nm) was simulated by using a stack of two Gaussian curves. The lifetime of the minority carriers in p-type wafer material was taken to be 200 μs. Due to the high doping level of phosphorus and the high recombination rate in the emitter, the dependence on surface passivation is negligible, such that even surface passivation schemes with negative fixed charges, like shown by Hoex *et al.* [32], can be used to coat the surface. When reducing the doping density however, the sensitivity for the surface passivation increases and with negative fixed charges the V_{OC} will reduce again due to depletion at the interface. For a low doping density of 3×10^{18} atoms/cm^3, even inversion will occur, which will further reduce the solar cell efficiency due to inversion layer shunting. This effect of shunting cannot be simulated with PC1D.

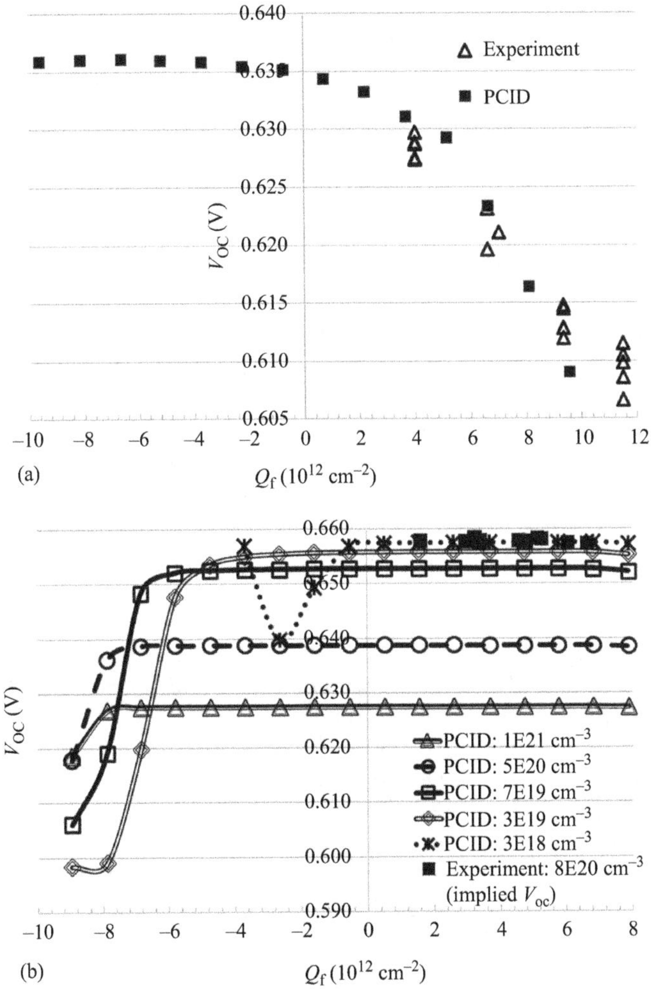

Figure 11.15 The V_{OC} of the n-type solar cells: (a) and p-type solar cell (b) as function of Q_f both modeled in PC1D and measured experimentally. In (b), the experimental data are given for a phosphorus surface concentration of 8×10^{20} atoms/cm^3 and the modeling results are shown for different phosphorus surface concentrations [12]

11.6 Conclusions

Over the recent decades, surface-passivating properties of PECVD a-SiN$_x$:H have been investigated extensively. These properties depend on Q_f and D_{it} and are in turn related to bulk and interface properties. In general, there is a positive correlation between Q_f and D_{it}, which have opposite effects on the surface-passivating properties: a higher Q_f improves the so-called field effect passivating quality whereas a lower D_{it} positively affects the chemical passivation. For an optimal effective surface passivation, an optimization is therefore necessary.

Q_f finds its origin in the variation in volume density of the so-called K-defects over the interface between Si and a-SiN_x:H. Higher Q_f and D_{it} are found to correspond to a higher and deeper insertion of N in the c-Si thereby increasing the concentration of these defects. This insertion is related to NH_3 flow and temperature during plasma treatment. This is shown when separate NH_3 plasma pretreatments prior to a-SiN_x:H were done to examine the effect. This technique for forming defect layers has been experimentally validated by EELS as well as high-resolution TEM. MD simulation and *ab initio* DFT studies demonstrated the formation of defects and higher DOS in the bandgap for the same layer composition. Introduction of hydrogen was simulated with DFT as well and showed a lower DOS in the bandgap resulting in a better surface passivation. The effect of Q_f at the interface on the passivating properties of a-SiN_x:H applied to completed solar cells was proven as well and confirmed by PC1D device simulations.

References

[1] I.G. Romijn, W.J. Soppe, H.C. Rieffe, A.R. Burgers, and A.W. Weeber. In *Proceedings of the 20th European Photovoltaic Solar Energy Conference and Exhibition*, page 1352, Barcelona, Spain, 2005.

[2] H.F.W. Dekkers, L. Carnel, G. Beaucarne, and W. Beyer. In *Proceedings of the 20th European Photovoltaic Solar Energy Conference and Exhibition*, page 721, Barcelona, Spain, 2005.

[3] H.F.W. Dekkers. *Study and Optimization of Dry Process Technologies for Thin Crystalline Silicon Solar Cell Manufacturing*. PhD thesis, Katholieke Universiteit Leuven, Leuven, 2008.

[4] J. Hong, W.M.M. Kessels, F.J.H. van Assche, *et al. Progress in Photovoltaics: Research and Applications*, 11:125, 2001.

[5] C. Boehme and G. Lucovsky. *Journal of Vacuum Science and Technology A*, 19:2622, 2001.

[6] M.W.P.E. Lamers, K.T. Butler, P.E. Vullum, J.H. Harding, and A.W. Weeber, *Physica Status Solidi A*, 210, 658–668, 2013.

[7] A.G. Aberle. *Crystalline Silicon Solar Cells – Advanced Surface Passivation and Analysis*. Centre for Photovoltaic Engineering, University of New South Wales, 2000.

[8] T, Lauinger, J. Moschner, A.G. Aberle, and R. Hezel, *Journal of Vacuum Science and Technology A*, 16:530, 1998.

[9] H. Mäckel and R. Lüdemann, *Journal of Applied Physics*, 92:2601, 2002.

[10] J.D. Moschner, J. Henze, J. Schmidt, and R. Hezel, *Progress in Photovoltaics: Research and Applications*, 12:21, 2004.

[11] S. Duttagupta, F. Lin, M. Wilson, M.B. Boreland, B. Hoex, and A.G. Aberle, *Progress in Photovoltaics: Research and Applications*, 22:641–647, 2014.

[12] M.W.P.E. Lamers, K.T. Butler, J.H. Harding, and A.W. Weeber, *Solar Energy Materials and Solar Cells*, 106:17–21, 2012.
[13] M.W.P.E. Lamers, L.E. Hintzsche, K.T. Butler, *et al.*, *Solar Energy Materials and Solar Cells*, 120:311–316, 2014.
[14] D.E. Aspnes. *American Journal of Physics*, 50:704, 1982.
[15] W.L. Warren, P.M. Lenahan, and J. Kanicki. *Journal of Applied Physics*, 70:2220, 1991.
[16] L.E. Hintzsche, C.M. Fang, M. Marsman, *et al. Physical Review B*, 88:155204, 2013.
[17] W. Soppe, H. Rieffe, and A. Weeber. *Progress in Photovoltaics: Research and Applications*, 13:551, 2005.
[18] M.W.P.E. Lamers, K.T. Butler, I.G. Romijn, J.H. Harding, and A.W. Weeber. In *Proceedings of the Materials Research Society Fall Meeting*, page 1423, Boston, USA, 2011.
[19] F. Demichelis, F. Giorgis, and C.F. Pirri. *Philosophical Magazine B*, 74:155, 1996.
[20] S. Wolf. *Journal of Applied Physics*, 97:063303, 2005.
[21] M. Dai, Y. Wang, J. Kwon, M.D. Halls, and Y.J. Chabal. *Nature Materials*, 8:825, 2009.
[22] T. Takakura, R. Imai, Y. Okamoto, and H. Taniguchi. *Japanese Journal of Applied Physics*, 49:046502, 2010.
[23] N. Tanaka, J. Yamasaki, K. Usuda, and N. Ikarashi. *Journal of Electron Microscopy*, 52:69, 2003.
[24] J. Park, S. Heo, J. Ghung, *et al. Ultramicroscopy*, 109:1183, 2009.
[25] N. Ikarashi, K. Watanabe, and Y. Miyamoto. *Journal of Applied Physics*, 90:2683, 2001.
[26] L.E. Hintzsche, C.M. Fang. T. Watts, *et al. Physical Review B*, 86:235204, 2012.
[27] S. Dauwe. *Low Temperature Surface Passivation of Crystalline Silicon and its Application to the Rear Side of Solar Cells*. PhD thesis, University of Hannover, Hannover, Germany, 2004.
[28] M.W.P.E. Lamers, I.G. Romijn, M. Gagliardo, M.N. van den Donker, C.J.J. Tool, and A.W. Weeber. In *Proceedings of the 23rd European Photovoltaic Solar Energy Conference and Exhibition*, page 1708, Valencia, Spain, 2008.
[29] A.W. Weeber, A.R. Burgers, N. Guillevin, *et al.* In *International Workshop on Science and Technology of Crystalline Silicon Solar Cells*, Taipei, Taiwan, 2010.
[30] W. Shockley and W.T. Read. *Physical Review*, 87:835, 1952.
[31] P.A. Basore. In *Proceedings of the 25th IEEE Photovoltaic Specialists Conference*, page 377, Washington, USA-DC, 1996.
[32] B. Hoex, M.C.M. van de Sanden, J. Schmidt, R. Brendel, and W.M.M. Kessels. *Physica Status Solidi RRL*, 6:4, 2012.

Chapter 12

Microwave PE CVD reactor and process for industrial high throughput fabrication of aluminum oxide layers for solar cell applications

Hermann Schlemm[1], *Hans-Peter Sperlich*[1], *Gunnar Köhler*[1], *Thomas Große*[1], *and Egbert Vetter*[1]

12.1 Introduction

First concepts for backside passivated solar cells from [1,2] did suggest silicon oxide layer for backside passivation. In 2003, the concept of laser-fired contacts [3] was published. Following this, a screen-printed full area backside contact on silicon oxide, capped with a silicon nitride layer, was contacted by local laser-contacting process.

Another concept developed some years later did use the same backside passivation layers followed by a local opening of the passivation layer stack by laser ablation and applying after this a screen printed backside aluminum contact, which forms locally an aluminum back surface field inside the contact holes [4].

At this time, PE CVD reactors working with linear microwave plasma [5] already did work well in industry. Between 100 and 200 equipment from Roth&Rau (now Meyer Burger (Germany)—MBG) had already been produced for the growing solar industry for front-side passivation with silicon nitride.

So, it was a simple step to extend the existing inline PE CVD reactors with microwave plasma sources from MBG (called "SiNA") by an additional PE CVD process chamber, equipped with the same plasma sources, to produce backside passivation layer stacks first of silicon oxide and later of aluminum oxide and silicon nitride as a capping layer.

In 2008, Kessels *et al.* [6–8] published a backside passivation process with aluminum oxide, deposited by atomic layer deposition (ALD). This uses field effect passivation, caused by fixed charges incorporated in aluminum oxide, which is a much more robust process and well suited for industry.

In 2010, Roth&Rau and Fh ISE published successful industrial PE CVD process [9] by using microwave plasma and enabling the combined PE CVD of aluminum oxide and capping silicon nitride in one cost effective machine called

[1]Meyer Burger (Germany) GmbH, Hohenstein-Ernstthal, Germany

Multiple Application inline Apparatus (MAiA). Using the experience from microwave PE CVD of silicon nitride, Roth&Rau could demonstrate good homogeneities and passivation properties of the backside passivation stack [10].

In 2013, [11] first successful industrial production of PERC cells with 19.7% efficiency, produced on this microwave plasma PE CVD reactors combined with laser ablation, was published.

Today, more than 250 PE CVD equipment's with microwave plasma work worldwide in industry for backside passivation with aluminum oxide, which corresponds to a market share of >85% for this plasma process.

This chapter gives an overview of industrial microwave PE CVD reactors starting at details of the used linear plasma sources and showing then, that a system of modular equipment can be constructed, covering a wide range of wafer throughput and customized backside passivation layer systems.

12.2 PE CVD reactor development leading to high throughput tools

12.2.1 Principle of plasma deposition with linear microwave plasma sources

Figure 12.1 shows the principle of plasma deposition with linear plasma sources.

A carrier containing e.g. $4 \times 6 = 24$ solar cells is moved with the carrier transport velocity v_c in the x-direction under an linear plasma source. The linear plasma source in Figure 12.1 is arranged over the moving carrier with the solar cells, so that the direction of the plasma deposition is downwards. The opposite case, that a linear plasma source makes the plasma deposition from button up onto solar cells lying in open pockets on a moving carrier also can be carried out (see

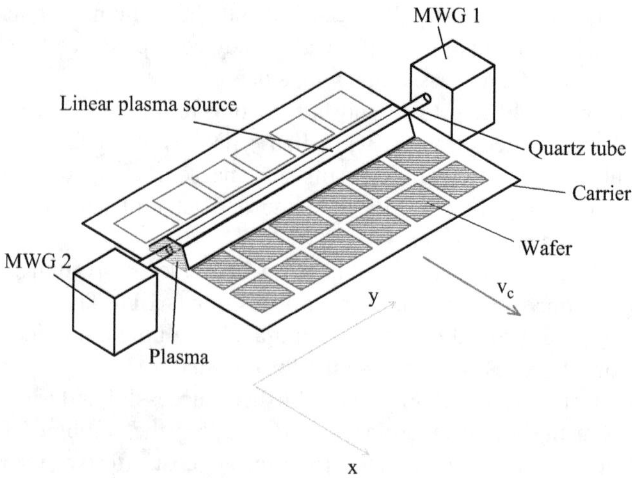

Figure 12.1 Principle of plasma deposition with linear plasma sources

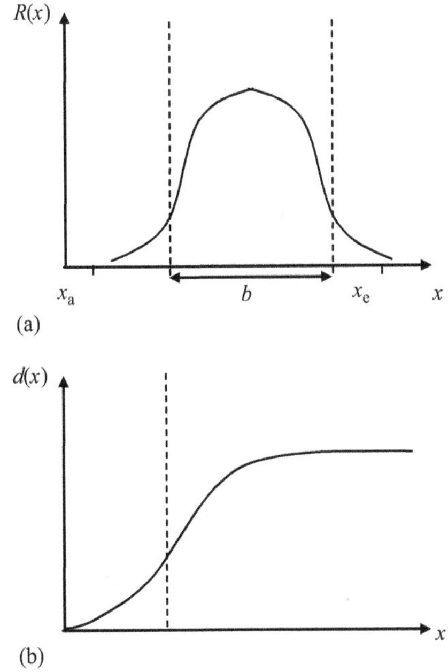

Figure 12.2 Principle of dynamic deposition at a linear plasma source: (a) Deposition rate profile R(x) in carrier movement direction (x-direction), (b) Accumulated layer thickness d(x) for a substrate, moved with constant velocity under the linear plasma source in the x-direction

Figure 12.15). After passing the linear plasma source the solar cells are coated with a PE CVD layer of a defined thickness.

The linear plasma source can be operated by microwave power [12] but also radio frequency plasmas can be applied [13]. In case of linear microwave plasma source, two microwave generators (typ. 2.45 GHz, mean microwave power up to 3 kW) generate a homogeneous microwave plasma in the y-direction outside the quartz tube, which connects the two microwave generators MWG 1 and MWG 2 in Figure 12.1. The generated microwave plasma is a traveling surface wave plasma [14] around the quartz tube. More details of this plasma are shown in Figures 12.3 and 12.4.

The principle behavior of a plasma deposition is shown in Figure 12.2. In the x-direction, the deposition rate $R(x)$ has a deposition profile, which mostly can be approximated by a Gaussian profile. The layer thickness accumulation $d(x)$ on a substrate moved in the x-direction under the plasma source has the typical increase like shown in Figure 12.2.

The accumulated layer thickness can be calculated from the deposition rate profile and the carrier velocity by

$$d(x_e) = \int_{x_a}^{x_e} \frac{R(x)}{v_c} dx = \frac{b_{\text{mean}} R_{\text{mean}}}{v_c} \tag{12.1}$$

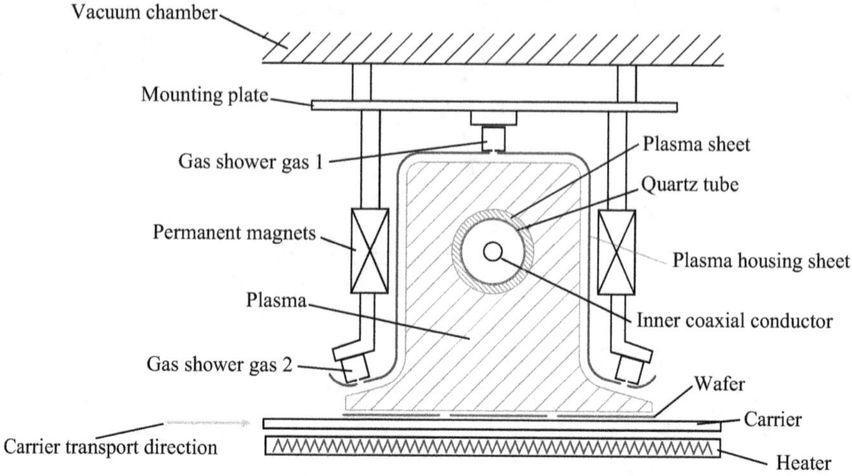

Figure 12.3 Principle cross section of the linear microwave plasma source MWPS 1, used for PE CVD of aluminum oxide

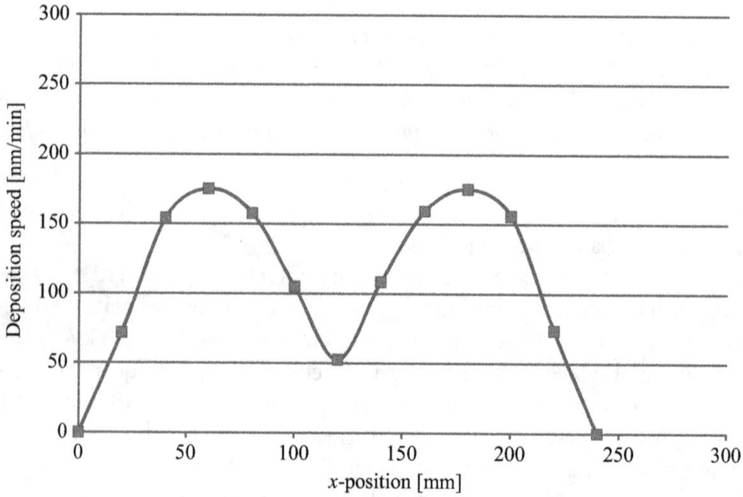

Figure 12.4 Deposition rate profile of microwave plasma source MWPS 1 for PE CVD of aluminum oxide (0.2 mbar, 1,700 W microwave power)

with

b_{mean}—mean deposition width of the deposition rate profile,
R_{mean}—mean deposition rate,
x_a—start coordinate plasma deposition,
x_e—end coordinate plasma deposition.

The wafer throughput of PE CVD by linear plasma sources can be simple scaled to higher values if several linear plasma sources are arranged like shown in Figure 12.1 over or under a moving carrier at a defined distance, so that each source (operated at the same process parameters) gives the same contribution to the growing layer. For n_{PS} plasma sources, the film thickness is then

$$d(x_e) = n_{PS}\int_{x_a}^{x_e} \frac{R(x)}{v_c} dx = \frac{b_{mean} R_{mean}}{v_c} * n_{PS} \quad (12.2)$$

With this type of linear microwave plasma source, several PE CVD- and plasma etch processes can be carried out like:

- PE CVD of silicon nitride
- PE CVD of aluminum oxide
- PE CVD of amorphous silicon
- Plasma etching of crystalline silicon.

12.2.2 Linear microwave plasma sources for PE CVD of aluminum oxide

Figure 12.3 shows details of the linear source (cross section in the x-direction) for the microwave plasma source MWPS 1 from MBG, which is used for PE CVD of aluminum oxide from precursors like trimethylaluminum (TMA) and dinitrogen-monoxide (N_2O). The stoichiometric overall equation of this process can be written like (12.3):

$$2Al(CH_3)_3 + 20N_2O \rightarrow Al_2O_3 + 2CO_2 + 4CO + 9H_2O + 20N_2 \quad (12.3)$$

N_2O is fed to the plasma by gas shower 1 and TMA comes from gas shower 2, so that it is added to the plasma very close to the substrates.

The microwave plasma generation is carried out outside the quartz tube in a circular plasma sheet of typically 5 mm thickness. Very high plasma densities up to 5×10^{12} cm^{-3} can be reached inside this sheet [14]. From this sheet, the plasma diffuses by ambipolar diffusion away from the quartz tube. This diffusion is controlled by a magnetic field of the permanent magnets, shown in Figure 12.3. The magnets stabilize the microwave plasma generation and increase the plasma density at substrate position.

The plasma source is complete mounted on a mounting plate for internal mounting at a vacuum chamber. It can be easy dismounted and changed if necessary. The distance from the gas shower 2 to the substrate position is typically 20 mm for this source. Under the moving carrier (at a distance smaller than 10 mm) a heater is arranged, to keep the carrier and the solar cells at the desired process temperature for the process, which is approx. 300 to 400 °C for aluminum oxide PE CVD.

Figure 12.4 shows the deposition rate profile for the microwave plasma source MWPS 1 (see Figure 12.3) for PE CVD of aluminum oxide in the x-direction.

Two deposition speed maxima near the gas showers for TMA are observed because of the high plasma decomposition of TMA at the microwave downstream

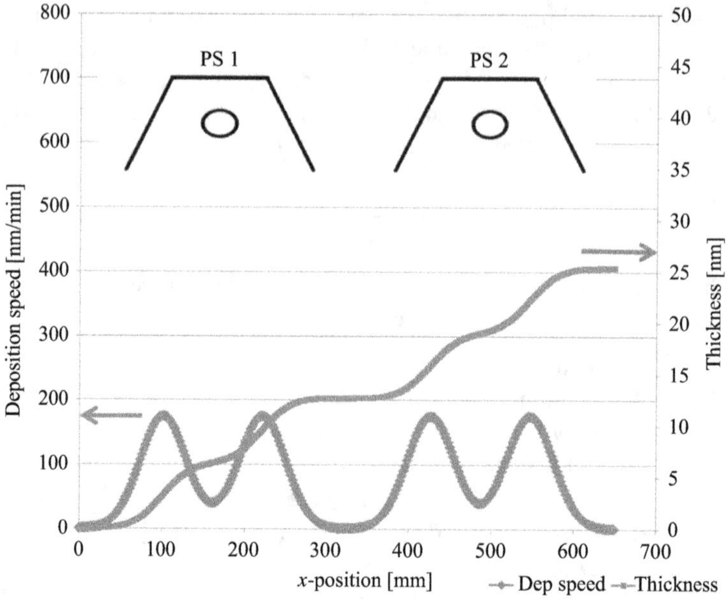

Figure 12.5 Simulation of dynamic deposition of 25 nm aluminum oxide by two linear plasma sources MWPS 1

plasma. After a penetration length inside the plasma of lower than 100 mm nearly all TMA is consumed.

A typical reactor for combined PE CVD of aluminum oxide and silicon nitride for backside passivation of solar cells has a wafer throughput of approx. 3.200 wafer/h corresponding to a carrier velocity of 200 to 250 cm/min (see Figure 12.14). A typical aluminum oxide layer for solar cell backside passivation should have a thickness of 15 to 25 nm. Figure 12.5 shows a simulation, where the accumulated layer thickness was calculated by (12.2) for two plasma sources and a carrier velocity of 200 cm/min, using the deposition profile from Figure 12.4.

Figure 12.5 shows both the deposition profiles and the plasma source positions (plasma source distance in the x-direction is 325 mm) and the accumulation of the aluminum oxide layer to a thickness of 25 nm.

The most interesting feature for industrial PE CVD processes is the homogeneity in the y-direction of an installed process. Figure 12.6 shows the measured values of thickness and refractive index for six wafers on a carrier like shown in Figure 12.1 for PE CVD of aluminum oxide. Homogeneities better 5% for thickness and better 2% for refractive index will be reached.

12.2.3 *Linear microwave plasma sources for PE CVD of silicon nitride*

The principle arrangement of a linear microwave plasma source for PE CVD of silicon nitride is the same like shown in Figure 12.1, but the cross section of the

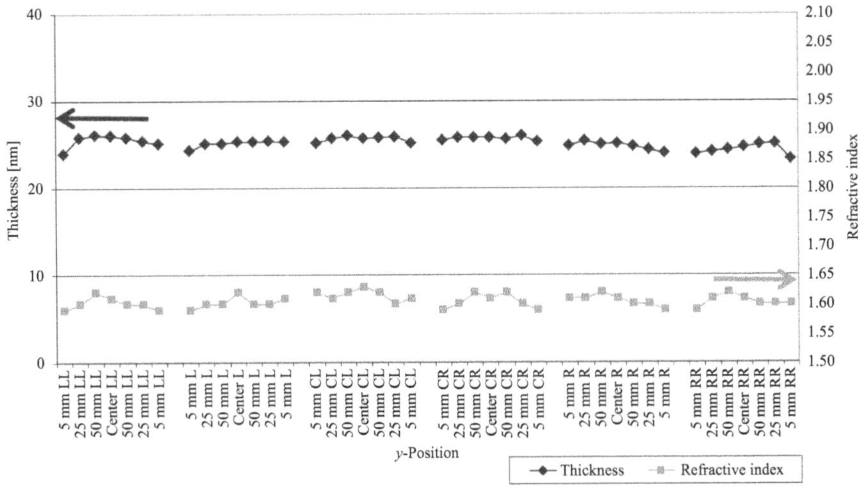

Figure 12.6 Homogeneity of deposition of aluminum oxide (layer thickness and refractive index) at 6 solar wafers in the y-direction

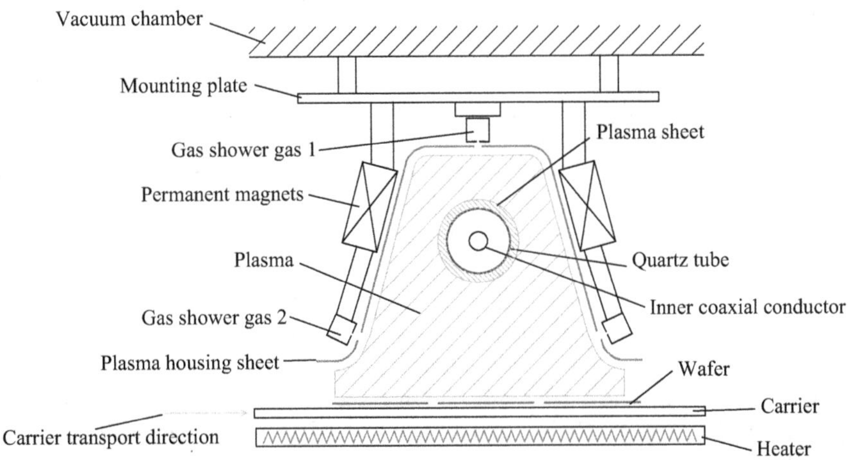

Figure 12.7 Principle cross section of the linear microwave plasma source MWPS 2 used for PE CVD of silicon nitride

source is different in details. Figure 12.7 shows details (cross section in the *x*-direction) for the microwave plasma source MWPS 2 [15], which is used for PE CVD of silicon nitride from precursors like silan (SiH_4) and ammonia (NH_3). The stoichiometric overall equation of this process can be written as (12.4):

$$3SiH_4 + 4NH_3 \rightarrow Si_3N_4 + 12H_2 \tag{12.4}$$

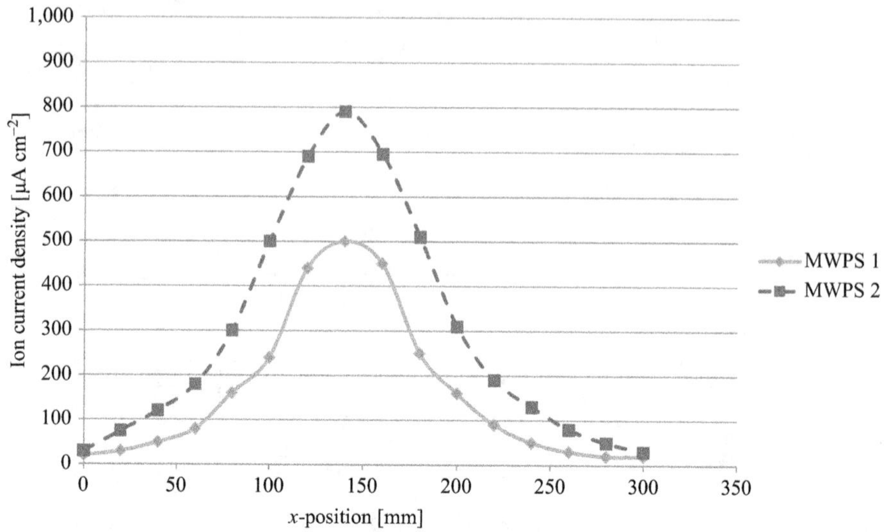

Figure 12.8 Ion current density profile at substrate level for both microwave plasma source versions in the x-direction (0.2 mbar nitrogen, 2,000 W microwave power)

NH_3 is fed to the plasma by gas shower 1 and SiH_4 comes from gas shower 2, so that it is added to the plasma at a typical distance to the substrates of 40 mm. The quartz tube is arranged at the center of the source like at version MWPS 1, but the magnetic field, produced by the permanent magnets is optimized to increase the plasma density on the substrates.

Figure 12.8 shows a measurement, where an array of 16 plasma probes at a distance of 20 mm had been arranged at substrate level under both microwave plasma sources in the x-direction. If a negative voltage of typically −25 V is applied to all probes, then the collected ion current represents the saturation ion current density, which is shown in Figure 12.8. Figure 12.8 shows that both plasma source versions generate a nearly Gaussian ion current density profile. The optimization of the magnetic field of MWPS 2 (see Figure 12.7) increases the maximum ion current density from 500 to 800 $\mu A/cm^2$ (at same microwave power).

Both microwave plasma sources have a nearly identical ion energy distribution at substrate level (Figure 12.9). To measure this, a retarding field analyzer (RFA) was mounted under the plasma sources at the x-position with the maximum ion current density (see Figure 12.8) and the ion energy spectrum had been measured. The mean ion energy is approx. 4 eV for both microwave plasma sources. A small tail of higher energetic ions up to 25 eV is also observed.

Figure 12.10 shows the deposition speed at the x-direction corresponding to the cross section of Figure 12.7 for the microwave plasma source MWPS 2. In maximum (under the quartz tube), a deposition rate of more than 400 nm/min is achieved, which is comparatively high [15].

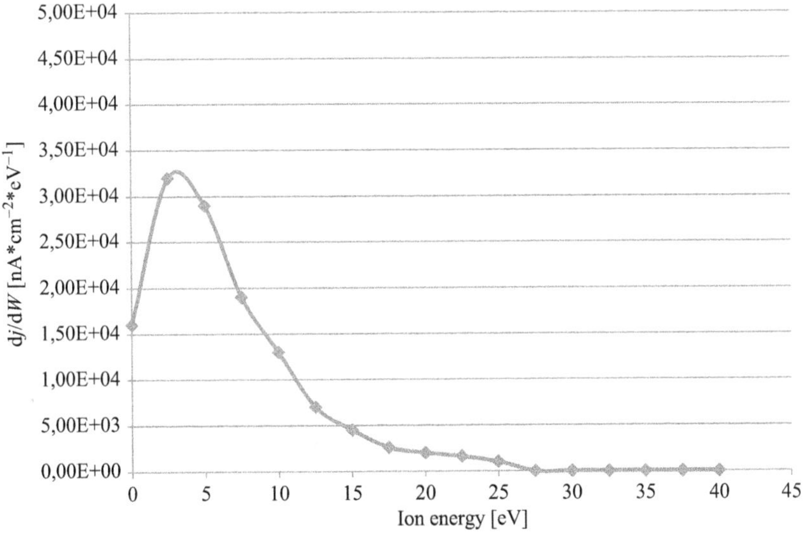

Figure 12.9 Typical ion energy distribution of both linear microwave plasma sources at substrate level (0.2 mbar nitrogen, 2,000 W microwave power)

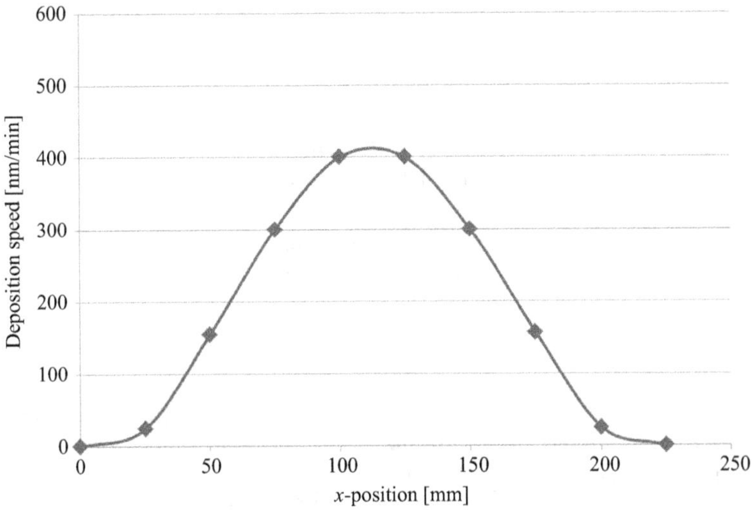

Figure 12.10 Deposition rate profile of microwave plasma source MWPS 2 for PE CVD of silicon nitride (0.2 mbar, 3,500 W)

Following the procedure of simulation of layer accumulation by moving a carrier under several linear plasma sources like shown in Section 12.2.2, and applying (12.2) for six plasma sources MWPS 2, with the deposition rate profile of Figure 12.10, a silicon nitride layer thickness of approx. 100 nm can be reached,

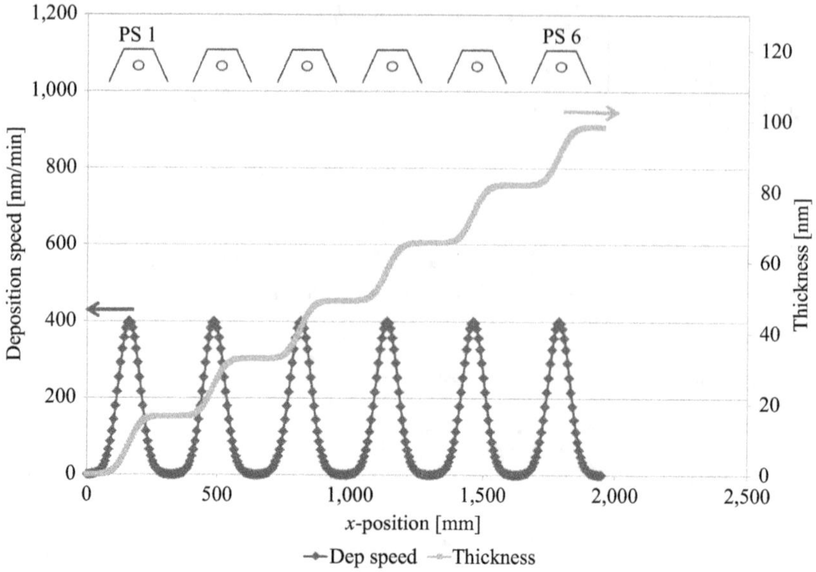

Figure 12.11 Simulation of dynamic deposition of 100 nm silicon nitride by six linear plasma sources MWPS 2 at same carrier speed like for deposition of aluminum oxide (200 cm/min)

like simulated at Figure 12.11. The distance of the plasma sources in the *x*-direction is 325 nm, which means that a process chamber can be kept lower than 2 m in the carrier-moving direction.

The most interesting feature for industrial PE CVD processes is the homogeneity in the *y*-direction of an installed process. Figure 12.12 shows the measured values of thickness and refractive index for six wafers on a carrier like shown in Figure 12.1 for PE CVD of silicon nitride. A homogeneity of better than 4% can be reached for layer thickness and better 2.5% for homogeneity of the refractive index.

12.2.4 Reactors for combined PE CVD of aluminum oxide and silicon nitride for backside passivation

A PERC-cell-line is established by only introducing two additional process steps into a standard aluminum BSF-cell-line. The first of this steps is the PE CVD deposition of a backside passivation layer system consisting of aluminum oxide covered with a capping silicon nitride layer. After the standard or improved PE CVD of silicon nitride for front side passivation and antireflection coating an additional laser step follows, which opens approximately one percent of the backside layer system by laser ablation of small holes for contacting the backside. Figure 12.13 shows this extension of the process flow of a standard aluminum BSF-line to a PERC-line. In total, three additional process steps are necessary. Because of the fact that PE CVD machines can deposit several layers in one PE CVD machine, the PERC process can be shortened, if PE CVD equipment

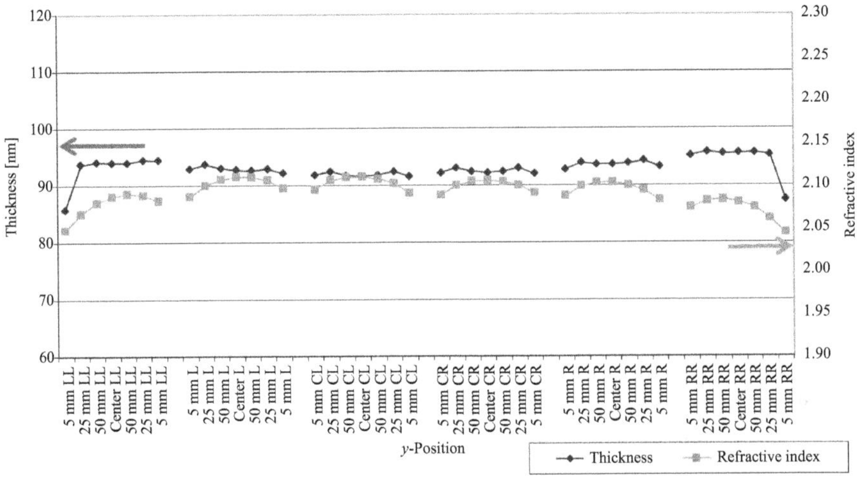

Figure 12.12 Homogeneity in the y-direction of silicon nitride deposition (layer thickness and refractive index) over six solar wafers

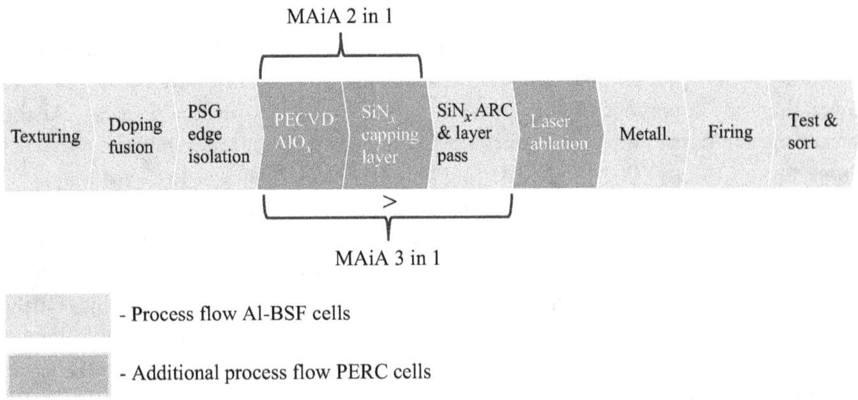

Figure 12.13 Overview process flow of PERC solar cells and usable PE CVD equipment for combined cost effective processes

like "MAiA 2 in 1" (combined PE CVD of aluminum oxide and silicon nitride at backside) or "MAiA 3 in 1" (combined PE CVD of aluminum oxide and silicon nitride at backside and frontside PE CVD of silicon nitride) are used. "MAiA" stands for "**M**ultiple **A**pplication **i**nline **A**pparatus" from Meyer Burger and means a modular system of PE CVD processes like shown in the following. So, by replacing at a BSF–cell-line, the old PE CVD machine for front-side silicon nitride by a MAiA 3 in 1 only one additional process step (laser ablation) is necessary.

Because of the excellent throughput scalability of PE CVD with linear microwave plasma sources (see Section 12.2.1), PE CVD reactors with small wafer throughputs of some hundred wafers per hour for R&D applications up to machines

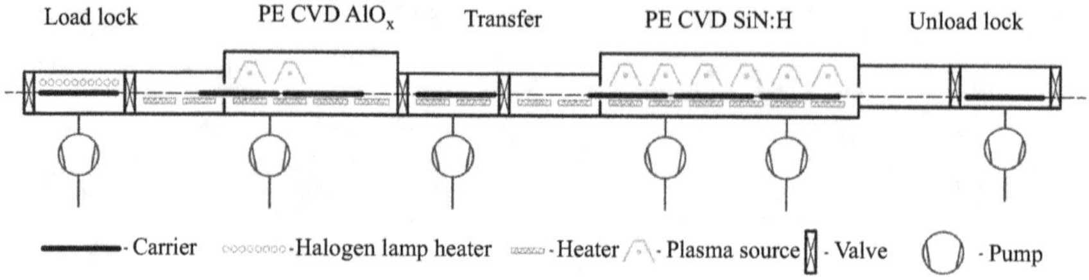

Figure 12.14 Schematically view on an inline equipment for combined PE CVD of aluminum oxide and silicon nitride for backside passivation

overcoming the level of 4.500 wafer per hour, which is the demand from the International Technical Roadmap for PV (ITRPV, [16]) for the years after 2017 can be realized.

Figure 12.14 shows in principle an inline PE CVD equipment ("MAiA 2 in 1") for backside PE CVD of approximately 25 nm aluminum oxide and 100 nm silicon nitride for a wafer throughput of approximately 3.200 wafer per hour.

The wafers are placed on a carrier with 24 wafers laying on the front side, so that the backside can be deposited from plasma downwards. The carriers move at the same level through the vacuum chambers, which are separated by vacuum valves.

The carriers enter at Figure 12.14 the equipment from left into the load lock chamber, which pumps fast down from atmospheric pressure down to a pressure lower than some percent of the pressure of the following plasma process. At the same time the carrier and the wafers are heated from room temperature to process temperature (300 to 400 °C) by a strong halogen lamp heater at the load lock.

The carrier movement at the load lock (and in same manner at the transfer lock and the unload lock) is discontinuously, while at the PE CVD process chambers under the linear plasma sources an accurate continuous carrier speed is necessary to achieve homogeneous layer deposition (see Section 12.2.1). To ensure this, buffer zones are arranged before and after the plasma source zones, where the carrier process speed starts or ends. Moreover, the carrier process speed has to be the same in both process chambers.

The PE CVD process for 25 nm aluminum oxide can be realized by two linear microwave plasma sources (see Section 12.2.2), with a deposition length in carrier movement direction of 700 mm. The residual length of this process chamber works as buffer zone before the carriers will be sluiced through the transfer lock and additional buffer zones to the process chamber for silicon nitride, which is equipped with six linear microwave plasma sources for depositing of approx. 100 nm silicon nitride like simulated at Figure 12.11.

By using the shown transfer lock to separate the oxygen-containing process gases from the aluminum oxide PE CVD (see also (12.3)), from the process gases of the silicon nitride PE CVD process (see (12.4)), a process gas cross contamination of smaller than 0.1% is reached. Moreover, both PE CVD processes can operate at different process pressures, enabling optimum processes for both layers without any restrictions.

All vacuum chambers up to here are heated to process temperatures of 300 to 400 °C. A moderate change in process temperature between aluminum oxide process (typical at 325 °C) and silicon nitride process (typical at 375 °C) can be realized. After passing the last linear microwave plasma source at the silicon nitride process, an unheated buffer zone and an unheated unload lock follows to cool carrier and wafers down to temperatures below 150 °C, when they leave the equipment at the last gate.

Depending on the type of backside printing process, PERC cells with 100 to 150 nm silicon nitride capping layer are produced today. Following the calculations from Sections 12.2.1 and 12.2.3, approx. 10 linear microwave plasma sources are required for a 150 nm capping layer of silicon nitride. This is simple realized at an inline equipment by only introducing an additional PE CVD process chamber with 4 plasma sources between the second PE CVD chamber and the unload buffer zone (see Figure 12.14) realizing a total process with 10 microwave plasma sources for silicon nitride PE CVD. All the sluices with their tact time etc. remain unchanged, so that the same wafer throughput is realized for 150 nm silicon nitride.

12.2.5 Reactors for front- and back-side PE CVD

Figure 12.15 shows the principle of a PE CVD equipment, where a maximum of plasma processes are integrated in one machine (concept of "MAiA 3 in 1", see Figure 12.13).

The first part of the inline equipment is identical to "MAiA 2 in 1", shown in Figure 12.14. After PE CVD of silicon nitride for the capping layer of the cell backside, an additional process chamber with typical 6 microwave plasma sources is introduced. After that the unload lock combined with the buffer zones realize the output of the carriers to atmosphere.

Because of the capability of flexible inline mounting of the microwave plasma sources (see Sections 12.2.1 to 12.2.3) a bottom-up deposition like shown in Figure 12.15 simple can be realized. The wafers must lie now in a complete open pocket at the carrier, only haltered by some very small hooks. That means that the solar cell backside is deposited top-down and the front side is deposited bottom-up. To avoid cross depositions from front side to back side, the gap between open pocket and wafer should be as small as possible and not bigger than 0.5 mm. That means that wafers processed with "MAiA 3 in 1" should have a wafer dimension tolerance not bigger than 0.5 mm in both dimensions.

The process chambers for back- and front-side silicon nitride PE CVD operate at the same process pressure because they are not separated by a transfer lock, which would increase price and footprint of the equipment too much. But that is in this case not a disadvantage. Back side and front side silicon nitride PE CVD operate both at the optimum process pressure of approximately 0.25 mbar. Different front- and back-side silicon nitride layer properties can be simple realized by optimized gas flows for each process chamber, because each chamber has its own vacuum pump.

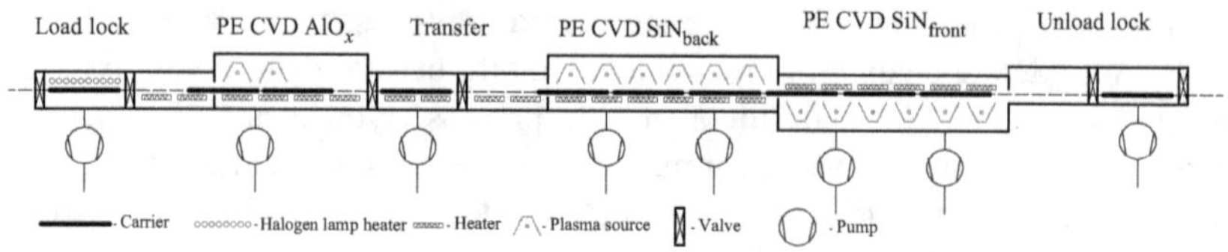

Figure 12.15 Schematically view on an inline equipment for PE CVD of aluminum oxide and silicon nitride for backside passivation and combined front side passivation with silicon nitride

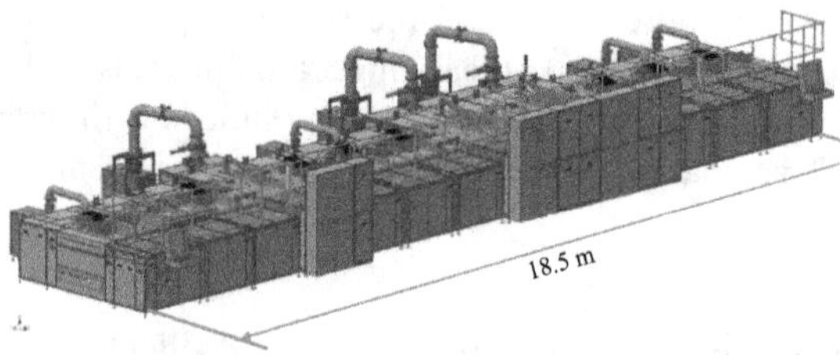

Figure 12.16 PE CVD equipment "MAiA 2 in 1" concept study for a wafer throughput of 4.800 wafer per hour

12.2.6 Future high throughput reactors—limits and advantages

In the near future (year 2020 or earlier), PE CVD equipment for PERC lines should have a wafer throughput in the range between 4.000 and 5.000 wafer/h (see [16]). The MAiA—concept, shown in Sections 12.2.4 and 12.2.5 with a wafer throughput of 3.500 wafers per hour simple, can be extended into this range by:

- A bigger carrier with 36 wafers instead of 24 (+50% wafers per carrier),
- Additional linear microwave plasma sources at the process chambers in accordance to (12.2), combined with a higher carrier speed at the PE CVD process.

Figure 12.16 shows a "MAiA 2 in 1" equipment for 4.800 wafer per hour equipped with three linear microwave plasma sources for PE CVD of aluminum oxide and 12 linear microwave plasma sources for approx. 100 nm capping layer of silicon nitride.

Also PE CVD reactors following the "MAiA 3 in 1" concept can be constructed for this wafer throughput range. It can be shown, that the growth of equipment and production cost from a 3.200 wafer per hour PE CVD reactor to a 4.800 wafer per hour reactor is smaller than the wafer throughput gain (+50%). By this way, a further cost reduction is possible (see Figure 12.19). The length of equipment like the one shown in Figure 12.16 is 18.5 m.

12.3 Experimental results

Some characteristic results of the aluminum oxide films, deposited like described at Section 12.2.2, should now be presented. Figure 12.17 shows the infrared spectrum of a 25 nm thick aluminum oxide layer at the initial state after deposition and after an annealing step (400 °C, 5 min) and after firing ($T_{peak} = 850$ °C) of the corresponding PERC cell. The hydrogen content of the layer after PE CVD deposition is in the range between 0.5% and 3%, which is understandable following the chemistry of (12.3). Figure 12.17 shows, that with increasing temperature O–H groups and water will be baked out and the corresponding stretching (ν O–H at 3.500 to 3.200 cm^{-1}) and scissoring modes (δ H_2O at 1.600 to 1.400 cm^{-1}) disappear at the FTIR spectrum.

The fix charge density after firing, caused by residual O–H groups is still at a sufficient level of 2 to 5 × 10^{12} cm^{-2} [10], which is enough for excellent surface passivation. At the float zone, wafers carrier lifetimes at a level of some milliseconds can be reached [10].

While Figure 12.17 gives an impression of the structural changes at the PE CVD deposited aluminum oxide layers by thermal treatment, Figure 12.18 shows the electrical improvement of 15 fabricated PERC cells with increasing thermal treatment (after PE CVD, after annealing with 400 °C, 5 min and after firing ($T_{peak} = 850$ °C)).

The initial state after PE CVD with charge carrier lifetimes of about 100 μs is already improved by the annealing to 200–300 μs. After firing, all cells reach a stable lifetime of about 300 μs, corresponding to an implied V_{oc} of 700 mV. On Czochralski Wafers mean PERC-cell efficiencies of 20.4% correspond to the data of Figure 12.17 and Figure 12.18.

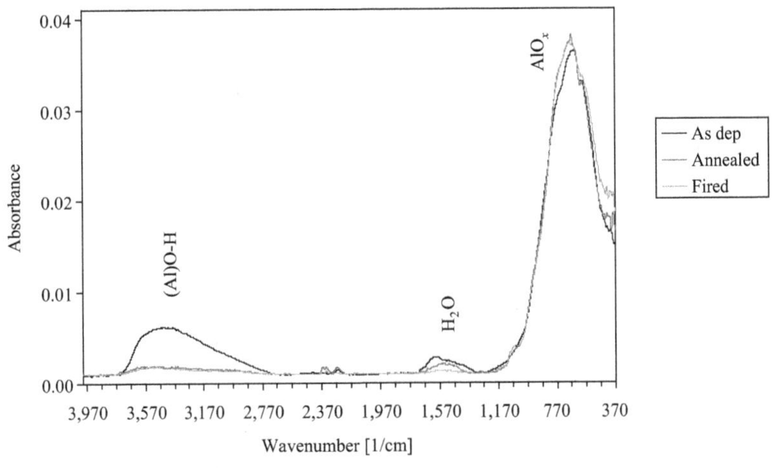

Figure 12.17 Fourier transform infrared spectra of 25 nm aluminum oxide layer after PE CVD deposition, after annealing (400 °C, 5 min) and after firing process ($T_{peak} = 850$ °C)

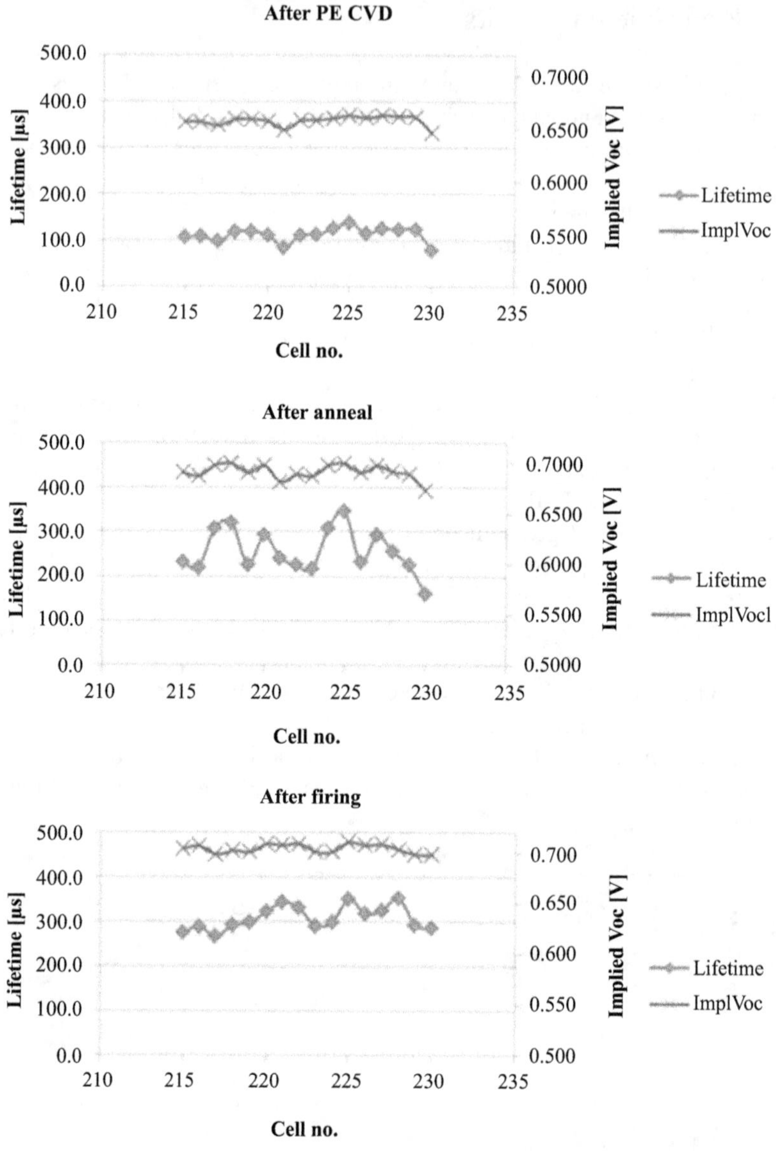

Figure 12.18 Lifetime and Implied V_{oc} of 15 PERC – cells after PE CVD, after post annealing (400 °C, 5 min) and after Firing ($T_{peak} = 850$ °C)

A significant advantage of PE CVD reactors with combined aluminum oxide and silicon nitride deposition (see Figures 12.14 and 12.15) is, that a post anneal process (at typically 375 °C over 4 to 8 min) is already carried out after aluminum oxide PE CVD without any vacuum brake while depositing silicon nitride. Not additional annealing after PE CVD is required, and excellent cell data can be

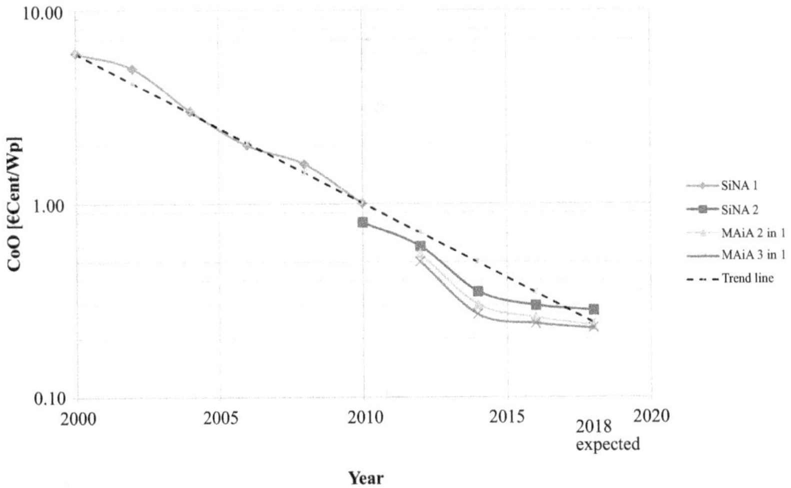

Figure 12.19 Overview on mean Cost of Ownership of one PE CVD process at microwave PE CVD reactors over last 16 years (calculated for producer at Asia)

reached after firing. Top efficiencies of best monocrystalline PERC cells fabricated with low cost industrial process using microwave PE CVD [17] are at the end of year 2016 between 21.0% and 22.5%.

The mean cost of ownership, calculated for one PE CVD process with linear microwave plasma sources in the equipment shown in Figures 12.14, 12.15, and 12.16, did dramatically decrease in the last 16 years (see Figure 12.19).

Starting at several €Cent/Wp in the PV beginning (year 2000 to 2005 at SiNA 1 equipment's from former Roth&Rau), today process costs between 0.25 and 0.40 €Cent/Wp are reached. PE CVD processes at equipment's with two or three integrated processes (reactors like Figures 12.15 and 12.16) moreover have additional smaller production cost than a single process (like SiNA 2). Very high throughput reactors (like shown in Figure 12.16 for a wafer throughput of 4.800 wafer per hour) will further decrease the production costs, following the trend line in Figure 12.19.

12.4 Summary

PE CVD reactors with linear microwave plasma sources have a very successful history over the last 16 years. More than 750 inline equipments from MBG work worldwide for front side and backside PE CVD. Some thousands of linear microwave plasma sources of two optimized types had been produced therefore. With combination of this linear plasma sources (see Section 12.2.1), a wide range of wafer throughput up to 5.000 wafer/h can be realized.

Backside PE CVD of aluminum oxide and capping silicon nitride starts in PERC production in year 2012. By the end of 2016, more than 20 GWp yearly production capacity of PERC cells is installed in the world. Microwave plasma reactors have a market share of more than 85% for backside passivation layer deposition.

References

[1] Green M.A., *Silicon Solar Cells Advanced Principles & Practice*, Publ. Centre for Photovoltaic Devices and Systems, Sydney, 1995, p. 218–219.

[2] Blakers A.W., Wang A., Milne A.M. *et al.*, 24% efficient silicon solar cells, *Applied Physical Letters*, **55**, 1363, (1989).

[3] Schneiderlöchner, E., Preu, R., Lüdemann, R., Glunz, S. W., Laser-fired rear contacts for crystalline silicon solar cells, *Progress in Photovoltaics*, **10**, no. 1, p. 29–34, (2002).

[4] Glunz, S. W., Preu, R., Schaefer, S., *et al.*, New simplified method for patterning the rear contact of RP-PERC high efficiency solar cells, *Proceedings of the 28 Photovoltaics Specialists Conference*, Anchorage, 19–22 September 2002.

[5] Schlemm H., Mai A., Roth S., Roth D., Baumgartner K.-M., and Muegge H., Industrial large scale silicon nitride deposition on photovoltaic cells with linear microwave plasma source, *Surface & Coatings Technology* **174–175**, 208–11, (2003).

[6] Schmidt J., Merkle A., Brendel R., Hoex B., van den Sanden M.C.M., and Kessels W.M.M., Surface passivation of high-efficiency silicon solar cells by atomic-layer-deposited Al_2O_3, *Progress in Photovoltaics: Research and Applications*, **16**, 461–466, (2008).

[7] Hoex B., Gielis J.J.H., van den Sanden M.C.M., and Kessels W.M.M., On the c-Si surface passivation mechanism by the negative-charge dielectric Al_2O_3, *Journal of Applied Physics*, **104**, 113703, (2008).

[8] Hoex B., Schmidt J., Pohl P., van den Sanden M.C.M., and Kessels W.M.M., Silicon surface passivation by atomic layer deposited Al_2O_3, *Journal of Applied Physics*, **104**, 044903, (2008).

[9] Rentsch J., Sperlich H.-P., Kohn N., *et al.*, *25th European Photovoltaic Solar Energy Conference and Exhibition, Proceedings*, Valencia, Spain, 1715.

[10] Sperlich H.-P., Decker D., Saint-Cast P., Erben E., and Peters L., High productive solar cell passivation on Roth&Rau MAiA MW-PE CVD inline machine – a comparison of Al_2O_3, SiO_2, and SiN_x-H process conditions and performance, *25th European Photovoltaic Solar Energy Conference and Exhibition, Proceedings*, Valencia, Spain, 1352.

[11] Tjahjono B., Yang M.J., Wu V., *et al.*, Optimizing CELCO cell technology in one year of mass production, *28th European Photovoltaic Solar Energy Conference and Exhibition, Proceedings*, Valencia, Spain, 775.

[12] Schlemm H., Fritzsche M., and Roth D., Linear radio frequency plasma sources for large scale industrial applications in photovoltaics, *Surface & Coatings Technology*, **200**, no. 1–4, p. 958–61, (2005).

[13] Decker D., Schlemm H., Mai J., Wolf P., and Mehlich, H., Inline a-Si:H PECVD using linear microwave and new linear RF plasma Sources, *27th European Photovoltaic Solar Energy Conference and Exhibition, Proceedings*, 2012, p. 1624–9 of lxv+1388 pp.

[14] Petasch W., Raeuchle E., Muegge H., and Muegge K., Duo-plasma line – a linearly extended homogeneous low pressure plasma source, *Surface and Coatings Technology*, **93**, no. 1, 112–18, (1997).

[15] Mehlich H., Wolf P., Kohler G., *et al.*, PECVD processes with increased deposition speed by linear microwave plasma sources, *Proceedings of the 25th European Photovoltaic Solar Energy Conference and Exhibition and the 5th World Conference on Photovoltaic Energy Conversion*, 2010, p. 2659–2662,

[16] http://www.itrpv.net/Reports/Downloads/2016/

[17] http://taiyangnews.info/technology/new-trina-perc-cell-record/

Chapter 13

Spatial atomic layer deposition of Al_2O_3: Levitrack, a one-pass ALD system with throughputs exceeding 6,000 wafers/h

Ernst Granneman[1] and Wil van Velzen[1]

13.1 Introduction

With the introduction of PERC solar cells [1,2], Al_2O_3 films established a solid position as a passivation film for B-doped silicon. Various types of deposition technologies were developed: plasma-enhanced CVD (PECVD). Atomic layer deposition (ALD), atmospheric pressure CVD (APCVD), and sputtering (PVD). For PERC cells, PECVD gained the largest market share [3], followed by various types of ALD processes [4]. It is expected that with the gradual penetration of n-type cells, the penetration of ALD films will gradually increase. An important reason for this is that for n-type cells thinner ALD Al_2O_3 films, in the range of 4–6 nm, are preferred above the standard PECVD films with thicknesses in the range of 10–20 nm.

Basically two types of ALD Al_2O_3 systems are commercially available: temporal and spatial ALD. In both cases, the wafers are exposed to two precursors, TMA (tri-methyl-aluminum) and water. In temporal ALD, the wafers are located at a fixed position while being exposed to precursor pulses separated in time. In contrast to that, in spatial ALD tools the precursor gas flows are continuous and separated in space while the wafer is transported from one precursor location to the next. In both techniques, an inert gas, usually nitrogen, is used to prevent mixing of the two precursors in time and space, respectively. In case of temporal ALD, large throughputs can only be realized in batch-type tools with extensive automation. An overview of various types of spatial ALD (SALD) systems is given in [5].

A number of variations on spatial ALD tools exist, see Figure 13.1. In the most straightforward way, the wafers are placed on a belt. During transport, subsequent zones of TMA and H_2O are passed. The TMA and H_2O zones are separated by an inert gas zone that is designed such that, ideally, TMA and H_2O do not mix. It

[1]Levitech BV, R&D Department, The Netherlands

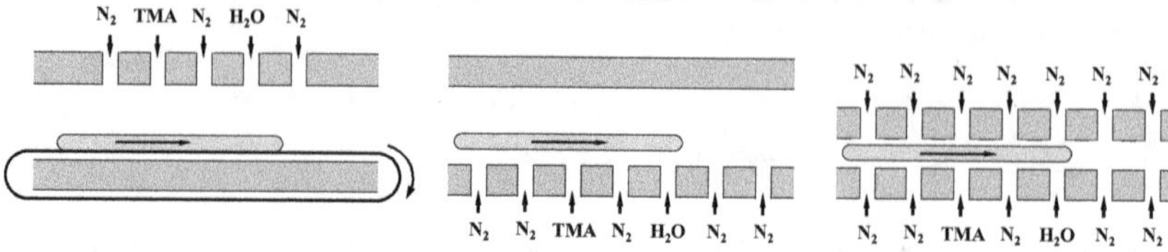

Figure 13.1 Various layouts of spatial ALD tools: (a) wafer on belt; (b) wafer floating on single side gas bearing; and (c) wafer floating in double-side gas bearing

should be noted that mixing of TMA and H_2O results in large amount of Al_2O_3 powder that will clog the system rapidly. In principle, when the gas separation is ideal, no deposition on the wall of the system takes place.

However, as the belt moves underneath the gas injection points, deposition will also occur on the belt. Further, when the distance between the gas injection plane and the belt is large, separation of the precursors is not trivial, and large N_2 flows are required. Deposition on a belt can be avoided when the wafers are transported on a gas bearing, see Figure 13.1(b). Also in this case, no deposition on the walls takes place. The disadvantage of a single-side gas support system is that the gas bearing is not very stiff, the wafer tends to shift sideways easily, and also the vertical position is not well controlled. This can be avoided by supporting the wafer in a double-side gas bearing. When the gas flows are sufficiently large, such bearings are very stiff. The forces in vertical direction are generally considerable, much larger that, for example, the gravitational force pulling the wafer down. Now wafers can be transported in channels with very small wafer-to-wall distances, in the range of 20–300 μm. With such small gaps between the wafer and both wall, the flows are basically laminar, and relatively small N_2 gas flows suffice to separate the TMA and H_2O precursors.

The gas bearing shown in Figure 13.1(c) can be applied in two different ways. When the wafer passes the TMA and H_2O gas injection points once, one (sub-) monolayer of Al_2O_3 is deposited. Additional monolayers can be deposited by adding additional TMA/H_2O injection points in transport direction, or by moving the wafer back and forth passing the single (or multiple) set(s) of gas injection points as many times as is required to build up a desired film thickness. In case of ALD, Al_2O_3 0.12 nm is added to the film for each ALD cycle (at 200 °C). This implies that the wafer needs to move 21 times in and out to build up a film of 5 nm (or proportionally fewer times when more than one ALD precursor arrangement are placed in series). When high throughputs are needed, high acceleration/deceleration forces are required to make the wafer go back and forth several times per second. In order to get a sufficiently high throughput, many ALD modules are processing in parallel [6].

Alternatively, one can place a large number of ALD gas injection sets in series such that the wafer passes each ALD cell only once. In this case the wafers move forward at a constant, low speed. No large acceleration/deceleration

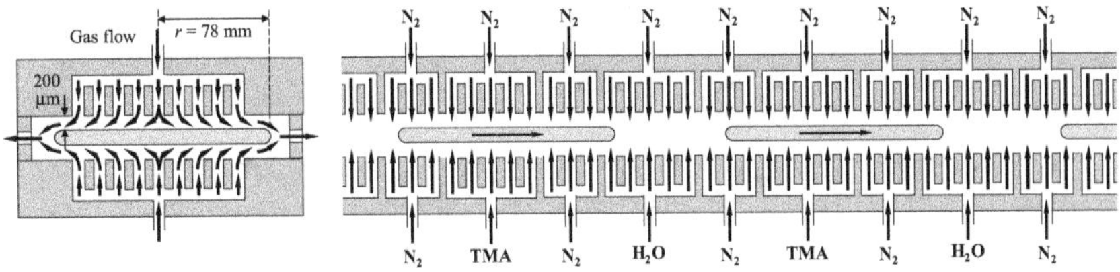

Figure 13.2 Layout of the Levitrack SALD system. The wafers are injected into the system continuously, and travel with a speed of 200–350 mm/s. One (sub-) monolayer of Al_2O_3 is deposited each time a pair of TMA/H_2O gas injection lines is passed

is required, and wafers can be introduced in the tool close to each other. The forces on the wafer are small and constant, resulting in very little, if any, wafer breakage. Levitech decided to develop a system based on the latter SALD principle, see Figure 13.2. The high-level design specs are summarized in Section 13.2.

The result of the project, the so-called Levitrack system, is described in Section 13.3. Details are given on wafer transport mechanism, lateral stability of wafer transport and wafer heating and cooling. In Section 13.4, the fundamentals of the underlying physics are derived; it is shown that wafer velocity can be accurately described as a function of track angle, gap between wafer and walls, wafer thickness and system temperature. In Section 13.5, a comparison is made with experimental results. Section 13.6 deals with the lateral stability during wafer transport. In Section 13.7, a recent addition to the tool is described, the so-called moving side wall, a feature that allows a considerable increase of throughput. In Section 13.8, examples are given on the deposition uniformity and the efficiency realized in state-of-the-art PERC cells. It should be noted that the primary objective of the paper is system design philosophy and not so much solar cell results. Details on the passivation of solar cells can be found in other papers given in this book. Finally, in Section 13.9 conclusions are given.

13.2 High-level system specs

This paper focuses on the design philosophy underlying the Levitrack. Prior to developing the system, a number of high-level design objectives are defined as follows:

- Wafer transport based on gas bearings confining the wafer in vertical as well as in the lateral direction
- No or limited deposition on the walls of the system
- Zero wafer breakage
- Uniform deposition; for passivation, this implies a uniformity <5%
- Film thickness tunable between 0 and 10 nm (during the development process, it turned out that 6 nm suffices)

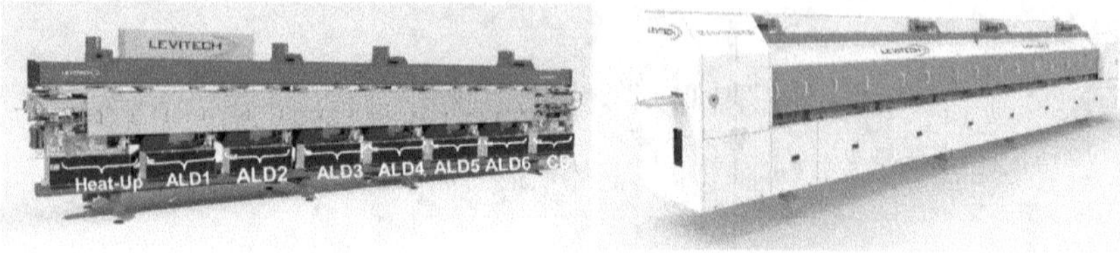

Figure 13.3 Levitrack ALD system. Left: System without automation and skins, heat-up, ALD1 to ALD6, and cool-down segments are shown. Right: system with outside skins. On the entrance side, one can see the wafer feed-in belt feature onto which the loading automation is connected (not shown). All segments have a length of 1 m

- Deposition on one side of the wafer only
- Throughput exceeding 6,000 wafers/h
- Interval between maintenance activities >500,000 wafers
- Competitive, low cost-of-ownership.

13.3 Levitrack system

There are basically two ways to drive the wafers forward. The first one is to apply a pressure gradient across the entire length of the tool. The disadvantage of this method is that it will result in flows in transport direction (potentially) resulting in mixing of precursors. The second method is to place the system at a small angle (1–1.5°, see Section 13.4). The main advantages of this "gravity drive" are simplicity and the fact that gravity works on each wafer individually, rather than on the full train of wafers present in the track. The wafer velocity is linearly proportional with the track angle. As the rates at which the precursors react with the wafer surface are high [4], the resulting throughput is only limited by wafer transport irregularities (see Section 13.6), and the capabilities of the loading and unloading automation to keep up. The result is a high throughput, combined with zero breakage. As very limited deposition takes place on the wall of the system, the precursor consumption efficiency is high, of the order of 25%–30%.

The length of an ALD cell in the Levitrack is currently 12 cm. This implies that 1 nm of Al_2O_3 is deposited for each meter system length. The disadvantage of this approach therefore is that the final film thickness is linearly dependent on the length of the system. Fortunately, the range of required ALD Al_2O_3 film thicknesses in the PV industry is rather small, typically 2–6 nm (2–4 nm for n-type and 5–6 nm for p-type PERC). This implies that a maximum overall ALD process length of 6 m suffices for all currently known applications.

A picture of the system is shown in Figure 13.3.

The six ALD segments, kept continuously at process temperature, typically 200 °C, are preceded by a heat-up segment in which the wafers are gradually heated up to process temperature (see Section 13.4). At the end of the tool, the wafers are cooled

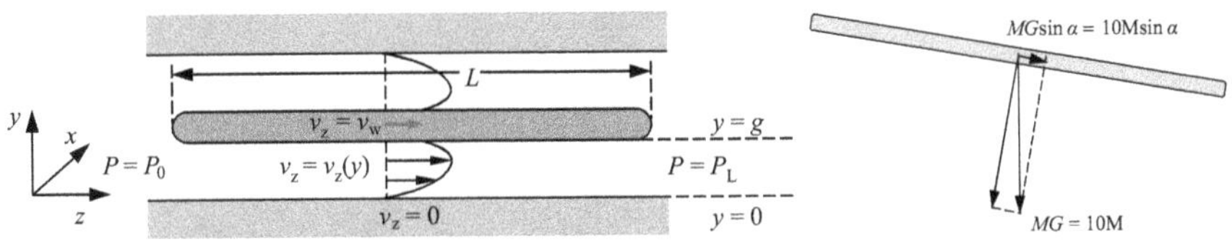

Figure 13.4 Arrangement used in the calculations. It has been assumed that edge effects can be neglected. The track can be placed at a small angle α

down in a cool-down segment in which the wafer temperature is gradually reduced to (slightly above) room temperature. The six ALD segments all operate independently, and can be switched on and off individually. This implies that with 8 ALD cells per segment of 1 m, the Al_2O_3 thickness can be set at 1, 2, 3, …, or 6 nm [7,8].

13.4 Basic wafer transport theory

From a theoretical point of view, the transport of a thin wafer inside a transport channel is close to ideal for the following reasons:

- The width of the gaps between wafer and top and bottom walls is much smaller than the length of the wafer. It is safe to assume that irregularities of wafer flow at the location of the wafer leading and trailing edges penetrates the area between wafer and wall over a distance of at most a few time the gap width (i.e. <1 mm). As this is much smaller than the length of the wafer (156 mm), such edge effects can be neglected in the calculations.
- The gap between the wafer edge and the side wall is of the order of 1 mm. As this is much smaller than the wafer width (156 mm), also the effects of gas flows on the wafer sides can be neglected.

Under the assumption that edge effects can be neglected, we can formulate the following, see Figure 13.4.

A pressure gradient $\Delta P = P_L - P_0$ is assumed to be present between the leading and the trailing edge of the wafer. This induces a flow between the wafer and the walls. In case of laminar flow, this causes the well-known parabolic velocity profile between the wall ($y = 0$) and the wafer ($y = g$). When we assume that the wafer has a velocity v_w, the parabolic profile is slightly distorted, see Figures 13.4 and 13.5. The dimensions of the wafer are L and L_x in transport and lateral directions, respectively.

Several forces act on the wafer and are discussed in detail in Sections 13.4.1, 13.4.2, and 13.4.3.

13.4.1 Shear force F_Q

The gas flow over the wafer surface exerts a shear force [9]:

$$\frac{d\tau_{yz}}{dy} = \frac{(P_0 - P_L)}{L} = \frac{\Delta P}{L} \quad \text{in which} \quad \tau_{yz} = -\mu\frac{dv_z}{dy} \tag{13.1}$$

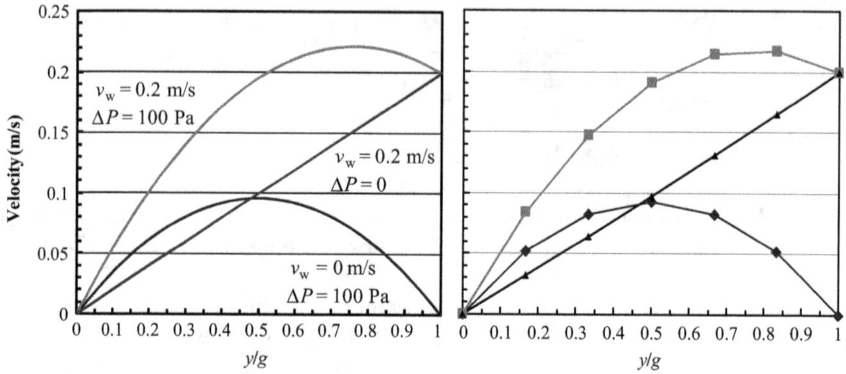

Figure 13.5 Velocity profiles for a number of typical conditions. The right-hand picture is a numerical calculation carried out in Fluent. The system temperature is 20 °C and the gas ambient N_2

μ is the viscosity of the gas; as the concentrations of the precursors are a few percent only, the viscosity is dominated by that of nitrogen. Solving (13.1) with boundary conditions $v_{z,gas} = 0$ and v_w at $y = 0$ and $y = g$, respectively, results in the following velocity profile:

$$v_z = \frac{-\Delta P}{2\mu L} y^2 + \left\{ \frac{v_w}{g} + \frac{\Delta P}{2\mu L} g \right\} y \tag{13.2}$$

In Figure 13.5, typical examples are shown for a number of different conditions. For the viscous drag force on one side of the wafer, we find

$$F_Q = L_x L \left(-\mu \frac{dv_z}{dy}\bigg|_{y=g} \right) = \frac{\Delta P g L_x}{2} - \frac{\mu v_w L_x L}{g} \tag{13.3}$$

Note that the viscosity is a function of temperature and the type of gas: in case of N_2 at 20 and 200 °C, we have: $\mu = 1.88 \times 10^{-5}$ and 2.51×10^{-5} Pa · s, respectively.

13.4.2 Gravitational force F_G

When the track is placed at an angle of $\alpha°$, we can write (see Figure 13.4):

$$F_G = MG \sin \alpha \tag{13.4}$$

where M is the weight of the wafer and G is the gravitational constant.

13.4.3 Pressure force ΔP

The pressure gradient ΔP acts on the leading and trailing edges of the wafer with thickness d:

$$F_{\Delta P} = d L_x \Delta P \tag{13.5}$$

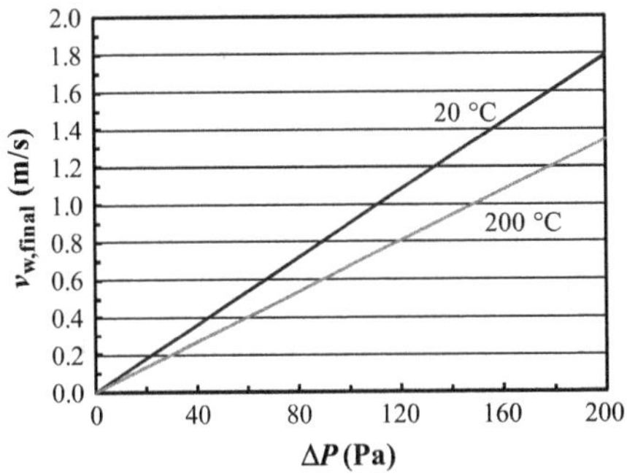

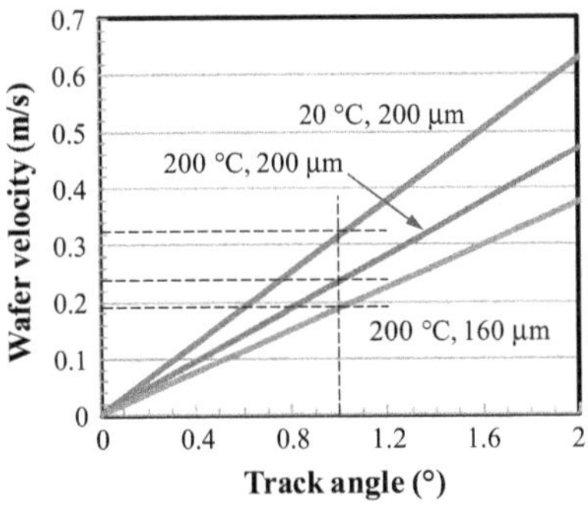

Figure 13.6 Left: Wafer velocity as a function of pressure gradient across the wafer at two different temperatures. The track angle is zero. Right: Wafer velocity at varying track angle with zero pressure gradient across the wafer. Various combinations of temperature and wafer thickness are shown. In all the cases, the gas ambient is N_2

The total force F_{tot} is the combination of (13.3), (13.4), and (13.5):

$$F_{\text{tot}} = dL_x\Delta P + MG\sin\alpha + 2\left(\frac{\Delta PgL_x}{2} - \frac{\mu v_w L_x L}{g}\right) \tag{13.6}$$

The factor 2 was added to the third term because of the fact that the viscous forces act on both sides of the wafer.

When wafers are injected into the track, the driving forces caused by either a pressure gradient across the wafer, and/or a gravitational force (when the track is placed at an angle) will continuously accelerate the wafer; the associated velocity will increase up to the point where the viscous friction forces (increasing linearly with wafer velocity) will balance the driving forces. At this point, there will be stable transport with constant wafer velocity, and $F_{\text{tot}} = 0$.

One can now consider various limiting cases:

- The wafer is accelerated by only a pressure gradient, i.e. $\Delta P \neq 0$, $\alpha = 0$. When we now assume $F_{\text{tot}} = 0$, we find

$$v_w = \frac{g(d+g)\Delta P}{2\mu L} \tag{13.7}$$

- Track placed at an angle α, and no pressure gradient across the wafer: $\Delta P = 0$. This results in

$$v_w = \frac{MG\sin\alpha}{2\mu L_x L} = \frac{dg\rho G\sin\alpha}{2\mu} \tag{13.8}$$

where ρ is the substrate density.

Numerical examples for both cases are shown in Figure 13.6.

Typical equilibrium wafer velocities that are realized are in the range of 0.2–0.6 m/s in case of track angles 1–1.5° and/or pressure gradients 20–100 Pa. It is worth noting that in case of a track angle of, e.g., 1°, the difference in height between the entrance and the exit of an 8 m tool is ~14 cm.

So far we based the calculations on equilibrium conditions, i.e. the wafers have reached a constant velocity. It is of interest to investigate what time (or travel length in the tool) it takes to reach this equilibrium condition. In order to calculate this, we use the force equation (13.6) and combine this with the time-dependent wafer acceleration:

$$F_{\text{tot}} = M\frac{dv_w}{dt} \tag{13.9}$$

This results in

$$\begin{aligned}\frac{dv_w}{dt} &= \frac{L_x\Delta P}{M}(d+g) + G\sin\alpha - \frac{2\mu L_x L}{gM}v_w \\ &= \frac{\Delta P}{\rho L}\left(\frac{d+g}{d}\right) + G\sin\alpha - \left(\frac{2\mu}{\rho dg}\right)v_w\end{aligned} \tag{13.10}$$

When we assume the boundary condition $v_w = 0$ at time $t = 0$, we find as a general solution

$$\begin{aligned}v_w &= \frac{a}{b}(1-\exp(-bt)) \quad \text{in which} \\ a &= \frac{\Delta P}{\rho L}\left(\frac{d+g}{d}\right) + G\sin\alpha \quad \text{and} \quad b = \left(\frac{2\mu}{\rho dg}\right)\end{aligned} \tag{13.11}$$

In Figure 13.7, a typical example is shown.

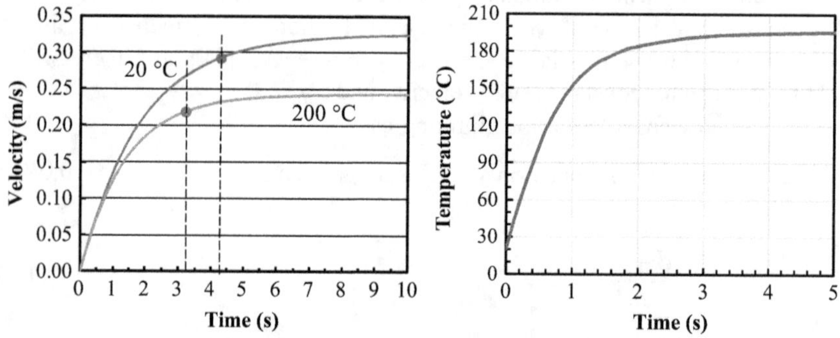

Figure 13.7 Velocity and temperature entrance effects for 200 μm wafers in track at 20 and 200 °C. The gap between wafer and walls is 150 μm. Left: acceleration of wafers in a track at an angle of 1°, injected with initial velocity zero. In this case the (asymptotic) equilibrium velocities are v_w = 0.235 and 0.314 m/s, respectively. The dots indicate the time at which 90% of the final velocity is realized. Right: heating of wafer in a track with a constant wall temperature of 200 °C

From Figure 13.7, we conclude that at 20 and 200 °C, 90% of the final (constant) velocity is realized after 4.3 and 3.2 s, respectively. This corresponds with distances of 0.85 and 0.48 m travelled. The reason why at 200 °C the equilibrium velocity is reached at a much shorter traveling length in the track is because at higher temperatures the viscosity is much higher. This results in a lower equilibrium velocity that is realized at a shorter distance.

It is important to make sure that the wafer is properly heated to the desired process temperature before it reaches the first ALD cells in the system.

It should be realized that no special heating sources (such as radiation) are required to heat the wafer. It suffices to heat the walls surrounding the wafer. As the gap between wafer and walls is very small (150–200 μm) conduction of heat from the hot walls to the wafer is very efficient. This so-called conduction heating has been demonstrated earlier in the so-called Levitor system, an RTP system in which wafers are heated uniformly to temperatures as high as 1,200° in a matter of seconds [10,11].

When we carry out a calculation on the heat-up of wafers in a track at 200 °C, assuming a gap of 150 μm between the wafer and the walls, we find a heat-up curve as shown in Figure 13.7, right. In this calculation, it was assumed that the wall temperature is constant, and that only conduction of heat takes place (i.e. radiation and convective heating are neglected).

From the figure, it can be concluded that it takes ~2.3 s to heat the wafer to a value that is off by <10 °C of set point. When this is combined with typical wafer velocities in the range of 0.2–0.4 m/s, this indicates that the heat-up length is in the range of 0.45–0.9 m.

The conclusion is that the velocity and temperature equilibration times (and therefore lengths) are very similar and both <1 m.

13.4.4 Accuracy of the analytical model

An alternative approach is to carry out 3D fluid (CFD) simulations of the system. However, depending on the complexity of the model such simulations may take days to run, and are therefore not very attractive to use. In addition to that, such simulations do not provide the user with parametric relations, and are therefore not very useful for predicting scaling of specific parameters of interest. However, these simulations are useful to verify the analytical results of selected sets of parameters. This was done, and it turns out that the differences between the analytical results discussed in the previous sections and the fluid simulations are identical within 1%–2%. The excellent agreement is easy to understand: first of all the edge effects that are included in the simulations and not in the analytical models are very small, see comments made at the beginning of Section 13.4. In addition to this, it should be realized that the physical laws applied in the analytical calculations and in the fluid simulations are identical.

13.4.5 Design impact

Based on the results obtained with the analytical model, the system was designed to meet the high-level specs defined in Section 13.2. The modeling resulted in a detailed design of a track operating within the following range of system parameters.

- Wafer velocity: variable between 0.2 and 0.5 m/s
- Track angle: variable in the range of 1–1.5°
- Pocket in which wafer is transported: 500–700 μm. Depending on the wafer thickness, this results in a wafer/wall gap to be selected between 150 and 250 μm. The pocket depth can easily be changed by replacing top plates present in the bottom block of the track.
- Heat-up and cool-down segments: length 1 m.

The length of the heat-up segment is based on the calculation of the typical system entrance lengths that are required to reach the equilibrium (constant) velocity and process temperature, see Figure 13.7. Likewise, at the exit side of the tool wafers should cool-down sufficiently such as to be able to handle the wafers safely with the unloading automation.

13.5 Comparison with experimental results

Various experiments were conducted to verify the validity of the calculations. In the Levitech engineering department a short, 4 m, version of the tool is used for engineering tests. The velocity of the wafers injected into the system can be varied by changing the speed of the entrance belts outside the system.

The velocity of the wafer is determined by monitoring the pressure sensors located at different positions along the track. The measurement is based on the fact that higher pressures are observed when wafers are passing such sensors. For details, see [7]. Figure 13.8 shows the wafer velocity at various positions inside the

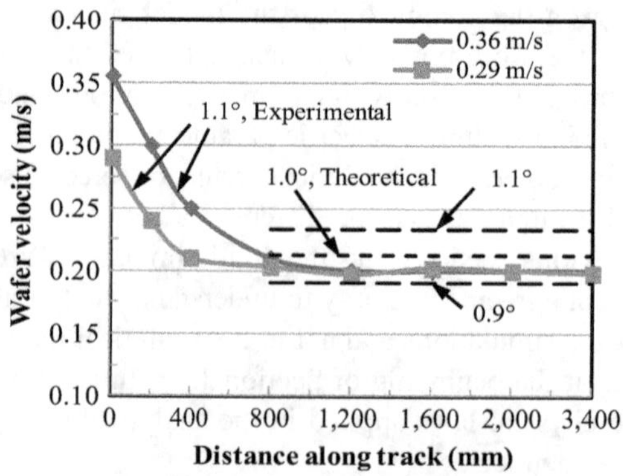

Figure 13.8 Wafer velocity measured at various positions inside a 4 m track. Wafer thickness is 180 μm, track angle 1.1°, temperature 200 °C, $\Delta P = 0$. The wafers enter the track with two different injection velocities, 0.29 and 0.36 m/s

track. Two different wafer injection speeds were used. Note that it is difficult to inject wafers at zero or low velocity.

From these measurements, the following can be concluded.

- The equilibrium wafer velocity under the conditions indicated in the figure is $v_w = 0.20$ m/s.
- The theoretical equilibrium wafer velocity is 0.232 m/s. This is ~15% higher than the experimental value.
- The equilibrium wafer velocity does not depend on the wafer injection speed. It takes ~0.5–0.7 m before the final velocity is reached. This is consistent with the typical numbers calculated analytically in Section 13.4.

The origin of the deviation between the calculated and experimental velocities is, most likely, related to the fact that the pressure build-up in front of the moving wafer is not taken into account. The resulting force will cause a decrease in wafer velocity.

It is important to be capable to predict wafer transport speed when system settings/parameters are changed. For this reason, the scaling that is predicted by the relation (13.8) was verified experimentally. The parameters that were varied are wafer thickness (100–200 μm), depth wafer transport pocket (0.7 and 0.7 mm) and track angle (1.3° and 1.5°). In all cases, $\Delta P = 0$.

This should be compared with the maximum variation of a factor ~3 in wafer velocity shown in Figure 13.9. In Figure 13.10, the deviations from the predictions given by the scaling relation (13.8) are shown. It turns out that the maximum deviations from the theoretical scaling is in the range of roughly +10 to −5%. This is sufficiently accurate for a proper operation of the tool.

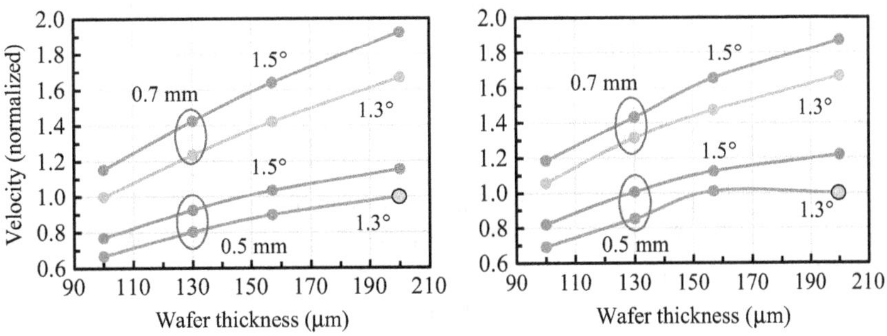

Figure 13.9 Left: theoretical velocity scaled relative to the velocity calculated for track at 1.3°, pocket depth (gap) 0.5 mm and wafer thickness 200 μm. Right: experimentally obtained wafer velocities normalized at the same point. The velocity is measured at the location of the one of the sensors present in the second segment in the 4 m system (ALD1 segment). All data are normalized at 1.3° track angle, 200 μm wafer thickness and 0.5 mm gap (150 μm between wafer and walls)

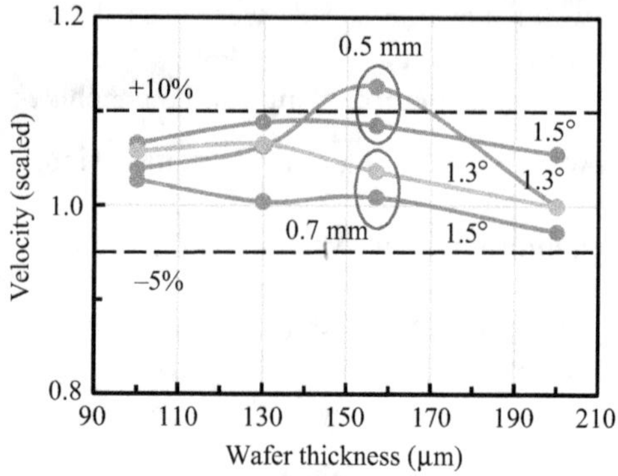

Figure 13.10 Verification of the scaling relation (13.8). The (scaled) velocity shows the mismatch between theory and experimental wafer transport data shown in Figure 13.9

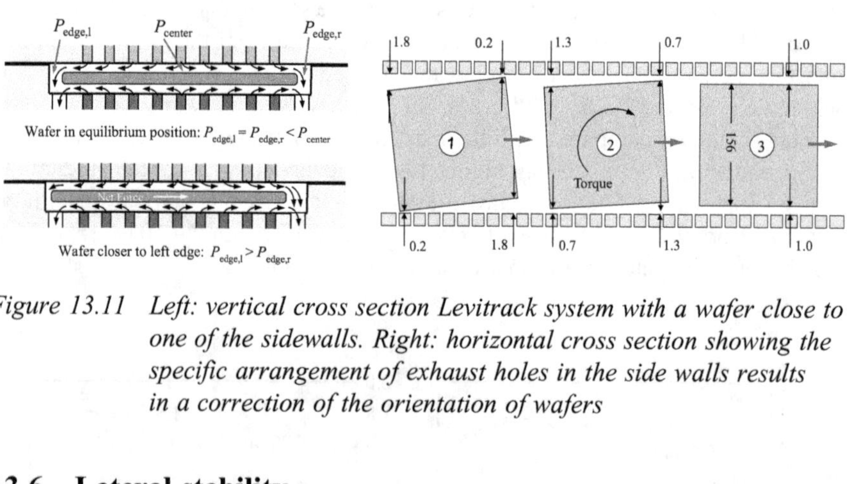

Figure 13.11 Left: vertical cross section Levitrack system with a wafer close to one of the sidewalls. Right: horizontal cross section showing the specific arrangement of exhaust holes in the side walls results in a correction of the orientation of wafers

13.6 Lateral stability

The forces in the vertical gas bearing are very strong; during normal operation, it is virtually impossible for the wafer to touch the top and bottom walls. When wafer is warped/bowed, the wafer is flattened by these forces, preventing such warp/bow to cause contact with top and bottom walls. Unfortunately, this does not prevent the wafers to touch the side walls. For that reason, the tool was designed to have a lateral stabilization mechanism as well. This is illustrated in Figure 13.11.

When the wafer moves laterally and comes close to one of the side walls, the exhaust opening in the left side wall is (partially) closed off, resulting in a higher pressure on that side. Likewise, the exhaust hole on the right side is more exposed, leading to a local decrease in pressure. This results in a force moving the wafer back to the center of the transport pocket. This force is a combination of a pressure-related force on the wafer sides and a viscous force caused by the pressure-induced

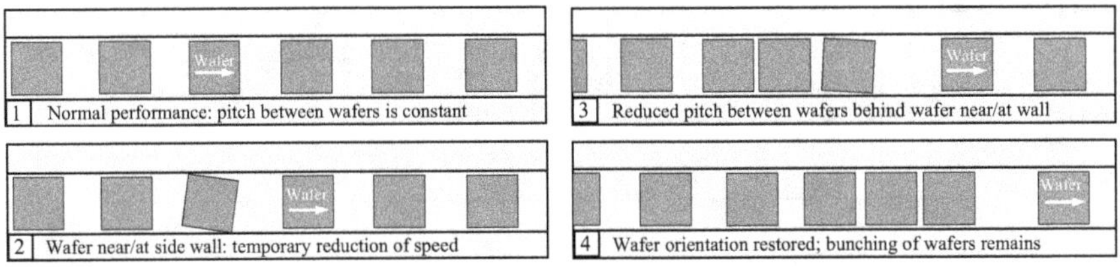

Figure 13.12 Local bunching of wafers: (a) regular, ideal, wafer transport; (b) one wafer comes close and/or slightly touches the side wall; (c) later wafers catch up, causing slow down, and bunching; and (d) the lateral stabilization corrects the orientation of the wafers; some bunching in transport direction remains

flow across the wafer, in between the wafer and the wall. As this force is highly local, it not only pushes the wafer back to the center of the track, but it also corrects the wafer orientation.

The design details of the exhaust hole arrangement are based on extensive numerical fluid modeling [12].

13.7 Throughput enhancement

As there is a strong desire for a continuous reduction of the cost-of-ownership of PV production tools, a number of steps were taken to increase the throughput of the system.

One of the main limitations of the throughput is the inability of the unloading part of the automation to handle to wafer output reliably. This has to do with the fact that the timing of wafers leaving the tool is not completely identical to that of the wafers entering. Contrary to belt-type transport systems, wafers in a gas track have some freedom to move back and forth in transport direction. This freedom is quite limited. When two adjacent wafers approach each other, the pressure in the empty space between the wafers increases, pushing them away from each other. However, some oscillations in transport direction are unavoidable. In addition to this, there are also slight oscillations in lateral direction. This causes some of the wafers to come very close to one of the side walls. Although the lateral stabilization forces described in the previous section tends to correct this, it cannot be avoided that slight (near-) collisions with the side walls slow down these wafers in transport direction, see Figure 13.12. Wafers following this wafer will catch up, resulting in local bunching of a number of wafers. Note that it is highly unlikely for the wafers to touch each other. When two wafers come very close to each other, this will result in a large increase in pressure in the open space between the wafers. The consequence is that these wafers will be pushed away from each other again. The resulting oscillation will damp out over a distance of the order of typically 0.5–0.8 m.

The phenomenon does not really cause problems inside the track; no wafers are permanently blocked and/or damaged. However, it creates a problem with the

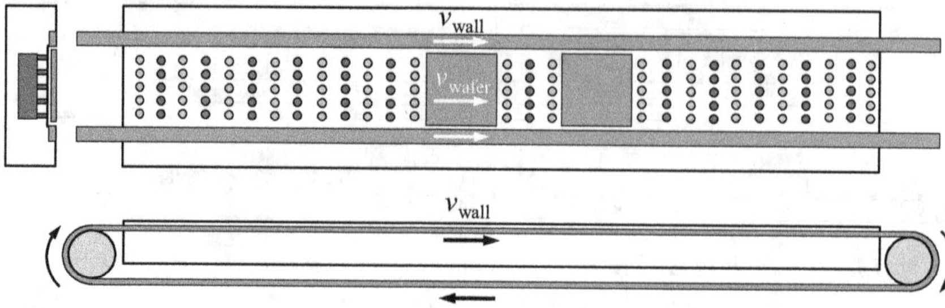

Figure 13.13 Moving side wall arrangement. The belts are made of fiber-enforced teflon, with a width (in the plane of the wafer) of a few mm and somewhat thicker than the wafer. The belt can handle temperatures >200 °C

unloading automation: when wafers exit the track at high speed, and are close together, the wafer handling robotics is no longer capable of reliably unloading adjacent wafers from the unloading belt. This limits the effective throughput to roughly 3,200 wafers/h. Consequently, higher throughputs can only be realized when the above-mentioned bunching is eliminated. This can be done by making sure that wafers are not slowed down when coming close to the side wall. This was accomplished by adding a so-called moving side wall feature to the system, see Figure 13.13.

The solid walls are replaced by a thin belt moving along with the wafer at the same speed. In the ideal case when the belt moves with exactly the same speed as the wafers, the velocity of a wafer does not change when it touches the side wall. As mentioned earlier, the vertical gas bearing is very stiff and wafers never touch the top and bottom wall. This being the case, the consequence of this arrangement is that the wafers basically travel in an environment that moves forward at the same speed. The belts are inexpensive and are designed such that they can be replaced within a few minutes.

The main disadvantage is that ALD of Al_2O_3 will take place on all components that move with the wafer, passing the subsequent precursor cells. In due course, this will cause buildup of Al_2O_3 on the belt eventually causing trouble with wafer transport. Fortunately, as the Al_2O_3 deposits adhere only loosely to the teflon surface, these can easily be removed continuously with brushes placed outside the track.

With this arrangement, it appeared to be possible to increase the throughput to >6,000 wafers/h, a 85% increase [13]. As the associated additional hardware and maintenance costs are very limited, the cost of ownership is reduced considerably. Increasing the throughput from, typically, 3,200 to 6,000 wafers/h results in a CoO reduction of 35%–40%. Note that as the increase in throughput partially comes from a closer packing of the wafers, the percentage open area between adjacent wafers is reduced, leading to a more efficient use of the precursors (precursors are continuously injected everywhere along the track, i.e. in areas with wafers as well as in empty areas between wafers).

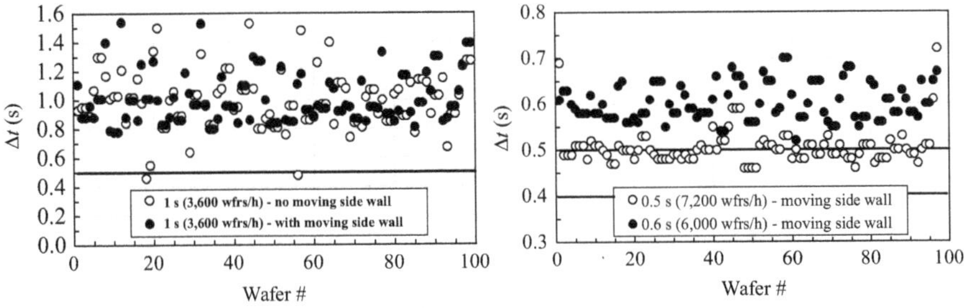

Figure 13.14 Variation of wafer arrival intervals (wafer pitch in seconds) for various throughputs, and with/without activation of the moving side wall. Solid line: spec current automation system. Dashed line: spec of automation system under development. For details, see text. Note the difference in vertical scale in both figures

In Figure 13.14, a comparison between wafer transport with and without the moving side wall is shown. In Figure 13.14, the differences in arrival time of successive wafers at the unloading automation unit are shown for a standard cassette of hundred wafers. These measurements were done with a fast, simple automation unit that can operate at very high throughputs. Unfortunately, it can only do so for one cassette at the time. The current high-volume manufacturing automation system is able to handle much larger quantities of wafers, including switching cassettes automatically, without loss of time. However, it can only handle wafers that are separated in time by 0.5 s or more. This implies that during continuous operation for days/weeks, there should not be a single event such as is illustrated in Figure 13.12, bunching of wafers. In Figure 13.14, left, transport of a cassette of 100 wafers at a throughput of 3,600 wafers/h. is shown, with and without activation of the moving side wall feature. It is observed that in case the moving side wall is not operational, some of the wafers arrive with intervals <0.5 s, which would cause serous automation problems. In contrast to that, when the moving side wall is operating with a speed equal to the velocity of the wafers inside the track, the wafer arrival intervals are always >0.8 s. Therefore, it is now safe to operate the tool at this throughput. It should be noted that in a HVM factory, several hundreds of cassettes are processed daily, and consequently, it is very critical that no arrival intervals shorter than 0.5 s are present.

In order to verify the moving side wall feature at higher throughputs, tests were done at 6,000 and 7,200 wafers/h, see Figure 13.14, right. From Figure 13.14, it can be concluded that it is safe to operate the system with 6,000 wafers/h. At 7,200 wafers/h, one can observe many wafer interval excursions <0.5 s. This is understandable: at this throughput, wafers are loaded at 0.5 s intervals. However, one can clearly see that no wafer intervals <0.45 s are present. Consequently, a new development project was started with the objective to be able to handle wafer arrival intervals of 0.4 s, enhancing the system throughput to 7,200 wafers/h. Note that the reaction rate of the ALD process is not limiting the throughput [14]. In case depletion effects are an issue, one can simply increase the precursor concentrations.

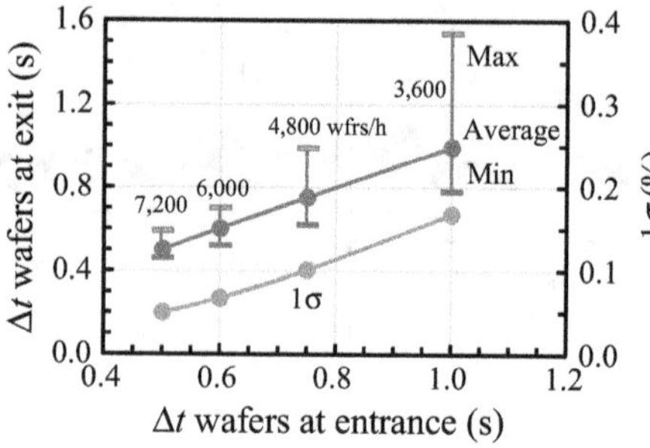

Figure 13.15 Wafer arrival interval as a function of the wafer-loading interval (throughputs from 3,600 to 7,200 wafers/h). Shown are the average wafer arrival intervals (which, ideally, should be identical to the wafer-loading intervals), and the variation in arrival time (maximum and minimum values). Also, the standard variation in arrival time is shown, as a percentage of the average arrival time. The first wafer and the last three wafers are eliminated from the calculations. These wafers always arrive at larger intervals, because the pressure build up in between successive wafers drives them forward, backward, respectively. This increases the associated interval times

Another interesting feature is that the spread in wafer arrival intervals becomes smaller at higher throughputs, see Figure 13.15. It turns out that the variation in arrival time decreases considerably when, at higher throughputs, the wafers are more closely packed. This is caused by the fact that when the separation in transport direction is reduced, higher pressures are built up in the empty space between the wafers. Whenever a wafer makes an excursion in forward or backward direction, it is pushed back more vigorously to the equilibrium position. Consequently, the "gas bearing" in transport direction becomes stronger. This is an attractive side benefit of going to higher throughputs.

13.8 Spatial ALD process results

As the main topic of this paper is the design and performance of the system, we will only give a few examples of process results. More details can be found in [5].

Figure 13.16 gives a typical example of a deposited film. In one pass of the wafers through a tool with six ALD segments a film with a thickness of 6.4 nm is deposited, with uniformity 3.2% (1σ). This is sufficiently good for passivation of solar cell surfaces. It should be noted that it is possible to improve the uniformity by increasing the TMA and H_2O flow. However, this reduces the TMA consumption efficiency. Further, except in a small region of 1–2 nm on the leading edge of the wafer, no deposition on the rear side takes place.

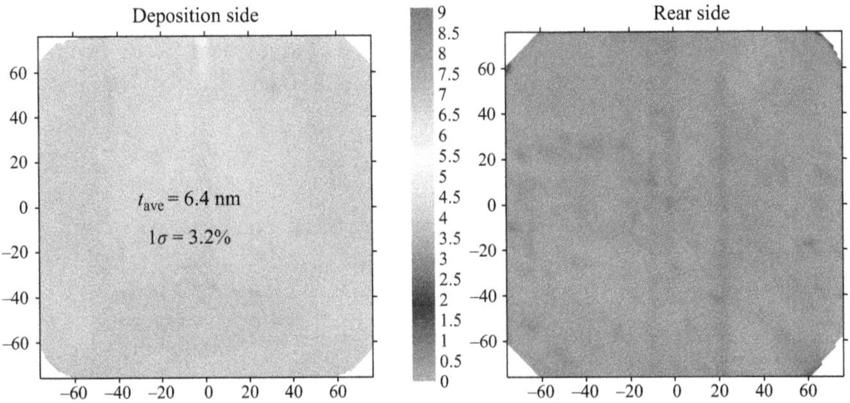

Figure 13.16 *Example of uniformity. Left: ellipsometer measurement of Al_2O_3 thickness after one passage in system with 6 ALD segments. Right: rear surface of same wafer. Both front- and rear-side start out with a native SiO_2 film with thickness ~1 nm*

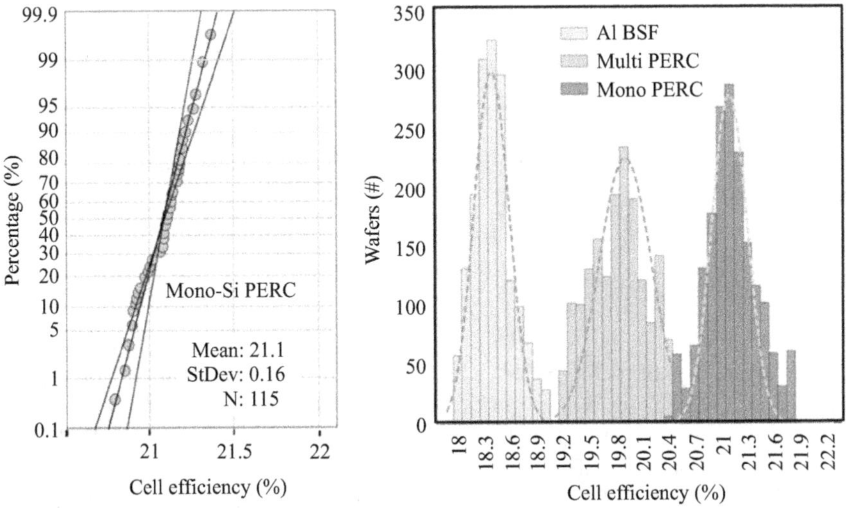

Figure 13.17 *Examples of passivation with ALD Al_2O_3 in manufacturing of PERC cells (customer data). Left: efficiency distribution for mono-Si cells. Right: comparison of multi- and mono-Si cells with a conventional Al BSF cells*

In Figure 13.17, typical cells results are shown. It turns out that efficiencies in the range of 19.5% and 21% are realized for multi- and mono-Si cells, respectively.

ALD Al_2O_3 is expected to grow considerably when n-type cells are being introduced into high-volume manufacturing. Levitech works closely with ECN to assist in the implementation of ALD Al_2O_3 in n-Pasha cells and n-IBC cells [15,16]. Further, it is worth mentioning that the Levitrack has already been introduced in HVM of n-IBC cells earlier (details are confidential).

13.9 Conclusions

It has been shown that Al_2O_3 deposition systems based on spatial ALD with wafer floatation are attractive for passivation of cSi solar cells in high-volume manufacturing.

- Passivation is excellent; it has been demonstrated that p-type PERC cells with efficiencies exceeding 21% can be realized;
- Uniformity is good, in the range of 2%–4%, sufficient for surface passivation;
- Very low breakage rate (in case of the Levitrack system described in this paper);
- The throughput in the Levitrack system was recently increased from 3,200 to 6,000 wafers/h, with little added system cost. This reduces the total cost-of-ownership considerably (~40%).
- In principle, the throughput can be further increased to 7,200 wafers/h; the limitation is not the ALD track, but the unloading automation. This is currently being addressed.

Acknowledgment

This work was partially funded by the Dutch Ministry of Economic Affairs within the TKI project IBCense.

References

[1] Deng, W., Ye, F., Liu, R., *et al.*, Invited paper, 22.61% Efficient Fully Screen Printed PERC Solar Cell, 44th IEEE PVSC, Washington, June 2017.

[2] Yoo, C.Y., Hong, K., Kim, J., Lee E., and Kim, D.S., Evaluation of ALD Passivation Layers for Industrial PERC Process, 44th IEEE PVSC, Washington, June 2017.

[3] For details of Al_2O_3 PECVD process, see relevant Mayer–Burger chapter in this book.

[4] Poodt, P., Cameron, D.C., Dickey, E., *et al.*, Spatial atomic layer deposition: A route towards further industrialization of atomic layer deposition, Vac. Sci. & Technol. A30, 2012, p010802.1.

[5] Macco, B., Van de Loo, B.W.H., and Kessels , W.M.M., Atomic layer deposition for high-efficiency crystalline silicon solar cells, in *Atomic layer deposition in Energy Conversion Applications*, (editor) Bachman, J., Wiley 2016.

[6] For details of ALD Al_2O_3 parallel process modules, see relevant Solaytec chapter in this book.

[7] Granneman, E.H.A. and Kuznetsov, V.I., Spatial ALD deposition of Al_2O_3 films at throughputs exceeding 3000 wafers per hour, Invited paper, Proc. ECS Conference, Orlando, May 2014.

[8] Vermont, P., Granneman, E., and Beijersbergen, J., Optimization of Al_2O_3 backside passivation into mass production, European Photovoltaic Solar Energy Conference and Exhibition, Amsterdam, Volume: 29th, September 2014.

[9] Bird, R.B., Steward, W.E., and Lightfood, E.N., 'Transport Phenomena', Wiley International, New York, USA, 1960.

[10] Kuznetsov, V.I., Storm, A.B., Snijders, G.J., *et al.*, Levitor 4000, An advanced RTP system based on conductive heat transfer, Invited paper Proc. Intern. Conf. on Rapid Thermal and other short-time processing technologies, Toronto, May 2000 ECS Edt. Roozeboom, F. *et al.* Vol 2000-9, p. 401.

[11] Granneman, E.H.A., Conduction heating in RTP fast, and pattern-independent, Mat. Sci. Forum, 573–574:375–386, January 2008.

[12] Granneman, E.H.A., Floating wafer track with lateral stabilization mechanism, Patent PCT WO 2011062490, May 23, 2011.

[13] Van Velzen, W.G., Van der Wal, M.G., and Kuznetsov, V.I., Moving side wall, patent submitted May 1, 2017: 'Substrate processing apparatus and to a method for processing substrates'.

[14] Poodt, P., Lankhorst, A., Roozeboom, F., Spee, K., Maas, D., and Vermeer, A., High-speed spatial atomic-layer deposition of aluminum oxide layers for solar cell passivation, Adv. Mater., 22, 2010, p.3564.

[15] Stodolny M.K., Lenes, M., Wu, Y., *et al.*, n-Type polysilicon passivating contact for industrial bifacial n-type solar cells, Solid Energy Materials & Solar Cells, 158, 2016, pp. 24–18.

[16] Mewe, A., Spinelli, P., Burgers, A., *et al.*, Mercury: Industrial IBC cell with front-floating emitter for 20.9% and higher efficiency, Proc. of 42nd IEEE-PVSC Conf., New Orleans, June 2015.

Chapter 14

Spatial Al_2O_3 ALD: from Lab to Fab

Roger Görtzen[1]

14.1 Time-based versus spatial

The atomic layer deposition (ALD) technique was invented by Suntola and Antson [1] in the 1970s. During that time, it was already know that different types of layers can be grown by dosing of process gases separately, and using a purging step in between. The two solutions of time-based and spatial ALD were already known in that period. But the time-based solution was the most commonly used. With ALD, layers are grown layer by layer on atomic level, delivering excellent layer thickness control and predictable layer properties. Since one atomic layer of aluminum oxide (Al_2O_3) has a thickness of only 0.1 nm, it takes 50 atomic layers to build a 5 nm passivation layer.

In time-based ALD reactors, deposition gasses are led through a process chamber in which the wafer is present (Figure 14.1). One atomic layer is formed in four steps: exposure to a first (metallic) precursor, purge with inert gas, exposures to a second (oxidizing) precursor, and purge with inert gas again. The two precursor gasses must be kept well separated to prevent contamination. This causes a minimum required time for the purge steps. Existing ALD tools can be divided in single wafer tools and batch tools. In a single wafer reactor, one full cycle takes 5 to 10 s.

In a batch reactor, one cycle is slower, typically 20 to 60 s, but a full batch of solar products (typically up to 1,000 wafers) is processed in parallel which makes it faster per wafer.

Single wafer tools are far too slow for the solar production lines, ~1 min process time instead of the required ~1 s! Batch tools may reach typical throughputs of 3,000 wafers per hour, but several batch reactors have to be clustered. On top of that, batch ALD has another disadvantage: it gives either wrap-around or both sides full coverage of the layer. Furthermore, the loading and unloading of wafers into the boats are in principle more complex and more expensive.

The spatial ALD (S-ALD) technology is based on spatial separation of precursor gasses instead of time-based separation (Figure 14.2). The different gases are confined in specific process areas while the wafers pass by. Every point Z on the wafer "sees" the gasses sequentially which induces the ALD effect.

[1]Co-founder of SoLayTec, Director Marketing and Sales, The Netherlands

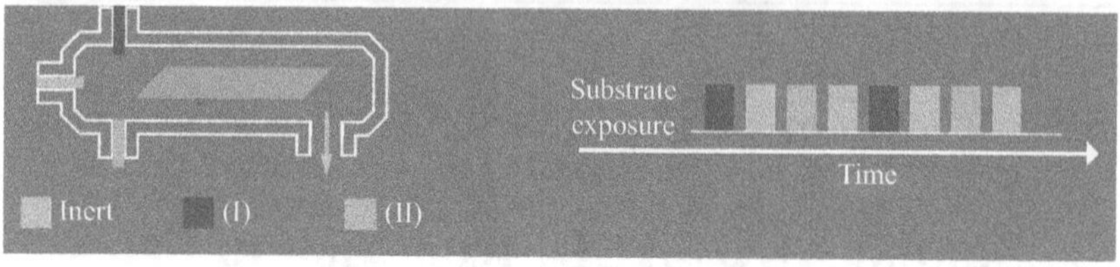

Figure 14.1 Illustration of time-based ALD

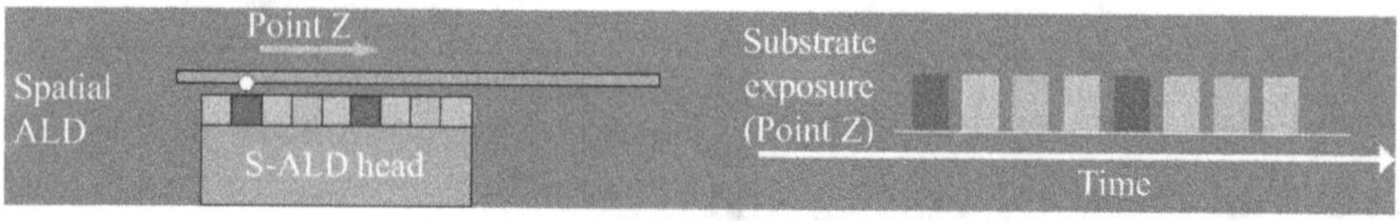

Figure 14.2 Illustration of spatial ALD

14.2 Proof of principle spatial ALD

In 2009, Levy *et al.* [2] published the results of the gas bearing S-ALD reactor, which was used to deposit oxides for TFT application. He performed with this setup that vacuum ALD was no longer needed and that large open atmospheric ALD was able to deposit large or continuous substrates. In this work, the gas and purge injectors were flat and both sides of the injectors an exhaust channel was applied. For example, Levy could reach a deposition rate of ~10 nm/min for Al_2O_3.

In that same period, the Dutch scientists at TNO were also developing their own S-ALD reactor, but their main objective was to reach an even higher deposition rate. In Figure 14.3, the several main parts of this S-ALD setup are shown:

(a) Presents a schematic drawing of the reactor head, in which two slot openings (inlets) are shown. Precursor tri-methyl aluminum (TMA) on one side and on the other side the reactant water was connected. On the outer side of each inlet, a cavity was formed, and that is used as the exhaust zone for the excess of chemicals for TMA and H_2O. The gas bearing plane of this circular reactor head is filled with small holes, which are connected with a nitrogen gas line. The function has two reasons. One it functions as a purge step in between the two materials and that prevents mixing of the two chemicals. Furthermore, the reactor head is floating on top of the wafer which is processed, so it functions as an air bearing.

(b) The complete setup is shown in a schematic way, the reactor head is placed in a fixed position at the top. At the bottom, a rotating/substrate table has been made, which is connected to a motor. In between the substrate table and reactor head, a 6″ wafer could be placed. The complete setup was built into a standard oven.

(c) A picture of a 6″ silicon wafer, that has been deposited by the proof of principle and shows a 100 nm thick Al_2O_3 ring pattern.

(d) Picture shown the actual S-ALD proof of principle at TNO (Figure 14.3).

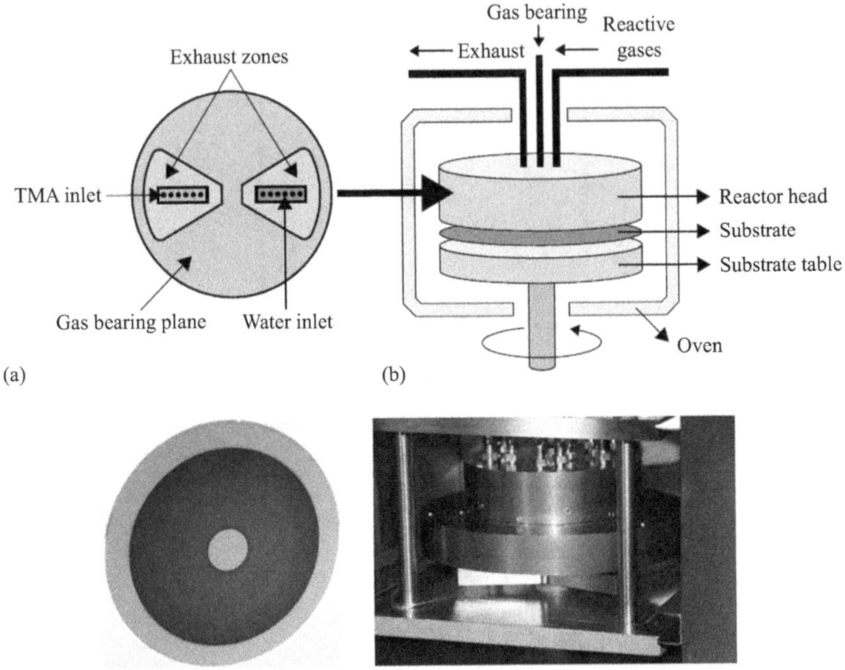

Figure 14.3 Proof of Principle at TNO

The motor speed of the substrate table was at a maximum of 600 rpm, and that gave a record deposition speed of ~1.2 nm/s (72 nm/min) [3] for this S-ALD setup, which was at that moment a factor 7× higher compared to Levy.

14.3 Spatial ALD application

After this successful proof of principle, the right market had to be addressed. During 2008–2009, the semicon market was not interested in this new kind of S-ALD. So, the scientists at the TNO looked for new markets. After a search, the development in the PV industry at IMEC, ISFH, and the TU/e came around. In 2009, Al_2O_3 was the next big thing for the PV industry, which was proven by Bram Hoex *et al.* when reading his thesis [4] in 2008. Beside the fact of the unique properties of Al_2O_3, it was also mentioned that, unfortunately, no viable ALD mass production tool was available yet.

This started the cooperation between institutes to proof that this S-ALD tool could deliver excellent passivation properties, which were needed for high efficiency solar cells.

Together with IMEC [5] and ISFH [6], several independent tests have shown that S-ALD was as good as the standard time-based ALD (thermal ALD), in terms

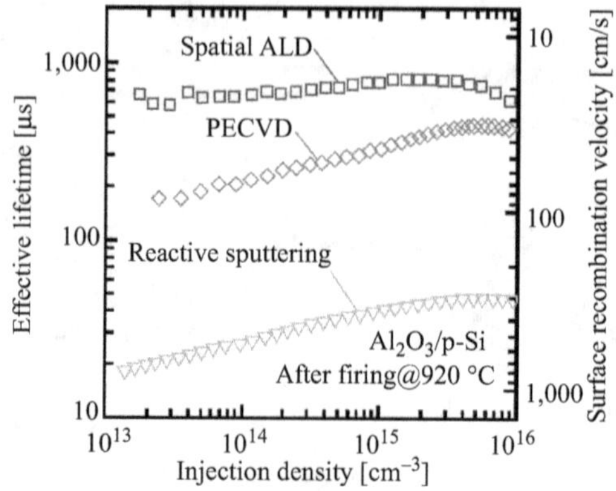

Figure 14.4 Effective lifetime eff (left scale) and upper surface recombination velocity (SRV) limit S_{max} (right scale) as a function of the injection density n measured on 1.3 cm p-type FZ-Si passivated by Al_2O_3 deposited by spatial ALD, PECVD, and reactive sputtering

of its passivation effect. The graph presented in chapter 4, figure 6 [6] in cooperation with ISFH shows clearly that the effective lifetime τ_{eff} as function of the injection density Δn is for time-based (thermal ALD) equal to S-ALD. So, the factor ~100 faster deposition rate did not had any negative effect on its passivation quality (Figure 14.4).

Next, the passivation comparison with state-of-the-art deposition techniques such as PECVD and reactive sputtering was also tested [6]. In the graph below, the effective lifetime after firing at 920 °C can be found. At the injection density 10^{15}, it can be clearly concluded that S-ALD has the lowest SRV. Which is needed for next generation high efficiency solar cells. Especially now the wafer producers, for example, Longi, are improving the bulk quality the surface passivation will be even more critical than before.

14.4 From the proof of principle to InPassion® LAB

With these process results, the next phase of machine development could be started. In 2010, the establishment of the company SoLayTec was a fact and the first real commercial S-ALD lab tools (later called the InPassion® LAB) had to be delivered early 2011. Now the circular proof of principle had to be translated into an inline solar cell R&D development tool. This tool was called the InPassion® LAB tool and it had the following three important technologies: (1) floating wafer transport, (2) S-ALD injector head, and (3) modular design for scaling to mass production.

The InPassion LAB tool features two H_2O inlets and one TMA inlet, as shown schematically in Figure 14.5(a). Between these two gasses, an inert gas is separating in space, to prevent mixture. The 6″ wafers are transferred from right to left by CDA pulses, as shown in Figure 14.5(b). In every movement from the left to right or right to left, one ALD cycle is completed and approximately one atomic layer of Al_2O_3 is deposited. The InPassion® LAB achieves wafer oscillation speeds

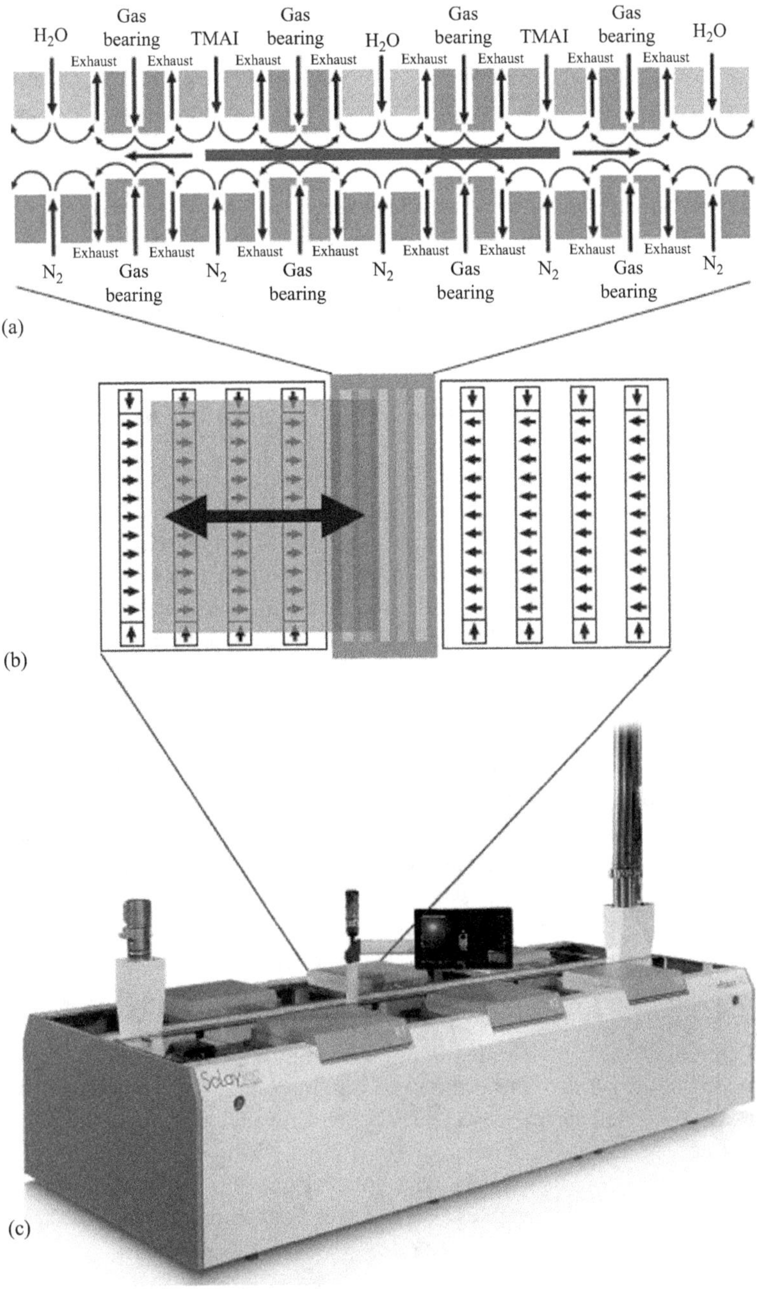

(a)

(b)

(c)

Figure 14.5 *(a) In spatial ALD, the TMA and H_2O steps are separated in space by an inert gas (in this case N_2); (b) in the deposition unit the wafer oscillates underneath the spatial ALD injector head; and (c) the InPassion® ALD 6 features two main conveyors that feed various ALD units at either side of the conveyor*

of 2 Hz, resulting in the deposition of four atomic layers of Al_2O_3 per second and thus a deposition rate of ~0.5 nm/s without compromising the ALD process.

The floating wafer transport was developed, so no carriers and less heating was required. In recent years, a lot of know-how has been developed, to make sure wafers are moving on a speed of 1 m/s from the left to right and backwards, in the reactor without touching the metal reactor parts. Because when a wafer would touch the metal parts of the reactor, breakage of wafers would occur. To accomplish this with an oscillation of 2 Hz, wafers had to move in a straight line through the reactor. The second large development was the design of the injector head to get a uniform gas flow and no mixing of gasses to prevent contamination. The third aspect will be addressed in the next paragraph.

14.5 Transform from Lab to Fab—InPassion® ALD

After the learning phase of the InPassion® LAB tool introduction and acceptance of those tools, the next phase of development was started. The building blocks of the InPassion® LAB tool could be reused. The only new add-ones were included: two ALD steps in one pass and a modular mass production platform is developed.

Doubling the gas inlets meant add one TMA and one H_2O more in the same injector head spacing. This effectively doubled the deposition rate to 1 nm/s.

There might be a need for the Al_2O_3 layer to be changed to another thickness to optimize cell efficiency or costs. This can be changed easily within the graphical user interface of the system. When you reduce the layer thickness, the throughput for InPassion® ALD 6 will increase up to a maximum of 3,600 wph (Figure 14.5(c)).

Depending on the capacity levels that are needed, SoLayTec offers two types of InPassion® ALD tools, the main difference being the number of deposition units in such a system. The basic two products which can be offered, are 4 or 6 deposition units, that result into 2,400 or 3,600 wph, respectively (based on 4.7 nm layer thickness).

In the case that one deposition unit (1 of 6) needs to be stopped for service, maintenance or caused by an error, the other five deposition units will continue production, whereas tools of competitors will be completely down. This gives the end user the highest possible uptime.

InPassion® ALD can be scaled in throughput by adding deposition modules up to a number of 4 or 6 delivering the maximum capacity. For your first pilot line, this means you order the InPassion® ALD platform including gas cabinet, abatement and choose the right number of throughput. With e.g. 4 modules you can reach a throughput of 2,400 wph. At a later stage, you can add throughput by adding more modules, this enables you to lower your initial capital investment.

14.6 Second generation—InPassion® ALD

After installing machines at various PV manufacturers, we started to learn about the pros and cons of our system. In this article, I would like to explore both topics and convince you to become one of our next ALD users.

14.6.1 How wafer thickness variation affected the adaptation of the InPassion© ALD product

If you use a thermal batch or wet chemical process, the wafer thickness variation does not seem to pose any problem in the industry; the wafers are loaded into a quartz boat and the total thickness variation (TTV) is fairly irrelevant. However, in the case of the new spatial ALD processing technology, where we use a reactor gap of about 280 μm and we process wafers of about 180 μm thick, there is only a 50 μm gap separating the wafer from the reactor wall. This makes the TTV/Bow specification an important aspect of feasible and successful production.

Let me explain the ALD process of SoLayTec. The Al_2O_3 is deposited atom by atom by moving the wafers on an air bearing and, at the same time, moving the wafer back and forth without any kind of mechanical contact. Every time the wafer is moved, it passes the patented injector-head design, which builds the layer in a very uniform way. For a 180 μm thick wafer, with a local bump of about 80 μm, there is not a lot of room left between the wafer and reactor wall to cause the movement to fail. Mass production experience suggested that wafers of different thickness qualities would not reach the ALD system and that the industry would have good quality inspection at the beginning of their production line. Of course, some customers do use the appropriate wafer quality, but not all. Where customers have their own in-house wafer production, it is essential for these customers that all the wafers, which are sawn, are eventually also finished into a complete solar cell. In other words, even the wafers that are not completely within specification can be successfully deposited by the ALD tool.

In the past, SoLayTec developed its own 3D thickness measuring tool and started to measure thousands of wafers that became stuck in our ALD system. It opened our eyes completely. One of the conclusions was that if we want to run mass production, the ALD tool has to be more flexible for a wider variation of wafer quality. Several wafers were measured and it was found that wafers of TTV and bow/warp of up to 80 μm were used for production. See Figure 14.6(a) and (b) for a 3D representation and Figure 14.6(c) for within-spec wafer representation.

14.6.2 Adapt the reactor gap to improve reliability of the ALD system

Based on this learning phase, SoLayTec developed a second-generation InPassion® ALD system designed with a much larger reactor gap. The gap spacing has been

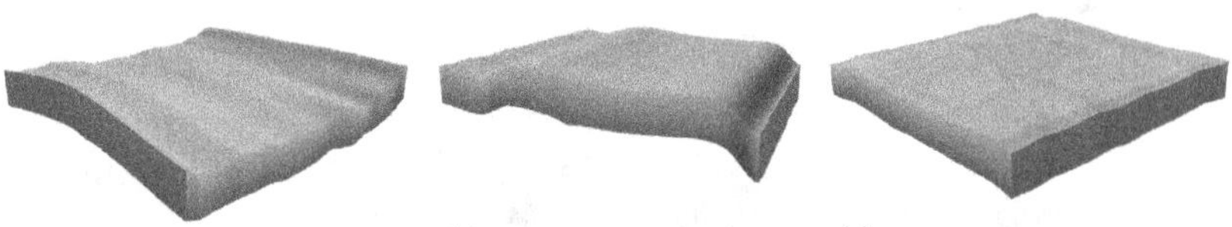

(a) (b) (c)

Figure 14.6 (a) and (b) The 3D wafer structure of wafers that were analyzed by SoLayTec inline 3D measuring equipment, showing that some production wafers are highly nonspec wafers; (c) within-spec wafer

changed from 280 to 380 μm. By doing this, the newest ALD reactor design is more reliable for the large-scale PV wafer geometry.

The main advantages of this second-generation InPassion® ALD in production are as follows:

- Increased robustness
- Less breakage
- Fewer operator assists
- Higher uptime >95% (previous design max. 92%).

In the meantime, this design has been proven and tested by different customers in mass production and some repeat orders are already signed.

14.7 Combining the stack layers

Since the beginning, we know, for example, PERC cells a stack of AlO_x–SiN_x is required. So, some PECVD tools manufacturers can deposit by either direct PECVD or MW-PECVD those two layers in one machine. But most recently leading PV cell manufacturer Hanwha Q CELLS published results of using the InPassion ALD compared with their MW-PECVD system [7]. In a comparison of the two technologies, the ALD offered an average gain of 0.15% better efficiency for multi-cSi PERC compared to the MW-PECVD AlO_x. The evaluation showed that this efficiency was achieved by "optimising the process flow and integrating an anneal into the SiN_x capping process," as stated by a Hanwha Q CELLS research paper.

In the past, to match the capacity of an ALD tool for a SiN_x direct PECVD capping layer, this required 6 or 8 tubes, which made the initial investment costs for a solar manufacturer to large and that is why the MW-PECVD had an advantage. Tempress Systems recently introduced the SPECTRUM PECVD, which has at least the same net capacity as the InPassion ALD (Figure 14.7). This system can load

Figure 14.7 Tempress Systems five-stack direct SPECTRUM PECVD system that can load 400 wafers/boat

400 wafers in a boat and contains five tubes. Now, capex wise SoLayTec and Tempress have a highly cost-effective solution available. Especially, if we take into account that the uptime, throughput, TMA usage and efficiency gain are even better than our main AlO_x PECVD competitor.

References

[1] T. Suntola and J. Antson, "Method for producing compound thin films," United States Patent 4058430 (1977).

[2] D.H. Levy, S.F. Nelson, and D. Freeman, "Oxide Electronics by Spatial Atomic Layer Deposition," Journal of Display Technology, vol. 5, p. 484 (2009).

[3] P. Poodt, F. Roozeboom, A. Vermeer *et al.*, "High-Speed Spatial Atomic-Layer Deposition of Aluminum Oxide Layers for Solar Cell Passivation," Advanced Materials, vol. 22, p. 3564 (2010).

[4] Hoex, B. "Functional thin films for high-efficiency solar cells Eindhoven:" Technische Universiteit Eindhoven DOI: 10.6100/IR634627 (2008).

[5] B. Vermang, A. Rothschild, J. John *et al.*, "High Speed Atmospheric pressure ALD for Industrial Scale Solar Cell Passivation," Proceedings of the 25th EUPVSEC, Valencia, Spain, 2010.

[6] J. Schmidt, F. Werner, B. Veith-Wolf, *et al.*, "Industrially relevant Al_2O_3 deposition techniques for the surface passivation of Si solar cells," Proceedings of the 25th EUPVSEC, Valencia, Spain, 2010.

[7] F. Kersten, I. Förster, and S. Peters, "Evaluation of Spatial ALD of Al_2O_3 for rear surface passivation of mcSi PERC Solar Cells," Proceedings of the 32th EUPVSEC, Munich, Germany, 2016.

Index

www.ingramcontent.com/pod-product-compliance
Lightning Source LLC
LaVergne TN
LVHW011752200926
848484LV00003B/681